SQL £40.00

The Market for Energy

The Market for Energy

Edited by
Dieter Helm, John Kay, and
David Thompson

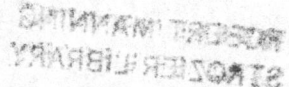

CLARENDON PRESS · OXFORD
1989

Oxford University Press, Walton Street, Oxford OX2 6DP
Oxford   New York   Toronto
Delhi   Bombay   Calcutta   Madras   Karachi
Petaling Jaya   Singapore   Hong Kong   Tokyo
Nairobi   Dar es Salaam   Cape Town
Melbourne   Auckland
and associated companies in
Berlin   Ibadan

Oxford is a trade mark of Oxford University Press

Published in the United States
by Oxford University Press, New York

© Institute for Fiscal Studies 1989

All rights reserved. No part of this publication may be reproduced,
stored in a retrieval system, or transmitted, in any form or by any means,
electronic, mechanical, photocopying, recording, or otherwise, without
the prior permission of Oxford University Press

British Library Cataloguing in Publication Data
The market for energy.
1. Economic aspects
I. Helm, Dieter   II. Kay, J.A. (John
Anderson), 1948-   III. Thompson, David
338.4'3621042
ISBN 0-19-828608-2

Library of Congress Cataloging in Publication Data
The Market for energy.
Includes papers from the Institute for Fiscal Studies
research project on energy.
Bibliography: p.   Includes index.
1. Energy industries—Great Britain.   2. Energy policy—Great
Britain.   3. Energy consumption—Great
Britain.   4. Energy industries.   5. Energy policy.
6. Energy consumption.   I. Helm, Dieter.   II. Kay,
J.A. (John Anderson)   III. Thompson, David.
IV. Institute for Fiscal Studies (Great Britain)
HD9571.5.M37 1989   333.79'0941   88-31252
ISBN 0-19-828608-2

Set by Colset Private Limited
Printed in Great Britain
by Biddles Ltd.
Guildford & King's Lynn

Preface

This volume has grown out of the Institute for Fiscal Studies energy research project, funded by the Economic and Social Research Council. The project initially focused primarily on the privatization and regulation of the British Gas Corporation and the analysis of the effects of the 1983 Energy Act on the UK electricity supply industry. It subsequently developed a wider interest in the energy utilities of western Europe, the rationale for help with energy costs to the poor, and the general question of energy policy itself.

Many of the papers in this book present the results of this research. However, we have endeavoured to produce an integrated volume which addresses the various aspects of our theme—the market for energy, and the appropriate role for State intervention and regulation. To that end, we have included additional papers by leading experts in the economics of the energy sector.

The development of energy policy in the 1980s has been rapid, and therefore many of the papers respond to particular events—the privatization and regulation of gas, the early proposals on electricity, and the development of tax and macroeconomic responses to the growth of oil production. They are therefore inevitably a product of the state of the policy debate at the time they were written, and this context should be borne in mind. Nevertheless, they all raise general questions of wide applicability, demonstrating the role and limits of the market in each energy industry.

The editors are grateful for the co-operation and efforts of the contributing authors, and to the Economic and Social Research Council for financial support under grant number B00232130. Special mention should be made of Judith Payne's extremely skilful and patient copy-editing. Secretarial help provided by Hayley Bell, Chantal Crevel-Robinson, and Monica Hyde is gratefully acknowledged.

Every effort has been made to trace the copyright-holders. If any have been inadvertently overlooked, the Institute for Fiscal Studies will make due acknowledgement at the earliest opportunity.

Contents

List of Tables ix
List of Figures xiii

Introduction: Energy Policy and the Role of the State in the Market for Energy 1
Dieter Helm, John Kay, and David Thompson

Part I. Energy Policy 21

1. Energy Policy 23
   The Rt. Hon. Nigel Lawson
2. Energy Policy Issues after Privatization 30
   David Newbery
3. Energy Policy, Merit Goods, and Social Security 55
   Andrew Dilnot and Dieter Helm

Part II. Energy in the Public Sector: Modelling and Performance 75

4. The Demand for Energy 77
   Department of Energy
5. Modelling Public Enterprise Performance 92
   Ray Rees
6. Performance of the Public Sector Energy Utilities between 1968 and 1978 115
   Richard Pryke
7. The Rationale for Marginal Cost Pricing 133
   Martin Slater

Part III. Electricity 155

8. Competition in Electricity Supply: Has the Energy Act Failed? 157
   Elizabeth Hammond, Dieter Helm, and David Thompson
9. The Potential of Incentive Regulation 178
   Richard Schmalensee
10. Regulatory Issues in the Electricity Supply Industry 188
    George Yarrow

Contents

11.	The Role of Public Service Commissions in Facilitating the Development of Combined Heat and Power Generation in the US *Alex Henney and David Thompson*	206
12.	Risk Analysis and Optional Investment in the Electricity Supply Industry *Ian Jones*	214
13.	Electricity Supply in Europe: Lessons for the UK *Dieter Helm and Francis McGowan*	237

Part IV. Gas — 261

14.	Gas Privatization: Effects on Pricing Policy *Catherine Price*	263
15.	Regulation of the Gas Industry *Elizabeth Hammond, Dieter Helm, and David Thompson*	279

Part V. Coal — 287

16.	Liberalizing the British Coal Industry *Colin Robinson and Eileen Marshall*	289
17.	The Economics of Coal *Bill Robinson*	313

Part VI. Oil — 347

18.	The Economic Implications of North Sea Oil Revenues *Peter Forsyth and John Kay*	349
19.	The Macroeconomic Impact of North Sea Oil *Christopher Allsopp and John Rhys*	377
20.	The British Experience of Taxing Oil Extraction *Michael Devereux*	411

References — 428

Index — 438

List of Tables

2.1	Forecast of primary fuel use	33
3.1	Percentage of households with various types of central heating, at different income ranges	64
3.2	Percentage of households with various types of central heating, by tenure type	65
3.3	Percentage of households with various types of central heating, by household type	65
3.4	Number of disconnections as a percentage of number of credit customers	68
3.5	Gains from alternative methods of reducing fuel costs for different household types	70
4.1	Demand effect of an energy price change	80
4.2	Sectoral energy demand elasticities	82
4.3	Other industry: price elasticities	84
4.4	Domestic energy demand: price elasticities	85
4.5	Public administration: price elasticities	85
4.6	Miscellaneous sector: price elasticities	86
5.1	Forecasts and out-turn of maximum electricity demand in England and Wales	105
5.2	Investment decisions and commissioning lags	106
5.3	Generation Development and Construction Division: estimated costs to completion of power-stations currently under construction	107
6.1	British Gas: output, productivity, and finances	116
6.2	British Electricity Boards: output, productivity, and finances	119
6.3	National Coal Board collieries: output, productivity, and finances	125
8.1	Electricity production by industrial and transport undertakings	159
8.2	Bulk supply tariff: system service charge	167
8.3	Bulk supply tariff: capacity charges	167
8.4	Bulk supply tariff: unit rates, 1985–6	168
8.5	Structure of private purchase tariffs	172
11.1	Private generation purchased by Boards	207

List of Tables

11.2	Sales by private generators to Area Boards	207
12.1	The CEGB's estimates of the net effective cost of future stations	220
12.2	Effect of a one-year deferment of Sizewell B on its net present costs as estimated by the CEGB	226
12.3	Conditions for zero cost of deferring Sizewell B by one year (scenario 'C')	227
12.4	Effect of a one-year deferment of Sizewell B on its net present costs for commissioning in 1995	229
12.5	Sensitivity test results on deferment	232
12.6	Subjective probability distributions for WMPO and X and associated values of R	234
12.7	Subjective probability distributions for WMPC and GA	234
16.1	British coal mining: supply and demand since nationalization	292
16.2	World primary energy consumption, 1964–84	292
16.3	National Coal Board financial results, 1980–5	293
16.4	Coal markets in Britain, 1957 and 1983	294
17.1	Who buys British coal?	316
17.2	NCB deep-mines: unit operating costs and revenues	324
17.3	Investment per employee	325
17.4	The scale of job subsidies in loss-making pits	332
17.5	Cost and price movements, 1981–2 to 1984–5, deep-mined coal	335
17.6	Effects on employment of changes in the coal terms of trade	338
17.7	Effect on employment of productivity changes	338
17.8	The coal industry: prices, wages, and unemployment	343
18.1	UK oil reserves	350
18.2	Long-term estimates of North Sea oil revenue	352
18.3	The structure of the UK economy, 1976	352
18.4	The structure of the UK economy, 1976 (by value added)	353
18.5	The post-oil economy	355
18.6	Production changes by sector	355
18.7	UK terms of trade in manufactures	358
18.8	Relative output, import, and export prices	359
18.9	The post-oil economy with changed relative prices	360
18.10	Output changes by sector	361
18.11	Gains from North Sea oil	362
18.12	Gains from North Sea oil	363
18.A1	Allocation of oil revenues, 1980–5	375
18.A2	The real exchange rate and the real price of oil, 1975–80	375

18.A3	The marginal rate of tax on North Sea oil revenue	376
19.1	UK crude oil production, consumption, and net trade	379
19.2	Oil price trends	380
19.3	Impact of North Sea oil	382
20.1	UK tax revenues from the North Sea	423
20.2	Tax rates under different systems	426

List of Figures

3.1	Annual household energy expenditures at different income levels	60
3.2	Annual household energy expenditures at different income levels	61
3.3	Fuel expenditure and food expenditure as share of total expenditure	62
3.4	Changes in pattern of expenditure: all households	63
3.5	Distribution of expenditures relative to the mean	72
4.1	Other industry: simulated and actual fuel demands, 1955–80	87
4.2	Domestic sector: simulated and actual fuel demands, 1961–80	88
4.3	Dynamic response of energy demand model: other industry	89
4.4	Dynamic response of energy demand model: domestic space and water heating	90
5.1	Percentage deviation of the average weekly earnings of CEGB NJIC staff from the national average for manual workers	102
5.2	Percentage deviation of gross average weekly earnings of CEGB NJB staff and electrical/electronic engineers from all manual and non-manual employees in all industries and services	103
7.1	Partial equilibrium analysis	138
7.2	The ideal pricing policy	145
12.1	The relationship between system demand duration curves, plant outputs, and system running costs	217
12.2	Running cost savings for an increment of nuclear plant on base load	218
12.3	Running cost savings for an increment of nuclear plant not on base load	219
12.4	Cost–benefit profile of a generating plant investment	225
12.5	The effect of deferment	226
12.6	Deferment with improved plant performance	231
12.7	Cumulative probability distribution for the effect of a one-year deferment on the expected net present cost of the Sizewell B project	235
16.1	UK coal production, 1900–84	291
17.1	Fuel prices and price ratio of coal to oil	318
17.2	Percentage shares of UK energy market, 1950–86	319
17.3	Percentage shares of UK energy market, 1970–86	320

List of Figures

17.4	UK coal production since 1900	321
17.5	Coal output since 1950 with projections to 2000	321
17.6	Relative fuel prices and energy ratio, 1950–84	322
17.7	UK energy consumption, 1950–86	323
17.8	Earnings in the coal industry relative to manufacturing	325
17.9	Deep-mined production: unit operating costs against cumulative output, 1981–2	326
17.10	Mining employment and profit/loss per man, 1981–2	334
17.11	Mining employment and subsidies: estimated cost, 1984–5	336
17.12	Mining employment and net revenue	336
17.13	Costs, prices, and profitable output	337
17.14	The supply curve for UK coal	341
17.15	Mining employment and profit/loss per man, 1986–7	342
17.16	Price ratio: coal/oil	344
20.1	Gross and net profitability of oilfields	424

Map

Major power-stations showing relationship to fuel sources 317

Introduction: Energy Policy and the Role of the State in the Market for Energy

1. Introduction

Energy policy is now in a state of transition. The policy consensus which emerged in post-war Britain—based on integrated public monopolies in gas, coal, and electricity, and central planning by government—was firmly rejected by the incoming Conservative administration in 1979. In its place was to be established a new market philosophy for the energy sector. The role of the public sector was to be reduced, first by encouraging competition in gas and electricity through the Oil and Gas (Enterprise) Act in 1982 and the Energy Act in 1983, and second by the privatization of the public sector utilities (starting with British Gas in 1986).

Yet it is clear that this market philosophy has so far been applied to the energy sector in a limited and partial way. Coal seems likely to remain in the public sector for some years, and it remains to be seen how far it will prove possible to introduce competition into electricity. Gas, although privatized, is still an integrated monopoly, subject to extensive regulation. The attempt to introduce competition in the gas industry has so far failed almost completely. Furthermore, the government retains an extensive network of controls over North Sea oil development and production.

Energy policy thus stands at a crossroads. Its future path could see further development toward a competitive 'market for energy' which Nigel Lawson set out as the policy goal in 1982 (and which provides the title for this collection of readings). Electricity privatization requires a clear choice between the claimed advantages of centralized co-ordination and management and the less certain outcomes of the free play of competitive market forces. If the competitive solution fails, or is not pursued, then British Gas may provide the model for privately owned integrated energy monopolies. A change in the political climate, or growing dissatisfaction with the results of liberalization, might subsequently lead to a return to planning and State control or ownership.

It is clearly an important time at which to take stock and to consider the various policy options and their comparative *merits* and *demerits*. That is the purpose of this book. The readings outline the results of analysis and empirical research which provide evidence on the likely effects of different policy choices.

In this introductory chapter, we provide an overview of the key issues and an organizing framework for individual chapters. We shall distinguish two contrasting policies for the energy sector. The first, which we shall call 'the Post-War Consensus', was pursued by governments of both parties in the 1950s,

1960s, and 1970s. This approach to policy, which we discuss in detail in Section 3, places emphasis on centralized planning of the allocation of resources in the energy sector. The second approach is 'the Market Philosophy' outlined by Nigel Lawson (Chapter 1, this volume). This approach emphasizes the role of market forces in the determination of prices and in investment and production decisions, in the energy sector.

In order to compare the merits of these very different policies for the energy sector, we shall use two familiar economic concepts—market failure and regulatory failure. Market failure arises when unregulated private markets fail to meet consumers' requirements with maximum efficiency. Section 2 of this introductory chapter considers the incidence of market failure in the energy sector and its relevance for policymaking. Regulatory failure arises when intervention by government in the operation of a market, perhaps to rectify a perceived market failure, has the effect of reducing efficiency. It will be clear that advocacy of the policies characterized by the Post-War Consensus requires a belief that market failure is endemic in the energy sector, and has serious consequences for efficiency, while the possibility of regulatory failure is viewed with equanimity. Conversely, the Market Philosophy assumes that market failure is not serious in the energy sector, or at least that it is less serious than the regulatory failures which have accompanied the scale and type of intervention implied by the Post-War Consensus. In Section 3 we consider regulatory failure and its importance in the publicly controlled energy sector of the Post-War Consensus. We go on to consider its likely incidence in a liberalized, and privately owned, energy sector. In Section 4 we consider the current policy issues in the light of our analysis of market and regulatory failure, and draw our own conclusions in Section 5.

2. Energy Market Failures

Most commodities in Britain are provided by private firms in competitive, unregulated markets. This has rarely been true of energy. A similar observation could be made in most countries in the western world. Why is it that governments have tended to think that markets which could be relied on to produce most other commodities could not deal adequately with the supply of energy? What is special about energy? In this section, we review the arguments in common, although often implicit, use. Some concern particular characteristics of the *demand* for energy; the majority, however, are aspects of energy *supply*.

There are, we believe, three main questions which those who are sceptical about the possibility of a market for energy would pose. First, energy is a particularly important commodity. Without it, individuals suffer acute physical distress and may even die; almost all industrial processes grind to a halt. The production of energy affects all other sectors of the economy to a degree that is characteristic of few, if any, other commodities. Can market forces give this adequate recognition?

Second, the time-scales associated with decisions about energy are exceptionally long. This is partly a result of the non-renewable character of many energy resources. It also follows from the sheer size and scope of energy projects—oilfield development and power-station construction are among the largest single investments in modern economies. Can market forces cope with the very long-term planning which these decisions require?

Third, even if a competitive market in energy were desirable, is it feasible? Many areas of energy supply seem to be natural monopolies—production by more than one firm is technically impracticable or would lead to wasteful duplication on a large scale. Ever since the Rockefellers sought to monopolize the US oil industry, energy production has been concentrated in the hands of a few major firms which have sought to influence the markets they face as well as to respond to them. No elementary textbook turns to the energy industry to illustrate the workings of perfect competition.

Can Market Forces Take Proper Account of the Importance of Energy to the Economy?

The analysis of competitive markets assumes that individual consumers are best placed to choose the goods and services they want. To do this, they must be well informed about the costs and characteristics of alternatives, capable of judging between them, and consistent in their behaviour. Each year, however, a number of elderly people die of hypothermia. It would be difficult for even the most fanatical admirer of the operation of market forces to argue that this outcome is the result of welfare-maximizing behaviour. Moreover, the fact that the market fails in this disastrous way in a small number of cases must raise the possibility that it works less than perfectly in many more.

There are two distinct problems here. Consumers with resources adequate for their needs may use insufficient energy because they are poorly informed —about the price of energy, how to operate appliances, or their heating requirements. More generally, consumers may lack appropriate resources to achieve a minimum standard of living. Since energy comprises a substantial proportion of the household budgets of the poor, the pricing policy of energy utilities is likely to have a considerable impact on poverty. It has often been argued that the poor should be protected directly, through lower fixed or standing charges for connection, and lower unit prices especially at the peaks in winter. But this is an expensive and inefficient means of helping a small subset of poor consumers, and lack of income is better dealt with, as Dilnot and Helm suggest (Chapter 3), through the social security system.

Security of energy supply is important to both domestic and industrial consumers. This requires that energy capacity should be available to levels in excess of normal requirements. But will profit-maximizing suppliers of energy provide capacity which will rarely—and may never—be used? If they do, how will the costs be recovered? In principle, the costs of a security margin could be recouped by imposing extremely high charges when the spare capacity was brought into operation, and firms might be induced to provide such a margin by the prospect of the revenues they could gain in these circumstances. It is easy to see political and practical reasons why this is unlikely to happen, at least on the scale required.

There are two possible solutions to the security problem. One is that some regulation is implemented to require firms to provide appropriate levels of investment. The other is that the public sector itself provides the spare capacity. However, public provision of capacity to meet supply shortages affects the incentives offered to the private sector. Since the government will always ensure excess supply, prices—and hence private investment—will be depressed. This phenomenon is described by Helm and McGowan (Chapter 13).

The output of the energy sector is a substantial proportion of gross domestic product (GDP), and oil trade in particular comprises a major part of the visible account of the balance of payments. Thus performance of the energy sector has a powerful effect on national economic performance. Energy policy in the North Sea oil and gas industries, the level of subsidy to British coal, and the cost of building new power-stations have unavoidable macroeconomic implications. The industrial relations problems of the coal industry have provoked repeated government interventions. The government has also been anxious to promote the development of nuclear power, partly in a (wholly unsuccessful) attempt to develop exportable advanced technology, partly with a view to weakening the bargaining power of the miners.

Energy may have effects on sectors of the economy other than through use. Externalities arise when the private costs of production and consumption are not equal to those of society, because costs or benefits spill over to those not directly involved. These social costs are typically considered to be large in the energy sector. They include the pollution effects of acid rain, the impact of nuclear risks on the general population, and the social consequences on miners and mining communities of declines in the coal industry.

There are two possible economic approaches to the existence of externalities, which we illustrate by reference to the acid rain example. The problem is that the Central Electricity Generating Board (CEGB) produces pollution for which it does not pay. Forestry and fishing in Sweden find that their costs of production are increased. Consequently output falls. The first solution is the tax/subsidy method. In order to reduce the output of pollution, we tax the output of the power-station. To compensate the fisheries and forestry firms in Sweden, subsidy is paid out of these receipts. Hence power-stations reduce output and the level of activity in forestry and fishing rises. The alternative approach is to view pollution as evidence of a missing market. On this view, there is no market in pollution or in clean air, but such a market could be created. For example, if Sweden owned the right to clean air, it could force the CEGB to pay compensation for acid rain. Thus if the property rights in clean air are defined, the optimal level of pollution can be attained as a result of the trading that takes place when the polluter offers to pay compensation. Except in a few cases, the practical problems of implementing these solutions are obvious and it is not surprising that direct regulation of processes or output levels is generally preferred.

Can the Market Cope with the Time-scales Implied in Energy Planning?

Many people are concerned by the use of non-renewable sources of energy. While minerals are also non-renewable, they are not destroyed in use: oil and coal, once burnt, are never available again. It follows from this that decisions about energy use foreclose options otherwise available to future generations. Do markets take this into account?

Current decisions *do* reflect the interests of future generations, because an alternative to using resources now is to retain them in order to sell them, at a higher price, in the future. Conserving resources is an investment for the future and, as with any other investment, private firms will undertake it if it is profitable. The view that energy depletion policy necessarily requires intervention because the interests of future generations will otherwise be ignored is certainly mistaken, but it does not follow from that that they are considered to an appropriate extent.

In a competitive market, non-renewable resources will command a price above the cost of production or extraction, and that difference will increase at the rate of interest, as shown by Newbery (Chapter 2) and Devereux (Chapter 20). If this did not hold, it would pay to deplete more now, rather than hold the resource back for future use. This rising price of natural resources generates a return on investment in conservation. It follows that if the market rate of return reflects the rate at which society would trade off present for future consumption—as it should and as it would in a competitive market—the competitive rate of resource depletion would also be the efficient rate.

Surprisingly, then, the fact of non-renewability does not, in itself, give rise to market failure. However, the assumptions of this model are sufficiently strained to preclude unqualified faith in the ability of markets to deal with resource depletion. The problems created for all public sector investment decisions by likely divergences between market interest rates and social time preference rates are well known. Current monopoly, or anticipated future monopoly, will distort depletion rates. Uncertainty about future ownership rights in the resource is a particular stimulus to rapid depletion.

Quite apart from the time horizons implicit in depletion decisions, many investment projects in the energy sector have their effects over an extended period. The longer the investment time horizon, the greater the degree of uncertainty over future returns. In itself, however, this is not an argument for intervention. Rather, the price of oil should reflect the additional risk. Indeed, future contracts should incorporate the risk premium in future prices of the output of the plan from the investment. In a perfect market, there would be a complete set of future contracts which perfectly incorporate these risk premiums. Failures in the investment decision arise either because futures markets are incomplete or because the social discount rate deviates from that of the market. Surprisingly, given the degree of concern expressed about the uncertainty of future energy prices, there are few futures markets in the energy sector. The oil futures market is short-term and small in size. Long-term contracts are usually concerned with the mechanisms of supply rather than the reduction of price variability. In the electricity industry, Hammond, Helm, and Thompson (Chapter 8) note that the absence of future contracts beyond the annual setting of the private purchasing tariffs can adversely affect private sector investment. In addition, the discounting procedures adopted in CEGB planning do not correspond to those of the private sector (see Jones (Chapter 12) and Helm (1987b)).

One of the primary activities of the Department of Energy (DoE) has been the monitoring and prediction of future levels of demand for energy. In Chapter 4 below, the main features of the DoE demand modelling procedures are set out, together with the major sectoral equations. The justification for the DoE taking on a demand-modelling role, apart from its own planning and regulatory function, is based on two premises—that the government is instrumentally better informed than the market and that a single consistent set of forecasts dominates a more pluralistic approach. The private sector may possess inferior information upon which to base its investment decisions, when compared with that of government. This may be because there are economies of scale in empirical research (for example, in the use of forecasts from a large-scale demand-forecasting model), because the quality of research staff is better in government, or because the government possesses relevant 'inside' information about its own demand, other developments in the economy, or the plans of firms.

The evidence for the above set of arguments is slim. The forecasting record of the British government has not been clearly superior to other attempts. An official view, if acted upon, is likely to be destabilizing if it is erroneous, since it will compound errors across the industries. Finally, if government failure tends to manifest itself in over-optimistic forecasting, the error is more likely to be an overestimate.

Can Energy Markets be Competitive?

The energy sector includes the largest monopolies in the UK economy—in coal, electricity, and gas. The oil industry is dominated by a few very large firms. The existence and apparent inevitability of monopoly have proved a motive for intervention. Nationalization was perceived as one solution to this problem, sufficient indeed to exempt these industries from the jurisdiction of the Monopolies and Mergers Commission up until the 1980 Competition Act.

Introduction

A monopoly can be either *natural* or *artificial*. A natural monopoly arises where technical cost conditions are such that the industry can support only one firm in the industry. Artificial monopoly exists where, despite the technical possibility of entry, the single incumbent firm is protected from entry either by strategic barriers or by statutory monopoly concessions.

Natural monopolies arise most frequently in networks—the electricity and gas grid, the North Sea oil pipes, the main telephone system, roads, and railways. They can be either *local* (like Area Boards) or *national* (like the national gas and electricity grids and transmission system). When these exist, it is wasteful to duplicate provision and hence, for cost reasons, one producer is better than two or more. The problem that arises is that, in the absence of competitive pressure, the monopoly can exploit its dominant position by marking up prices. Furthermore there is little direct pressure to minimize costs.

Natural monopoly does not, however, necessarily coincide with the industry, and it tends to change over time. The national transmission system of the British Gas Corporation is a natural monopoly, but the retailing of domestic appliances is not (Hammond, Helm, and Thompson (Chapter 15 and 1985)). The electricity grid is a natural monopoly, but household wiring and sales of appliances are not. Thus the natural monopoly problem is typically confined to parts of the industry and not the whole. Consequently, the regulatory problem is not coextensive with the industry. (As we shall see below, this lack of coincidence gives rise to serious regulatory failure problems if a simple rule is applied to the industry as a whole.)

By contrast, artificial monopoly arises where dominant firms erect barriers to inhibit competition from rivals. Apart from the gas and electricity local and national networks, much of the energy industry has been characterized by this second type of monopoly, through a combination of statutory provisions and other entry barriers. There is little or no evidence of natural monopoly in coal and oil extraction and delivery, electricity generation, and gas production. Auxiliary services, such as servicing, billing, and appliance sales, are similarly potentially competitive.

These two different types of monopoly require different regulatory solutions. Natural monopoly is addressed through direct regulatory control of prices, output, or rate of return, whilst artificial monopoly is remedied by competition policy, directed at reducing barriers to entry. These are tackled in Parts III and IV of the book respectively.

3. Regulatory Failure

Regulatory Failure in the Energy Market

In the previous section we identified the incidence and impact of market failure in the energy sector. The existence of market failure is, however, common to all markets. For intervention to be justified, these failures must not only be large, but they must also be greater than those which result from government intervention. In this section we consider the problems associated with government intervention to rectify these market failures in the energy sector.

The policies which can be followed by a government which wishes to intervene where private markets fail can effectively be divided into two broad groups. In the first, private markets are replaced by public enterprises which are set the objective of directly following welfare-maximizing policies. In the second, private firms are subject to regulatory constraints which aim to ensure that the pursuit of profit-maximizing policies will yield efficient outcomes.

'Regulatory failure' arises where interventionist policies fail to remedy the market failures which they seek to correct or where intervention has unintended, adverse consequences for efficiency. In reality, the contrast between nationalization and regulation can be drawn too strongly, and the problems which arise are not very different in the two cases. The underlying causes of regulatory failure are related to objectives and to the availability of information (see Kay and Thompson (1987) for a more detailed discussion). Should public enterprises or regulatory bodies choose not to follow welfare-maximizing objectives, then efficient outcomes will not be achieved. However, the successful specification of regulatory constraints to ensure the achievement of efficiency depends critically upon the information available to the regulated enterprise (whether public or private) and to the regulator. The policy problem can be characterized in a 'principal–agent' framework in which the principal (the regulatory authority or government department) relies on an agent (the utility) to achieve its objectives in circumstances where the objectives of principal and agent diverge and in which the two partners' access to information is asymmetrical (see, for example, Crew and Kleindorfer (1979) for an elaboration).

It can be seen that the problems of objectives and information interact. If the objectives of principal and agent coincide (that is, if the public enterprises and public regulatory bodies choose to follow welfare-maximizing objectives), then the principal is likely to face good access to information but to have little need for it. If, alternatively, objectives diverge, then information is required to set regulatory rules but will not be (easily) available to the principal from the regulated enterprise. Where, as is the case for the UK public energy utilities, the enterprise is a monopoly, with a corresponding dominance in information and technical expertise, the information asymmetry can be acute. In the sections that follow, we shall use this framework to assess the development of energy policy in the UK. We begin by looking at the performance of the nationalized energy utilities. Next, we consider liberalization and attempts to introduce competition with the State monopolies. Finally, we consider privatization in the energy sector and the regulation of privatized energy utilities.

Nationalized Industries in the Energy Sector

Policy and performance in the nationalized energy sector closely parallel those of the nationalized industries as a whole, not surprisingly given the importance of energy industries in the nationalized sector. In this section we trace the evolution of nationalized industry policy in outline only, before considering its application to the energy sectors. (For more general consideration of nationalized industry policy and performance, see National Economic Development Office (1976), Pryke (1981), and Molyneux and Thompson (1987).)

Nationalization by the Labour government between 1945 and 1951 was implemented on the basis of what is often called 'the Morrisonian model' (after Herbert Morrison). This established corporations which were publicly owned but which were separate from government and were intended to operate at arm's length from day-to-day political intervention. These corporations were typically given a national monopoly in the supply of goods and services.

The proposed solution to the perceived failure of private markets was to replace them with public corporations which, it was assumed, would seek to implement welfare-maximizing policies. The crucial determinant of the success or failure of policy was thus whether, in the absence of any explicit constraints or incentives, the public corporations would choose to follow the efficiency rules which provided the rationale for their existence. The development of nationalized industry policy to the present day can be caricatured, not altogether unfairly, as a progressive recognition of the inherent improbability of this outcome.

This recognition is reflected in a sequence of White Papers which attempted to prescribe regulatory rules which would constrain the industries to act efficiently. The White Papers in 1961 and 1967 focused upon allocative efficiency. The level and structure of prices were to be related to marginal costs. The benefits of proposed investments, discounted by the opportunity cost of capital, were to be compared with the costs of the project. In contrast, the main focus of the 1978 White Paper was on productive efficiency—establishing the primacy of financial controls and introducing performance targets.

From the mid-1960s, then, the nationalized fuel industries were instructed on the steps to be followed to implement welfare-maximizing policies. The procedure is admirably summarized in Posner (1973) and can be characterized as follows:

(a) prepare medium-term forecasts of the demand for energy and, within this, for component fuels;
(b) identify the investment paths required to meet this demand in each fuel industry;
(c) identify the efficient pricing policy for each fuel (using the marginal cost

principles outlined in the White Papers and discussed in Chapter 7 by Slater) and check the consistency of the planned investment path;
(d) check the path of relative prices against the demand forecasts, and iterate the process until the forecasts, prices, and investment paths are consistent.

The function of prices, in this framework, is to provide a medium-term signal to consumers and a bench-mark against which to evaluate investment plans. Unanticipated short-term mismatches between demand and supply are remedied not through adjustment to prices but through the under-utilization of capacity or through rationing (in practice usually the former for reasons discussed further below). Under this framework a medium-term view is taken by government of the likely future path of comparative advantage of the different fuels, and it is assumed that an orderly substitution can be implemented.

If this is implemented, then the consequences should be that consumers face prices which are stable in the short term and which give appropriate medium-term signals to inform consumer investment (in purchasing appropriate heating systems, cookers, etc.). Energy supply is provided by the most efficient mix of fuels produced using the most efficient technology, and prices are set at efficient levels. This requires, however, that demand forecasting is effective and the industries have incentives to implement the successive steps. In particular, they must produce the required level of output at efficient cost levels and set prices in relation to these costs.

Newbery (Chapter 2) assesses the success of energy planning by comparing forecasts prepared in 1972 for the year 1980 with the actual out-turn in that year. The differences between forecast and out-turn are striking and lead Newbery to conclude that policies built upon forecasts so prone to error are unlikely to be efficient. It is clear from Chapter 4, however, that the technical standard of the Department of Energy's forecasts is high and that the forecasting models are able to track closely the relationship between the demand for energy and the underlying determinants of demand, in particular economic growth. The forecasting failure identified by Newbery reflects both the sensitivity of energy demand to GDP growth (and the difficulty in accurately forecasting growth over the medium term) and a failure to achieve planned substitutions between energy sources (in particular, nuclear power for coal).

These problems are compounded by the probability that nationalized industry managers have concerns other than the maximization of social welfare. There are many objectives which nationalized industry managers may choose to follow in preference to welfare maximization, in particular output maximization (Rees (1984a)), expense preference (Williamson (1963)), and managerial slack.

Rees (Chapter 5) compares the performance of the electricity industry with that predicted by a model in which the firm maximizes output subject to constraints, set by governments, concerning the minimum level of profits and the maximum level of capital expenditure, and where the allocation of labour resources constrains labour bargaining power. Key features of performance accord with the predictions of the theory: wages are above average (for relevant skill groups), reflecting the exercise of union bargaining power in a monopoly industry, and investment plans have been consistently, and substantially, over-optimistic.

More generally, a specification of the behavioural objectives of public enterprise managers in the framework of output maximization or expense preference, subject to a government-imposed profit constraint, suggests that in periods when such constraints are weak, enterprises will fail to achieve both productive and allocative efficiency. In periods when the financial constraint is binding, however, this suggests that the achievement of technical efficiency will not be a serious problem and that the main failures will be allocative—both in relation to choice of production technique and in relation to pricing.

Pryke's assessment (Chapter 6) of the energy utilities' performance in the 1970s confirms that the weakening of financial constraints which resulted from the counter-inflation policy introduced in 1972 was followed by deteriorating performance in relation to productive efficiency. Analysis of developments since the 1978 White Paper, which elevated financial constraints to the centre for nationalized industry regulation, shows a sharp upturn in performance in relation to productive efficiency (see Molyneux and Thompson (1987)). The various chapters in this volume suggest that the most important failings in the public sector (or recently public sector) energy utilities are allocative. In the case of gas, prices are set below efficient levels in a way which is consistent with output maximization (see Hammond, Helm, and Thompson (Chapter 15 and 1985)). The over-ambitious investment plans of the electricity supply industry (see Rees (Chapter 5) and Jones (Chapter 12)) are also consistent with this goal. The output path in the coal industry, in which production has, until recently, continued at individual locations with supply costs well above current or prospective prices, also indicates significant allocative inefficiency (see Robinson (Chapter 17)). In these industries, however, there is little evidence of technical inefficiency (see Molyneux and Thompson (1987)). Most of these industries are characterized by uniformities in the structure of prices (between different geographic regions or between time periods) which fail to reflect variations in economic costs.

Thus regulation of public energy utilities in the UK has failed in a number of ways. The Morrisonian model—in which public enterprises are assumed to act as welfare maximizers—is obsolete. It is also clear that the system that replaced it—which essentially imposed rules or objectives but not constraints—also failed to ensure that productive or allocative efficiency was achieved. It was this system which effectively guided energy policy in the 1960s and 1970s, and the adverse consequences for efficiency are documented in various of the papers in this book. It does seem, however, that the redirection of nationalized industry policy in the 1978 White Paper—in which financial constraints and related efficiency targets were introduced—has had a beneficial impact, at least in relation to the achievement of productive efficiency.

The main efficiency failures in the energy sector are now allocative. The reasons for this relate both to the absence of product market competition and to specific regulatory failures. Particularly important are the inefficiencies in pricing and output paths in the coal industry which have been sustained from the earliest days of nationalization because of the bargaining strength of the miners and because British Coal and the CEGB, its main customer, are both public sector monopolies.

Cross-subsidization between energy industries (in particular from electricity to coal) has been accompanied by uneconomic cross-substitution within industries. Asymmetries in information between government regulators and monopoly industries have meant that requirements to relate the structure of prices to relative costs have been in some cases largely irrelevant. This summary thus suggests a fairly well-defined mix of successes and failures, and one which can be related directly back to the likely incidence of regulatory failure. This provides a frame of reference for analysing the recent reform of policy.

4. The New Market Philosophy

The *laissez-faire* economic philosophy of the incoming Conservative administration in 1979, and increasing dissatisfaction with the performance of public enterprises, led to the explicit attempt to opt for a market solution which abandoned the main features of the Post-War Consensus on energy policy. In 1982 Nigel Lawson, the then Secretary of State for Energy, set out the new objective—to create a market for energy (see Chapter 1).

What would a Market for Energy Look Like?

In a competitive energy market, *production* is carried out by many separately owned firms. There are no statutory restrictions on market entry other than general environmental planning requirements and those related to health and safety. A competitive market in the production of fuel is thereby established.

However, the infeasibility of competition in the *distribution* of some fuels (because natural monopoly exists, especially in electricity and gas) provides the opportunity for prices to be raised above efficient levels. Local distribution networks are separately owned, however, and a regulatory ceiling is fixed on the distribution charges for each area. This ceiling is set in a systematic relationship to existing charges in all other areas, thereby preventing significant exploitation of consumers, but also providing opportunities for distribution companies to increase profits by beating the average level of performance of all companies.

Distributional concerns relating to the energy sector are dealt with directly through the tax and benefit system. The development of natural resources is left to the market, but resource taxes are used to achieve the desired distribution of resource rents between producers, consumers, and the government. External costs and benefits are dealt with through regulation (for example, on emissions) or through specific taxes and subsidies.

Under this framework, prices perform a short-run allocative function. Consumers with a high preference for price stability can achieve this through purchasing of futures via long-term contracts. The path of output is determined by the investment decisions of the various market participants, which are based on their perceptions of the future path of costs and prices. The mix of fuels supplied is therefore essentially market-determined. Because these decisions are formulated in private markets, this policy framework is assumed to provide incentives which ensure that efficiency is achieved rather than simply planned for.

Evidently such a market approach is very different, in appearance and probably in consequence, from the structure which prevailed in 1980. It is also very different, in appearance and probably in consequence, from the structure which prevails in 1988.

The major legislative components of the new Lawson policy have been:

(a) the 1982 *Oil and Gas (Enterprise) Act* which ended British Gas's statutory monopoly in gas supply and distribution, forced the British Gas Corporation (BGC) to dispose of its oil interests (as Enterprise Oil), and provided for common-carrier provision and hence competitive entry;
(b) the 1983 *Energy Act* which extended the principles of the 1982 Act to the electricity supply industry. Most notably, it provided for the compulsory publication of the prices that Area Boards would pay for privately generated electricity (the private purchase tariffs) and tariffs for rent of the network;
(c) the 1986 *Gas Act* which privatized BGC, transferring it to the private sector as a single company subject to regulation by a newly created authority, the Office of Gas Supply (OFGAS).

We will consider how, and in what ways, these legislative initiatives have fallen short of the creation of a market for energy in two stages. First we consider the attempts to introduce competition into the production of energy, and then we consider the regulation of privatized energy utilities.

Competition and Regulation in Practice

The central policy dilemma in creating a competitive market was and is the selection of an appropriate liberalizing strategy. Much of the energy sector remains characterized by natural and artificial monopoly. Whilst the former type of monopoly requires careful regulation, the latter demands attention to the enhancement of competitive pressure: markets need to be liberalized and the terms of entry for rivals set to prevent artificial obstruction from barriers to entry.

Thus, in order to appreciate the impact of the 1982 Oil and Gas (Enterprise)

Act and the 1983 Energy Act, we need to look at the underlying entry-preventing strategies.

Any liberalizing strategy must be based on a prior view as to the degree of potential entry and the sorts of entry barriers which may inhibit it. As we noted above, the energy sector industries combine elements of both natural and artificial monopoly. Competition is possible only in the potentially competitive segments of the industry, and the natural monopoly elements should therefore be separated and subjected to regulation.

Non-natural monopoly elements are open to competitive entry. The dominant incumbent can, however, employ a number of strategies designed to create and sustain an artificial monopoly. In addition to the statutory barriers to entry, recent industrial theory has highlighted a number of strategic activities by which dominant firms may deter rivals. As applied to the electricity supply industry in the UK, Hammond, Helm, and Thompson (Chapter 8) show how these strategies may deter entrants and hence may have undermined the liberalization intentions of the 1983 Energy Act. The principal barriers they identify are excess capacity, entry costs, the bankruptcy constraint, and the strategic choice of objectives.

Where an industry is characterized by excess capacity, the potential entrant alters its expectation about the response rivals will give to entry. It will decide to enter only if the rivals' response is sufficiently muted to leave profitable opportunities for the entrant. If the incumbent has spare capacity, the cost to it of a loss of market share will be greater than if it were at full capacity. Thus entry is more likely to produce retaliation, and hence make entry less attractive.

Entry costs are of two forms—fixed costs and sunk costs. Fixed costs were dealt with under natural monopoly. Sunk costs are ones which are irrecoverable by the entrant, should it subsequently decide to leave the market. As these increase, the incentive to enter declines because the size of the initial investment at risk, and hence the costs of failure, rise. In the energy sector, the principal sunk costs relate to planning the entry decision, acquisition of energy skills, and the imperfection of the second-hand market for pipes, generating capacity, and so on. Much investment in the energy sector is specific.

The bankruptcy constraint is of considerable importance in the energy sector, because of the considerable presence of the public sector, and thus the fact that debt is underwritten. The potential entrant perceives that in a price war, the public sector incumbent is better able to withstand the short-term loss of profitability and hence is more likely to win. The existence of financial resources and a weak bankruptcy constraint is therefore a credible threat to the entrant that the incumbent would find it worth while to challenge the entrant, and hence reduces the entrant's expected return. This financial or bankruptcy barrier to entry is reinforced by another aspect of State ownership, namely the impact of managerial objectives on rivals' entry decisions. As we noted above, the separation of ownership and control encourages the presence of managerial rather than profit objectives. Following Rees (1984a and Chapter 5 below), we would expect an output-maximizer to lower prices and total profit to gain extra market share. This has a dual impact on potential entrants. The potential profit to the entrant is lowered because the nationalized firm lowers the general level of prices in the industry *and* is more likely to retaliate to loss of market share to the entrant, because market share more directly affects output than it does profits.

The 1983 Act did force the industry to publish the prices which would be paid to private sector producers (private purchase tariffs—PPTs) and the prices to be charged for the rental of the network (the common-carrier or network charge). The three problems with this liberalization were the setting of the tariff levels by the dominant incumbent, the absence of longer-term contracts, and the failure to set up an independent body to review the administration of these tariffs. The fundamental impact of public ownership was not altered by the 1983 Act: control of the tariffs enabled the excess capacity to remain, and its costs to be passed on to consumers as higher prices without encouraging entry, because entrants received only 'avoidable' variable costs. New entry is rarely promoted by letting the dominant incumbent set its rivals' price.

The dominance that the incumbents exercised could be addressed through restructuring. Such an opportunity presented itself in the privatization of BGC. The options for breaking up the industry are set out by Hammond, Helm, and Thompson (1985), but due to political pressure, concern for sale returns, and the crucial role of management, the industry was not broken up. The recent White Paper (Department of Energy (1988)), however, proposed greater restructuring of the electricity supply industry.

Enhancing competition through liberalization thus achieved very little in the UK electricity and gas industries, and government attention shifted towards improving the productive efficiency of existing dominant firms. The government argued, as we noted above, that changing the ownership of nationalized industries would in itself improve efficiency, subject to an appropriate regulatory structure. So far, a detailed regime has been developed only for British Gas.

The regulation of private energy utilities in the United States, where private ownership of energy production is most common, has been subject to well-known difficulties. The most usual regulatory instrument—rate-of-return, or rate-base, regulation—involves the specification of price ceilings which are based on the enterprises' actual costs and which provide for a pre-specified 'fair' return on the enterprises' capital assets. This system provides no additional profit in improving efficiency and reducing costs. Conversely, any increases in costs can be passed directly on to consumers in higher prices. The efficiency incentives usually associated with private ownership are therefore almost completely eliminated. Furthermore, because the price ceiling is determined to provide a specified return on capital assets, there is an incentive for profit-maximizing firms to adopt production techniques which are too capital-intensive (the Averch–Johnson (1962) effect). There may be also be adverse incentives for the efficiency of pricing policy with incentives to underprice capital-intensive (for example, peak) outputs (see Sherman and Visscher (1982)).

This regulatory failure is reflected in studies which have compared the costs and efficiency of regulated private utilities with those of similar publicly owned enterprises. Generally the findings of these studies show either no systematic differences in performance between the public and private sector or a differential which marginally favours the publicly owned enterprises (for reviews, see Millward (1982), Domberger and Piggot (1986), and Yarrow (1986)).

The underlying cause of this regulatory failure lies in the asymmetry in information between the regulatory authority and the regulated enterprise. If the regulatory authority knew what level of costs constituted efficient performance by the enterprise, then it would be able to structure the regulated price ceiling accordingly. In the absence of this information, prices are regulated in relation to achieved costs, with the adverse consequences for efficiency which have been noted. It can be seen that the nature of this regulatory failure —asymmetries of information in the achievement of efficiency between a monopoly enterprise and the relevant regulatory authority—is essentially the same as that discussed earlier in relation to UK nationalized industries.

The regulation of newly privatized enterprises in the UK has, however, been founded on an ostensibly different basis. The 'RPI $-X$ formula', as it has become known, was recommended in the Littlechild report (1983) for the regulation of the privatized British Telecom. It has subsequently been adopted in the cases of the British Airports Authority and British Gas.

The details of the British Gas formula ('RPI $-X+Y$') are complex but the essential elements are straightforward (see Price (Chapter 14)). The formula places a ceiling on the prices which BGC is permitted to charge its domestic customers. This ceiling allows BGC to pass on directly to customers any increase in the cost at which it purchases gas (Y) but requires that its non-gas costs (that is, labour costs, expenditure on materials, etc.) can only increase by the general rate of inflation (RPI) minus a factor X.

The basic principle underlying the 'RPI $-X$' system is that, with revenues constrained by the formula, the regulated enterprise's profitability will be determined by how effectively it controls its costs. There are thus rewards and penalties in relation to the achievement of efficiency which are largely absent in the case of rate-base regulation.

In theory, at least, the management of an enterprise which failed to achieve productive efficiency would find itself replaced (via take-over or at a shareholders' meeting through the actions of shareholders seeking to maximize their return).

The proper test of this regulatory system will lie ultimately in the performance (in terms of costs and efficiency) of the newly privatized enterprises. It will clearly be some years before any, even preliminary, verdict can be reached on this. Nevertheless, it is already clear that the practical implementation of

$RPI - X$ falls some way short of the idealized conditions necessary to generate the favourable incentives to efficiency discussed above (see Price (Chapter 14) and Yarrow (Chapter 10)).

These enterprises are not profit-maximizing in any usual sense. The limits to the enforcement of profit maximization are well known (see, for example, Helm (1988)) and in the case of the newly privatized utilities their size, market power, diffusion of shareholding, and immunity from take-over make them very different from the textbook model of the profit-maximizing firm.

Information necessary to assess efficiency appears, if anything, less readily available than before. If neither shareholders nor customers can monitor this, however, then the effectiveness of the $RPI - X$ system will turn on whether or not the regulatory authority can determine what constitutes an efficient level of performance, and may design the regulatory price ceiling accordingly. In capital-intensive industries, however, if the price ceiling is set too 'tightly' then the consequence is likely to be under-investment and, eventually, supply failure (see Yarrow (Chapter 10) for a discussion).

The regulator's task is thus not straightforward. Yet it is clear that unless the regulatory authority can form a judgement on the enterprise's efficiency which is largely independent of actual costs and performance, then the $RPI - X$ system starts to become very similar, in practice, to rate-of-return regulation, with all its associated weaknesses. This will be particularly likely if there is concern to avoid supply failure in capital-intensive industries.

One method of generating such an independent judgement is to make comparisons of the performance of particular activities in different geographic areas. Yarrow (Chapter 10) outlines a system in which each local distribution network for electricity supply (the 12 Area Boards) is under separate ownership. The regulatory price ceiling applied in each area can be established by reference to the costs and performance of the other Area Boards. Similarly shareholders will be provided with comparative information on their company's performance. Effectively the system establishes 'yardstick' competition (Shleifer (1985)) in which, although each distribution network is a natural monopoly, the regulatory framework requires each company to match the efficiency of other distribution companies if it is to maintain normal levels of profitability.

This 'regionalized' solution was not, however, adopted in the case of British Gas. Nor does it appear that the regulatory framework has established the information base upon which such comparative analysis could be carried out. There must therefore be very limited grounds for optimism that the regulatory system established for the privatized BGC will be effective in improving the efficiency of its activities.

The new Market Philosophy has thus far failed in its ambitions. The energy sector has not seen significantly enhanced competitive pressure, and the privatized British Gas has survived with its integrated monopoly intact, and subject to less control than when in the public sector.

It remains to be seen whether, following the recent White Paper (Department of Energy (1988)) on electricity, more effective competition can be introduced in that sector and whether the Monopolies and Mergers Commission's report on British Gas's pricing policy (MMC (1988)) will result in a more competitive gas industry.

The fact that the new policy has, to date, largely failed does not imply that it is infeasible. Many other factors have steered the privatization programme away from the competitive objectives (see Kay and Thompson (1986)). In the next section we suggest a number of factors which would enhance the liberalization strategy and form the components of a more competitive energy policy.

5. The Way Forward

In this introductory chapter, we have reviewed the institutional background and regulatory framework which have evolved in the post-war period, and we have examined the policy changes instituted since 1979. We have suggested that the rationale of energy policy lies in the identification of underlying market failures and the corresponding government failures that arise in corrective policies. The success of energy policy depends upon a proper analysis of such failures and a policy designed to intervene in the light of such failures.

In analysing the market for energy, we have focused on the two types of major failure—natural monopoly and artificial monopoly. The appropriate policy responses were then set out—regulation to prevent abuse of natural monopoly, and competition policy to ensure fair entry terms where competition is feasible. The creation of a market for energy depends upon the setting of these two policies.

Though it has been thought that nationalization 'solves' natural monopoly by replacing profit maximization by the pursuit of social welfare, it must now be relatively uncontroversial to claim that it in fact 'solves' very little in itself. The naïve view which dominated thinking in the early post-war period was clearly mistaken, and the painful attempts at control in the 1960s and 1970s reinforced this observation. Only effective regulation can mitigate the abuse of natural market dominance.

If ownership change via nationalization failed to 'solve' the monopoly problem, the lesson has hardly been learnt for privatization. Fundamentally, the privatization debate in the UK has focused on the wrong question—the appropriate competitive and regulatory structure is a more important determinant of performance than ownership. The structure of regulation and competition is the central issue, not a secondary consequence of privatization.

The important issue for monopoly is regulation. $RPI - X(+ Y)$, we noted, suffers from a number of serious drawbacks. The emphasis on prices rather than costs is more apparent than real, and freedom to set individual prices to exploit monopoly power or support predatory intentions remains. Performance would be improved by placing targets on monopoly sectors as separate cost centres, rather than on the industry as a whole.

This location of targets gives rise to two further regulatory questions—how much of what sort of information ought to be produced, and what should the optimal structure of the firm be? Cost and profit centres should be located according to the underlying characteristics of the industry. Natural monopoly elements should be separated from potentially competitive ones, and the latter broken up into competing units. Hence individual coal-pits, individual power-stations, and individual oilfields ought to be separated. This is the rationale behind the article on coal by Robinson and Marshall (Chapter 16), Hammond, Helm, and Thompson on gas (Chapter 15 and 1985), and Yarrow's contribution on electricity (Chapter 10), and, for the gas and electricity industries, the national network/grid should form a separate company, and the Area Boards each additional companies.

Information for regulation produced competitively gives regulators appropriate 'yardsticks' by which to compare differential performance. Thus the electricity regulators could compare Area Boards, pit costs could be contrasted, and gas showrooms ranked by performance. The regulator's need for information gives rise to the potential for his capture by the industry upon which he relies for his information. The wider the range of sources, the greater the independence and effectiveness of the regulator. If the regulator himself is independent of the industry, his position is enhanced. Allowing the Department of Energy to function as the sole monitor of performance is one of the central drawbacks of nationalization.

Thus a market for energy requires identified inescapable monopoly elements to be set up, ideally as separate firms, but at least with separate accounts. It employs targeted price regulation, set and monitored by independent regulators.

We have argued that monopoly regulation needs to be supplemented by appropriate measures to stimulate competitive entry, for it is competition, not ownership, which is most likely to stimulate efficiency gains. Competition does not, however, arise spontaneously. Where the incumbent retains substantial market power, regulation for entry is required. The regulator needs to set and referee entry conditions to ensure the incumbent does not create and exploit strategic advantages over rivals. This latter point is especially important where the incumbent is large—for example, in the cases of British Coal, the CEGB, and BGC.

Competition in the energy sector typically comes from three sources—within the industries, between the industries, and internationally. Current energy policy has failed to give adequate incentives and safeguards to entrants in each of these areas. Competition within the industry depends upon the actual and expected level, structure, and revision of prices. These in turn depend on the institutional method by which they are set and revised. As with monopoly regulation, independent rather than incumbent methods are more conducive to competition.

Competition between industries is typically unregulated, and argued to be intense in certain markets. Indeed, it is this sort of competition, or its absence, which has been the gauge of monopoly and the need for regulation. Thus the domestic gas market is regulated, but not the industrial one. However, a precondition for effective competition is consistent regulation of each of the industries considered separately. If, for example, the coal industry charges above marginal costs to the electricity industry (see Robinson and Marshall (Chapter 16)), the electricity industry ups marginal cost pricing taking the distorted coal price as exogenous, and the gas industry charges below marginal costs, then it follows that electricity/gas competition in the industrial market is unlikely to allocate resources efficiently. The criterion for inter-industry competition is consistency between individual industry regulation. Given, therefore, the framework devised for British Gas, the options for electricity regulation are heavily constrained.

Competition may also be international. Increasingly, electricity and gas are becoming traded European commodities. Firms and utilities would, in a liberalized energy market, choose the cheapest source of supply. In electricity, in practice, this implies that France and Scotland could provide powerful checks on monopoly profit for an English CEGB. The key to competition here is access to a network of common-carrier provisions.

A competitive energy policy thus has three components: a structure which follows the natural characteristics of an industry and its market failures; a regulatory system to control natural monopoly elements and enhance entry conditions; and an information system to allow the monitoring of performance. The choice of industrial structure strongly influences the required degree of regulation for competition and quality of information. The less competitive the structure, the greater the need for compensating regulation for competition. To date, restructuring has been largely avoided, $RPI - X$ suffers from a number of identifiable drawbacks, entry conditions have been inadequately addressed, and the dominant firms have effective informational monopoly. The government has a long way still to go to attain a competitive 'market for energy'.

Part I

Energy Policy

1 Energy Policy*
Text of a speech given in June 1982 by

THE RT. HON. NIGEL LAWSON

Unlike economic policy, energy policy clearly means very different things to different countries, depending on their individual resources and circumstances.

It means one thing to Saudi Arabia, quite another to Japan. Even within western Europe, the differences are more marked than the similarities. Norway, for example, with much of her electricity generated by hydropower and huge resources of oil and gas, has little concern over the cost of energy and even less over security of supply, but has to be very wary of allowing too rapid a development of oil and gas to dominate and distort the rest of the economy. Germany, with its massive dependence on imported energy, has to balance the strategic and economic risks of different levels of dependence on the Middle East and the Soviet Union. For the United Kingdom with its own indigenous supplies of all the fossil fuels, and a highly developed and diversified economy, the pre-eminent objective must be to ensure that the vitally important energy sector functions as efficiently and effectively as possible within the context of economic policy as a whole.

There are of course a limited number of agreed common objectives in the energy field. The Venice Declaration of 1980, for example, called on all seven Economic Summit nations to break the link between economic growth and oil consumption through conservation and by developing alternative energy sources. Although primarily motivated by consumer concerns, this Declaration was welcome to many of the oil-producing countries, particularly those that take a longer view. The measures agreed at Venice have subsequently been endorsed and built up both by Member States of the European Community and by the International Energy Agency. The UK has shown an impressive lead in this. Over the past two years, our oil use in relation to overall economic activity has fallen by 17 per cent.

But in general, as Secretary of State for Energy in the UK, I do *not* see the government's task as being to try to plan the future shape of energy production and consumption. It is not even primarily to try to balance UK demand and supply for energy. Our task is rather to set a framework which will ensure that the market operates in the energy sector with a minimum of distortion and that energy is produced and consumed efficiently.

* This chapter forms the text of a speech given by Nigel Lawson at the Fourth Annual International Conference, International Association of Energy Economists, Churchill College, Cambridge, on 28 June 1982. It is Crown copyright and is reproduced here with the permission of the Controller of Her Majesty's Stationery Office.

At the time of giving this speech, Nigel Lawson was Secretary of State for Energy.

Energy pricing is one key to this approach, in relation to both production and consumption. If energy prices are set too high, producers will be encouraged to invest in new capacity for which they may not be able to find a market. If energy prices are below economic levels, then energy will be used wastefully and consumers will be encouraged to invest in inefficient energy-intensive processes.

But what constitutes economic pricing of energy? Where there is a genuine market—as in oil—it is the price set by the market. Where there is no genuine market—as in electricity—prices will need to reflect costs of supply. Within this general concept there is clearly some room for flexibility, for example in response to the pressure of international competition on our industries. Hence the £250 million worth of concessions to industrial energy users, especially those with high load factors, in the last two Budgets.

Realistic pricing is a stimulus to efficiency. But its impact is muted because so much of the UK energy sector is composed of State-owned monopolies. How then can we improve the efficiency of the energy industries? The key lies in increasing the responsiveness of these industries to the forces of the marketplace.

We have made significant progress in this area over the past three years. The changes that are in prospect will further enhance the role of private enterprise and stimulate the action of market forces, thereby increasing efficiency and helping us to ensure that the supplies of fuel we need are available at the lowest practicable cost.

I shall return later to this subject. But there is something even more fundamental. This is to recognize, as governments have not always done in the past, that for the most part energy is a traded good.

Primary fuels can be imported and exported. For oil, the world market is well established, and although international trade in gas and coal is on a much smaller scale, it is building up. There is neither need, nor particular virtue, in having domestic production equal to consumption. The key to energy policy is flexibility. We should use our ability to import or export fuels at the margin to the best advantage in the context of an ever-changing world energy scene.

In seeking to achieve this, it does not help us very much to try to guess the unguessable—namely, what UK energy consumption will be in twenty, let alone fifty, years' time—and then aim to produce this amount judiciously divided up between the primary fuel sources.

We will do far better to concentrate our efforts on improving the efficiency with which energy is supplied and used, an objective that will remain valid and important whatever the future may bring. This means, among other things, that public sector energy investment decisions should in general be based not on a simple-minded attempt to match projected UK demand and supply, but rather, as in the private sector, on whether the investment is likely to offer a good return on capital. If these decisions are well based then the importing and exporting of fuels will match production to consumption on an economic basis. This does not mean that we can use imports and exports of primary fuels as a simple safety-valve. We cannot turn them on or off at will. But, as international trade develops, we can expect to see our energy supply industries acquiring an additional degree of flexibility in responding to changing market conditions. And at the same time the possibility of exploiting monopoly power to raise prices will be progressively reduced.

Within this overall approach, electricity poses special problems. With the development of appropriate infrastructures, coal, oil, and gas can be stored, or traded, to a sufficient extent to provide market disciplines and supply flexibility. This is not true of electricity. For many of its uses, there are no acceptable substitutes and, except for insignificant amounts at the margin, there is no flexibility for dealing with under- or over-supply through trade.

So the electricity supply industry, unlike the coal and oil industries, has a duty to ensure that there will be sufficient plant available to meet the top end of the range of most likely demand requirements.

But even in the case of electricity supply investment, reducing costs, to help improve the efficiency of supply and hence the efficiency of the economy, is at least as important as investment to meet projected demand. In this respect the electricity supply industry differs not at all from the industries involved in the supply of primary fuels. Diversification is also of particular importance for the electricity supply industry—a direct consequence of its limited ability to import. As many have found to their cost before, there are dangers in becoming too dependent on any one source of fuel.

In this context there is increasing public interest in the renewable sources of energy. The renewables undoubtedly offer considerable potential. They may well have a key role in the energy economy of the future, and the research and development work that the government is sponsoring is designed to evaluate this. But at their present stage of development, it is unlikely that they will be able to make a sizeable and economic contribution to energy supply this century.

So nuclear power is critical both to diversification and to reducing costs. It is significant that the Central Electricity Generating Board (CEGB), in applying for permission to build a pressurized water nuclear reactor at Sizewell, has explicitly based its case on cost and flexibility grounds. The specific decision on Sizewell has yet to be taken [as of June 1982]. But the government believes in general that if nuclear stations can be built to time and to cost, they can play a significant role in helping to keep down the price of electricity in the UK, thereby helping our manufacturing industries in particular to increase their competitiveness.

This is bound to take time. But we can see by comparing the UK with France what the potential advantages are. In this country nuclear power accounts for some 13 per cent of all electricity generated; in France the figure is about 40 per cent. This has given important parts of French industry a substantial advantage so far as its energy costs are concerned, and the disparity is unlikely to diminish for some considerable time.

In terms of meeting demand, the electricity supply industry has to look many years ahead. Power-stations take upwards of seven years to construct and have lifetimes of approximately thirty years. So the effects of decisions made now may still be apparent in forty years' time. However unknowable the future may be, these decisions still have to be taken—and taken on the most rational basis attainable, given the electricity supply industry's need to be able to meet economically the demand that will be made of it.

Let me say a word at this point about forecasts and their more modest and modern successors—scenarios and projections. In the last twenty years—the period in which forecasting has become a major industry—we have seen some startlingly wrong predictions. For example, the Electricity Council in the 1960s and early 1970s produced forecasts for maximum demand seven years ahead which were never less than 20 per cent too high and in one year were about 50 per cent out. It is only fair to add that the Electricity Council was by no means unique in this—forecasting errors are the rule rather than the exception—and that its error was in no small measure due to highly optimistic forecasts of economic growth provided by the government of the day.

By treating energy as a traded commodity, we greatly reduce the need for, and importance of, projections of UK demand and production. But for two reasons we still need to make such projections. First, for reasons I have given, the electricity industry has a special need to match capacity with likely demand, and the government, which provides the finance for the industry's capital investment programmes, needs to take an independent view of that likely demand. But it is only possible to form a sensible view of likely electricity demand in the context of demand for all the fuels. Second, while projections provide an unsure basis for planning, they can give rise to useful questions about the coherence of policy. It is for those two reasons that my Department is now preparing a revised set of energy projections which we expect to publish in the autumn [of 1982].

It is on the demand side of the equation that most of the errors in projections arise. The reason is a simple one. The demand side of the equation is an aggregation of decisions by millions of individual and corporate users. These range from the insulation of a domestic hot water tank to the installation of a large industrial boiler. Consumers have, since the first oil crisis of 1973-4, become increasingly aware of the need for using energy more efficiently, and conservation has become an important factor.

There is a tendency to talk of conservation as an alternative to supply. But this is misleading. Conservation is in no sense a source of energy. Rather, it is a lever on demand—a way for the consumer to cut his costs. But these decisions by consumers are often small, and always disaggregated, whereas supply investments are large and centralized. These two types of investment decision are taken by different sets of people using widely differing criteria.

The question for government is how these two sets of decisions can best be brought together. Certainly not by central planning. It is unlikely in the extreme that we would be better off if decisions about insulating millions of homes, building power-stations, operating oil refineries, and distributing gas and coal were all made within Whitehall.

Nor should the government seek to achieve overall and detailed control through the back-door methods of regulations and subsidies. The government's role is neither to induce the individual to take decisions against his better judgement, nor to waste public money in subsidizing investment that is already well worth while. The way to bring the two sides together and to ensure that they act consistently is to give them the same information and the same realistic signals. On the demand side, the UK government supplements the messages given by economic prices by improving the flows of information—for instance through advisory services and demonstration projects. The key here is not the amount of public money spent, but putting the message across properly. Initial results from the Extended Energy Survey Scheme (under which firms can get grants towards a survey of how to improve their use of energy, including the scope for combined heat and power) show energy savings equivalent to about £16 for every £1 invested by government. These modest schemes can help point up the messages given by the market and can accelerate the industries' responses. But above and beyond all this, the main spur must be competition, encouraging the consumer to cut his costs and the supplier to become more responsive to the customer's needs.

This is vital. The economy is now in a fundamentally healthier position than it has been for many years. If British industry is to make the most of this, it must, as an energy consumer, do all it can to increase the efficiency with which fuel is used.

I said earlier that I would return to discussing the efficiency of the energy supply industries. Most of the supply industries in the UK are State-owned. In part, State-ownership came about as a means of regulating natural monopoly. But this has not always been the case; and it is time to question both the extent of the natural monopoly and, where it can be shown to exist, the most effective means of regulation. State-ownership is neither a universal necessity nor the only means of regulation.

The Oil and Gas (Enterprise) Act, which reached the statute-book today, will enable us to establish Britoil, the greater part of the British National Oil Corporation (BNOC), as an independent private sector oil company in its own right, and will very significantly increase the competitive pressures to which the British Gas Corporation is subject. For the first time ever, there will be the prospect of competition for the custom of all gas consumers taking over 25,000 therms a year. This will provide a competitive spur not only for the Gas Corporation but also for all the potential suppliers in the private sector. We shall shortly be introducing legislation to encourage the supply of electricity by the private sector.

Where we can neither privatize nor introduce real competition, we have to do our best to simulate market disciplines. The external financing limits on the nationalized industries have a crucial role to play in this respect. We are now reinforcing control by setting clearer objectives for the State-owned industries and, wherever relevant, setting performance targets. For example, British Gas has been asked to reduce its costs—other than those represented by the purchase of gas—by 5 per cent before next April [i.e. April 1983].

We are also promoting external appraisals of the nationalized industries. Management consultants are being brought in for efficiency examinations of the Atomic Energy Authority and of British Gas. The Monopolies and Mergers Commission (MMC) report on the CEGB is already leading to much-needed changes in the way the Board evaluates investment projects. The National Coal Board and two Area Electricity Boards are now to be given a similar examination by the MMC.

There is, of course, one major energy industry that is fully subject to the disciplines of the market. For North Sea oil, where an eighth round of licensing has recently been announced, we have a genuinely free market approach. This is most unusual among the oil-producing countries of the world, among which the UK currently ranks fifth. But in fact it has always been the case, under successive governments, and our removal of BNOC's special privileges has merely reinforced this important fact.

The price of North Sea oil is determined not by government fiat but in response to market forces, and North Sea producers are free to produce as much oil as they wish. Even the much maligned North Sea fiscal regime is highly price-sensitive. We have made sparing use of our powers to control depletion and the rate of production, and have concentrated our regulatory intervention on minimizing wasteful losses through flaring.

But we know that our supplies of oil are limited. Sooner or later we shall have to make the transition from net oil exporter to net oil importer. The obvious question is whether government should act to defer some of the expected surplus of the next ten years or so, to help fill the gap that may start to emerge some time in the latter half of the 1990s.

At first glance the answer may seem equally obvious: of course government should ease the way. But behind the simple façade hide a host of complexities. Prospects for UK demand and supply, and for the world price of oil over the next ten or twenty years, are highly uncertain. We also have to consider whether action now would have any real economic justification. We have to consider whether it would damage either our more immediate prospects of general economic recovery or our longer-term objective of maximizing the economic exploitation of the North Sea over time.

We have to ask ourselves whether we are really so unenterprising as not to be able to put to good use the wealth which derives from oil, whenever it arises. That wealth is substantial. Last year UK Continental Shelf oil and gas production amounted to some £10 billion or 4 per cent of gross national product (GNP), without taking any account of the offshore supplies industry that rides on its back.

During the course of this short talk today, I have not sought to deliver a lecture on something known as energy policy. Rather, I have tried to explain how I see the development of our vitally important energy resources and our energy market fitting within the wider economic objectives that the government has set itself. This approach differs from that of previous administrations. It is one that many people with an interest in energy have perhaps been slow to understand. I am grateful to you for giving me this opportunity to explore my thinking with you.

2 Energy Policy Issues after Privatization*

DAVID NEWBERY

1. Introduction

Before 1973 all indigenous fuel was supplied by nationalized industries and the need for energy policy was seen as equivalent to the need for a policy for these nationalized industries. The present government has expressed the view that nationalized industries should, where possible, be transferred to private ownership, and is taking steps to achieve this for some of the fuel industries. The government has on occasion given the impression that this change of structure, from one of government control to one of market responsiveness, means that there is no longer any need for an energy policy as such. The purpose of this paper is to investigate how energy policy should respond to the profound changes in the organization of the energy sector.

I shall begin with a brief review of how energy policy was perceived before the 1973 oil shock, and the lessons learned from subsequent experience, both in terms of the issues that were thought to be important, and the theories that were argued to be relevant to its design. The period from 1973 to 1983 was a testing one for energy policy, but recent events in the oil market suggest that the future may be almost as unpredictable now as it was in 1972 (though we may now be more aware of the inherent uncertainties). How well did past energy policy fare in the face of unexpected shocks, and how well equipped is current policy to deal with similar shocks? This brings us to the present, and the need to rethink energy policy in the light of experience and given the changes in market structure. The four questions raised by these changes are: what the government's policy on competition in the energy sector should be; how the privatized industries should be regulated; how the remaining publicly owned industries should be instructed to behave; and finally, what changes in taxation are now desirable?

Economic policy consists of choosing whether and how to intervene in the economy to improve its performance measured by certain criteria. Energy policy is that part of economic policy that affects the production, supply, and use of energy. It follows that energy policy cannot be considered independently of economic policy, in particular the structure of taxation. The government has at various times implied that energy policy (as opposed to the lack of policy)

---

* This paper was presented at the conference 'Energy Policy' organized by the Institute for Fiscal Studies and the Economic and Social Research Council Industrial Economics Study Group in London on 1 May 1986. It was published as Centre for Economic Policy Research Discussion Paper 138 in 1986.

David Newbery is a Fellow in Economics at Churchill College, Cambridge.

The author is indebted to Jeremy Edwards and the participants at the conference for helpful comments.

involves the government taking action to alter the pattern of production and consumption away from the pattern that would arise from the free working of market forces. Indeed, the present government has sometimes claimed that there is no need for energy policy as such, meaning that decisions can be left to market forces. For present purposes I shall consider the 'market forces' option as one particular form that energy policy might take, for it still leaves open the important question of the choice of taxes which will modify these market forces.

2. Energy Policy in 1972

It is interesting to go back and see how energy policy was perceived before the dramatic oil price rise of 1973. In addition to the White Paper of 1967—*Nationalised Industries: A Review of Economic and Financial Objectives* (HM Treasury (1967))—and various reports of the House of Commons Select Committee, Michael Posner has provided just such an account in his *Fuel Policy* (Posner (1973)). Interpreting these various sources rather freely, one could argue that the main purpose of energy policy was seen as guiding and co-ordinating the investment decisions of the nationalized fuel industries (coal, gas, nuclear power, and electricity). The second aim was to design a price and tax structure for all fuels which would induce consumers to make efficient choices whilst at the same time meeting the government's revenue objectives. The case for an active energy policy (as opposed to leaving decisions to market forces) had two strands. First, it was felt that the coal industry in particular suffered from significant market failures, whilst inflation, incomes policies, and overvalued exchange rates further distorted market signals. Second, several of the industries were highly interdependent, and explicit co-ordination was felt to be needed. The size of the coal industry would depend on the demand for coal from the Central Electricity Generating Board (CEGB), and this in turn would depend on its use of oil, gas, and nuclear power. Given the long lead times involved in ordering power-stations and in adjusting employment and output levels in the coal industry, as well as the market failures already alluded to, and the considerable uncertainties on prices, costs, supplies, and construction times, this desire to improve co-ordination is understandable.

Given these aims, energy policy was then argued to be fairly straightforward in principle. The first part consisted in estimating the marginal cost of supply of the different fuels, correcting for the various market failures, for a sequence of dates in the future. The second step was to recommend setting prices at long-run marginal cost (which would be equal to some average of the estimated marginal supply costs over an appropriate time horizon) so that consumers, in making their investment decisions, would be led to choose the least-cost pattern of energy use. The final step was to estimate the levels of supply and demand at these prices and costs, and hence determine the rate of expansion (or contraction) of the various industries. The end result would be a desired investment plan, and this in turn would focus attention on the problems involved in achieving it. The potentially tricky issue of determining the appropriate long-run marginal cost for a depletable resource was side-stepped, for oil was imported and gas was to be depleted as fast as possible.

Seen from the perspective of the early 1970s, the main issues were fairly clear. Nuclear power appeared to be the least-cost method of producing base-load electricity, assuming no significant increase in the real price of oil, though allowing for the price rises of 1970–2. The cost advantage over alternative fuels was not large, but given anticipated technical progress in the nuclear industry and the likely rises in the marginal cost of indigenous coal, the main constraint was seen as the speed at which additional nuclear power-stations could be constructed. This in turn meant that nuclear power was infra-marginal and the main competition for fuels in the electricity generating industry was between coal, gas, and oil. The low extraction cost of North Sea gas suggested that the best depletion policy was to maximize the rate of extraction and use, which argued for using gas to displace oil, and possibly coal, in electricity generation. The price of oil was such that it was cheaper than a considerable fraction of British coal production. Consequently the main problem was seen as choosing the rate at which the coal industry should contract. Here, the main market failure was that the opportunity cost or shadow wage-rate of miners was argued to be significantly below their market wage, as their alternative employment prospects were poor. Thus energy policy consisted in large part in determining the rate at which oil should substitute for coal, given that oil was imported and the exchange rate was felt to be overvalued, whilst coal was domestically produced at an apparent cost felt to be higher than the true cost. Part of the answer consisted in the hydrocarbon duty, which raised the domestic price of fuel oil and hence made oil less attractive; part consisted in allowing the coal industry to run effectively at a (marginal) loss, by setting prices at average cost (well below marginal cost); and part consisted in preventing or discouraging the use of gas in power generation.

Table 2.1 reproduces the forecast figures for 1980 made by Posner (1973) towards the end of 1972, based on the assumption that the UK would supply energy at least cost by 1980, given the constraints on the rate of expansion of gas, given nuclear power, and given the estimated medium-run supply curve for British coal and for imported oil. It also gives the outcome for 1980.

Obviously, it was unreasonable to expect growth over the period 1972–80 to be so low (at 1.7 per cent per annum, compared with a figure that was assumed to be almost twice this rate). Nor was the sharp rise in all real energy prices foreseen, with the consequent reduction in fuel use per unit of gross domestic product (GDP). Nevertheless, several features stand out. First, the coal industry failed to contract as fast as considered desirable. Second, natural gas grew much faster than forecast. Both can be attributed to the sharp increase in the price of oil, as can the sharp drop in the 'fuel energy' oil use (i.e. non-transport use, in which oil competes with other fuels). The growth in nuclear power is surely

Table 2.1. Forecast of primary fuel use (million tonnes coal equivalent)

	1970 actual	'1980' forecast	1980 actual	Ratio of actual/forecast
Coal	157	95	121	1.27
Natural gas	18	50	71	1.42
Nuclear and hydro	12	50	15	0.3
'Fuel energy' oil	105	155	67	0.43
Total 'fuel energy'	292	350	274	0.78
Transport petroleum	41	73	54	0.74
Total petroleum	146	228	121	0.53
Total fuel	333	423	328	0.78

Sources: Posner (1973, Table 17.2, p. 336); *Digest of UK Energy Statistics 1984*.

deeply unsatisfactory, for although electricity demand stagnated over this period, it was still economic to replace existing high-cost plant by cheaper nuclear power. The slow growth in transport fuel use reflects the slow growth in real income as well as increased fuel efficiency.

3. The Experience of Post-1973 Energy Policy

What lessons can be learned with the benefit of hindsight? It was surely right to see the contraction of the coal industry as the main problem, though the dramatic oil price rise initially obscured this fact. But if the value of coal suddenly rose in 1974, so did the prospects for a growth in internationally traded coal, for the following reason. Transport costs for coal are significantly higher than for oil, and when oil was cheap, it was relatively unprofitable to export coal. When the price of oil rose, so did the demand price for coal delivered to ports in western Europe. The export value of coal in Australia, on the east coast of the US, and in South Africa (which was equal to the delivered price in Europe less transport costs) rose above the cost of mining it and delivering it to the port, and coal exports suddenly looked very attractive. It was reasonably clear that a significant fraction of British coal would be uncompetitive with foreign coal, even if coal became competitive with oil and gas. One of the factors which certainly did not help was the practice of pricing coal at average rather than marginal cost, so that the need to close marginal pits was obscured. The nuclear power issue was mishandled, certainly compared with the French programme. No doubt the Central Electricity Generating Board (CEGB) lacked the single-mindedness needed to push through an effective programme, as it was under pressure to take more coal than it wanted when the demand growth for electricity had disappeared.

The unforeseen rise in world oil prices and the subsequent exploitation of North Sea oil presented additional challenges to economic and energy policy. The main problem perceived by the government was to ensure that the country benefited from the oil wealth whilst consumers were encouraged to economize in the use of all, suddenly more valuable, fuels. In this we were probably lucky that oil became available only after the price rise, for it was then abundantly clear that a new system of rent taxation was urgently needed to retain the oil wealth for the national benefit (rather than for the benefit of the oil companies) and also to avoid inefficiently rapid depletion. For if oil companies were not confronted with what they believed was a credible level of taxation, they would expect future taxes on oil to be higher, and they would have an incentive to extract earlier.

It is instructive to compare Britain's experience with that of the US, which had a long-established and widely owned oil industry at the time of the first Organization of Petroleum Exporting Countries (OPEC) shock. In the US it was politically difficult to raise the domestic price of oil, as this would have generated large windfall gains to a visible and affluent section of society whilst at the same time cutting the real income of most consumers. If the US had had in place an efficient system of rent taxes, then the producers would not have gained, and the government could even have reduced other taxes as partial compensation to consumers. Given the political and constitutional power of the oil producers and oil states, it was not feasible to introduce suddenly a system of rent taxes, and the endlessly unsatisfactory compromise was to hold down the domestic price of 'old' oil to producers and consumers.

Compared with the US, and indeed many other oil-producing countries, Britain was arguably quite successful in devising an effective and efficient oil rent tax. On the one hand it succeeds in transferring a large fraction (nearly 90 per cent) of measured rent, whilst on the other it does not appear to have discouraged exploration or extraction, at least of the larger oilfields relevant in this period. The tax was criticized for its complexity and the large number of times it was adjusted. Certainly, instability in any capital or resource rent tax is a serious criticism, but in defence it might be claimed that the tax was being adjusted in the direction of greater efficiency and effectiveness, so that each adjustment reduced the fear of large future adjustments. The recent sharp fall in the price of oil provides an excellent test of the extent to which the fiscal system approximates a non-distortionary rent tax. A report in *The Sunday Times* (27 April 1986) revealed that the government is worried that the fall in oil prices will lead to the cancelling of exploration programmes and postponing of development. This may well be the efficient response to the reassessment of likely future oil prices, but it may also be induced by the fiscal regime, and the government made it clear on 25 April that it 'would not hesitate to make changes as soon as it became clear that developments were being frustrated by the fiscal regime.' 'One way for the government to help would be to forgo its royalty payments from producing fields; another would be to amend the tax regime and thus encourage incremental investment in existing oilfields.' (*The Sunday Times*, 27 April 1986, p. 65.) The fall in oil prices would seem to provide a good opportunity to remove some of the remaining distortions from the system of oil taxation, whilst the government's expressed willingness to continue to adapt the system towards neutrality is a good sign, though it would clearly have been better to have got it right first time.

The main problem with the fiscal regime is not so much that it may have discouraged the exploitation of marginal fields (or marginal investments in existing fields) but its possible inducement to wasteful expenditure. It appears that the very high tax on profits (and hence high rate of tax relief on allowable expenditures) inside the 'ring fence' has encouraged 'gold plating'—an excessive expenditure on extraction. Certainly, extraction costs increased dramatically in real terms over this period, and the price index of capital costs has systematically outpaced price indices of produced goods. It is interesting to speculate how this might have been avoided. Perhaps a lower rate of rent tax together with a more competitive auctioning of leases might have transferred the same amount of rent in a less distortionary way, though the great problem with low rent taxes is their credibility. The oil companies know that once they have discovered the oil and installed the platform, they are vulnerable to an increase in rent taxes. The only thing deterring the government from subsequently increasing the rent tax is the fear that it will discourage future exploration. But with a finite resource base like the North Sea, there comes a time when this deterrent evaporates.

If the government was forced to devise a reasonably efficient system of oil rent tax, it was under less pressure to do so in the case of natural gas, which was sold to the nationalized British Gas Corporation (BGC) under low and fixed price long-term contracts. As the Energy Committee pointed out, 'the fiscal regime applied to gas exploitation is highly complex and varies both with the timing of exploitation and the geographical location of the field.' (Energy Committee (1985, vol. I, para. 85).) Contracts signed before July 1975 pay a royalty of 12.5 per cent and corporation tax. After July 1975 contracts are also subject to petroleum revenue tax (PRT) at 75 per cent with various complicated allowances. After 1 April 1982 gas fields on the Continental Shelf outside the Southern Basin were exempted from the royalty. The problem with this fiscal regime was that whilst it was quite effective at taxing the oil companies that exploited the gas and brought it to the beach, it failed to tax the rent captured by the BGC on its existing long-term fixed price contracts. In other words, the value of the gas purchased by the BGC could (and did) diverge sharply from the contract price.

Valuing and pricing gas is more difficult than for any other fuel, as the high cost of transporting it makes its value at different places very different. In the case of oil, transport costs from one side of the world to the other are still small compared with its price, and landed oil has access to a unified world market. Its value can be immediately deduced from the price of comparable crudes. In the case of gas, the landed value can only be found by finding a margin at which this gas displaces some other fuel (other gas, or fuel oil), and then finding its value at that margin. The landed value is then its value at the margin of substitution less transmission costs, and is often termed the net-back value to indicate its method of determination. Given the size of British reserves of gas, the net-back value was determined by the price of oil, because at some future date oil would have to replace the depleted gas. (It could be argued that by then the UK could import sufficient foreign gas to defer that substitution until an irrelevantly distant date, but it is still probable that the price of the imported gas would be linked to the price of oil.) Consequently, when the price of oil rose sharply in 1973–4, the value of UK gas also rose sharply (though not by the same proportion). The demand for gas also increased sharply, as did the potential profits of the BGC. Part of these were realized, but part were passed on to gas consumers in the form of lower than justified prices, and some part may well have been dissipated in higher operating costs of the BGC and its high street showrooms. Given the considerable difficulty in calculating the value of gas, and the great secrecy about the contract terms on which the BGC negotiates to buy gas, these potential profits were obscured, and hence escaped the close attention that the tax experts in the government devoted to petroleum taxation. Belatedly, in 1980–1 the Gas Levy was introduced, and paid by the BGC on PRT-exempt gas purchases 'to secure for taxpayers a share of the benefits from early gas contracts.' (Energy Committee (1985, vol. I, para. 85, evidence from Department of Energy).) Nevertheless, the Energy Committee concluded that 'the fiscal system for UKCS [UK Continental Shelf] gas has evolved in an ad hoc fashion. . . . No case could be made for devising the current system from scratch . . .' (para. 93).

The government was concerned not only to tax the resource rents of oil and gas, but also that the prices paid for delivered energy (electricity, gas, and oil products in particular) should be set at appropriate levels, and intervened in a variety of ways to cause this to happen. The intent and success of these interventions will be discussed below, after setting out the theoretical economic argument for such interventions.

4. The Theoretical Debate on Energy Policy

In the 1970s, economists were reasonably confident of the principles needed to advise on energy policy, and again it is interesting to look back on those arguments with the benefit of hindsight. All indigenous energy industries were under nationalized ownership, and the government was required to specify how their prices should be set. All oil was imported, and its domestic price could therefore be altered by imposing excise taxes. The three main issues to address were how energy prices should be set, how investment decisions were to be made, and how to ensure the efficient management of the industries, bearing in mind the considerable uncertainties facing the industry. Welfare economics gave reasonably clear answers to the first two questions, but was silent on the third. The question of *whether* the industries should remain in public ownership was not seriously considered.

The clearest statement of the principles relevant to the first two issues was provided in Diamond and Mirrlees (1971). They argued that production and investment decisions in the public sector should aim at efficiency, whilst distributional issues should be addressed by the general tax system, and not by adjusting the prices of public sector outputs alone. In the absence of market failures elsewhere in the economy, and provided that after-tax private profits and rents were negligible (strictly, zero), the public sector should face the same set of prices that the private sector faced, there should be no taxation of intermediate goods, and, unless the country were large enough to affect the prices of imports and exports, there should be no trade taxes (i.e. no import duties, export levies, or export restrictions). Subsequent work showed that the government's redistributional objectives could, under not implausible conditions, be met by the system of direct taxes and transfers alone, and that differentiated indirect taxes would not be able to improve the income distribution further. If the government imposed value added taxes (as it does in the UK), then under these conditions the rates should be set at a uniform level, as argued in greater detail in Davis and Kay (1985).

The practical implications of this theory have been set out elsewhere (Newbery (1985)) and can be quickly summarized. Now that exchange rates float freely, the main market failures relevant to domestic energy policy are, in order of importance and difficulty: the existence of natural monopoly in electricity and gas distribution; problems of unemployment, labour immobility, and wage rigidity in the coal industry; the absence of futures and insurance markets; and problems of environmental pollution, primarily from power-stations. (I am ignoring the important international aspects of energy and economic policy which are directed to minimizing the disruptive effects of supply disruptions, as these do not directly affect domestic energy policy. Policies such as international agreements on the level of oil stocks to maintain, and the contingency sharing arrangements to make in the event of supply falling below an agreed trigger level, necessarily involve co-ordination with other countries and cannot usefully be considered in isolation. On the other hand, the residual risks that the UK faces from supply disruptions and oil price changes are relevant to domestic energy policy.)

The importance of natural monopoly in electricity and gas distribution is the main reason why these two industries are either publicly owned or regulated in almost every country. The problem of environmental pollution is conceptually straightforward—power-stations should be charged for the pollution damage caused, which will raise somewhat the cost of producing electricity. The second problem, of labour market imperfections in the coal industry, does not have much effect on the pricing of coal, for the domestic price of coal should be set equal to the world market price of coal, adjusted for transport costs. (Strictly speaking, the spot price of coal should be set at the world price level—either import or export parity, depending on whether the UK imports or exports coal. The actual price paid will typically be a contract price, which will be some average of expected spot price levels.) The reason for setting the coal price at the world market price is that coal, like oil, is now an internationally traded commodity, and its opportunity cost (or alternative value) is therefore given by the world price. It does, however, directly bear on the question of the size and rate of decline of the coal industry, or, putting it another way, on the determination of which pits are 'unprofitable'. The third problem, the lack of futures and insurance markets, means that pricing signals are blunter instruments than otherwise, as they have to signal future circumstances as well as current scarcities or gluts, and also avoid large unanticipated changes. It also means that substitutes for the missing insurance and futures markets are likely to be important, specifically that long-term contracts between suppliers and large users will need to be carefully drawn up to take account of possible contingencies, such as sharp changes in the world price of oil.

It is important to distinguish three types of prices in the energy sector—the prices offered to *suppliers* (e.g. to an oil company for North Sea gas, delivered to the beach-head), the prices charged to an energy-using *producer* of other goods (e.g. an aluminium smelter for electricity, or a pottery for gas), and prices charged to final *consumers* (e.g. households for heating and light). The ideal price system sets the supplier price and the producer price equal, extracts rent from suppliers via rent taxes or lease auctions, and sets the consumer price equal to the producer price plus the appropriate rate of value added tax (VAT). If, as could be argued to be the case in the UK, the system of commodity taxes does not afford the government redistributive instruments additional to the current system of direct taxes and transfers, then the appropriate rate of VAT is the uniform rate applied elsewhere.

There is another distinction which is useful when discussing prices—that between *efficient* or spot prices on the one hand and *contract* prices on the other. Efficient prices are the prices that would clear competitive markets at each moment, and, as for primary commodity markets which are the closest approximation available to competitive goods markets, one would expect the efficient price to fluctuate from moment to moment in response to changes in demands, supplies, expectations, and information. Contract prices would be the prices paid for the duration of a contract (either formal or implicit). One would expect contract prices to be an average of expected future efficient prices. The advantage of contracts are fairly obvious—they provide a surrogate for a futures market, with its attendant advantages of price insurance. Their limitation is that, unlike futures contracts, they cannot be freely traded and hence unwound. A futures contract has the great advantage of separating the functions of price insurance from the price-signalling role. If a coffee producer is worried that the price of coffee may fall, he can sell futures and insure himself against that outcome. If in fact the price subsequently rises and makes it profitable to apply extra inputs to increase output, the producer still has the incentive to apply those inputs, for he sells the coffee crop at the spot price whilst closing out the futures contract. A contract does not provide such an incentive, unless marginal purchases or sales take place at the spot price.

Much of the confusion on pricing principles stems from a failure to distinguish carefully between efficient prices and contract prices. Economic theory typically advises on the setting of efficient prices, whilst the fuel industries are frequently more concerned to set prices which are often best seen as contract prices. In the case of gas purchases from North Sea suppliers, the contract prices are just that, whilst for domestic gas consumers they are effectively contracts—implicit promises not to alter the price for a reasonably long period. Efficient prices are *short-run* marginal costs, whilst contract prices should be *long-run* marginal costs.

The producer prices of energy should then be set at the marginal social cost of production or the opportunity cost (the world price for traded fuels such as oil and coal, and the marginal extraction cost plus rent for gas, as explained in Newbery (1985, 1986)). One can then argue about the correct time period over which the marginal cost (or world price) should in practice be measured, and hence the extent to which the price actually charged approximates the efficient price or a contract price. If consumers respond quickly to price changes, and if future prices could be announced (and could be believed) then prices should be set at short-run marginal cost or spot world prices. If consumers respond more slowly, and future prices cannot be credibly announced, then some average of expected future short-run marginal costs or world prices, which balances errors in current consumption decisions against errors in fuel-using investments, will be preferable. This average will approximate an expected long-run marginal cost.

Investment decisions are then evaluated using a test discount rate (TDR). A project is accepted if the present discounted value of its incremental output, evaluated at the prices that will prevail given the pricing rules outlined above, less incremental operating costs, exceeds the incremental investment cost. Equivalently, capacity is expanded if the short-run marginal cost exceeds the long-run marginal cost, and capacity is optimal when all marginal costs (measured over all time periods) are equal. The only remaining problem is to specify the TDR, and here there is continuing debate. The Diamond–Mirrlees efficiency argument implies that the public sector TDR should be equal to the discount rate used by the private sector in its investment decisions, but it requires the perhaps unreasonable assumption that capital markets are efficient and do not suffer market failures. Credit rationing, itself the consequence of pervasive asymmetric information and moral hazard, casts considerable doubt on this assumption, and a variety of alternative methods exist for estimating the TDR, none fully satisfactory.

These principles, more or less those incorporated into the 1967 White Paper (with the possible exception of the principles for setting commodity taxes), have not changed significantly since the early 1970s, though the government has become increasingly disenchanted with them. The reason is simple: although the principles can guide pricing and investment decisions provided the nationalized industries are efficiently managed and minimize the costs of producing the desired levels of output, they do nothing to ensure that the industries are managed efficiently. Moreover, it is not difficult to argue that it is far more important to ensure that costs are minimized than that prices are set at the correct level. Suppose the elasticity of demand facing a nationalized industry is unity; then, roughly speaking, the welfare cost of setting the price 20 per cent too high or too low is no greater than the welfare gain of cutting production costs by 2 per cent; whilst if price elasticities are lower than unity, pricing errors are less costly. It is doubtful whether the disagreements over pricing amounted to more than 20 per cent, and I suspect that very few economists would have claimed that improved management, monitoring, or competitive pressure could not have been able to reduce costs by at least 2 per cent.

A variety of measures have been employed by the government to try to improve the efficiency of the nationalized industries, ranging from imposing additional profit targets to efficiency audits. (If the industries were operating efficiently and following the specified pricing and investment rules, these profit targets would either be redundant or inconsistent with the application of those rules.) None seems to have been very successful, though the past decade has been a turbulent one for the energy industries, and perhaps it is too soon to pass final judgement.

The growing evidence that deregulation and increased competition could dramatically lower costs in some industries has in any case changed the nature of the debate on the best way to control the energy industries. After some modest moves to permit private competition in electricity and gas supply, the government has privatized the oil industry and, at the time of writing, intends to privatize British Gas in the autumn of 1986. Next year the oil and gas industries will be in private ownership, competing with the still nationalized electricity and coal industries. It is not inconceivable that in due course coal, and perhaps even electricity, will also be privatized.

After this rather long introduction, it is now time to address the main topic of this paper—how should this change in ownership of part of the energy industry affect energy policy? This in turn can be broken down into a number of questions:

(1) What should be the government's policy on competition for the fuel industries, both nationalized and privatized?
(2) Should the prices of the privatized industries be regulated, and if so, how?
(3) What policies for pricing and investment should the remaining nationalized industries follow?

(4) Are there any implications for the taxation of energy?

In the following sections these questions are addressed in turn.

5. Competition Policy

Economists appear to be in reasonable agreement that firms are induced to operate efficiently less by the nature of ownership (private versus public) than by competition or the threat of competition. (See Bailey (1986), Kay and Thompson (1986), and the discussion by Brittan (1986).) The main success stories of deregulation have been in industries with few natural barriers to entry (and exit). When they were deregulated so that entry could occur, they were put under pressure to cut costs and reorganize to meet consumer demand at least cost. In the case of airlines in the US, the effects were dramatic, as competitive costs were sometimes less than half the inflated costs of the regulated companies. The British policy of privatization is not designed to achieve these benefits, since there appears to have been little concern to increase competitive pressure on the privatized industries, and indeed the government appears instead to believe that selling public industries as monopolies will be more profitable than the alternative of selling them after trying to increase the degree of competition. It is therefore important to realize that the fact that the government is planning to privatize parts of the energy sector is no guarantee that the problems of management efficiency will thereby be overcome. That objective will require an increase in the degree of competition within each fuel industry (as well as between them) and/or a carefully designed system of regulation. It is worth briefly examining each industry to see what options are available, and to what extent they are being adopted.

Gas

Gas is the most urgent case to look at as it is currently being considered by Parliament for privatization in autumn 1986. The most obvious option, that of selling the twelve Area Boards and the national transmission system as separate entities, has been foreclosed. Had this been done, the number of companies on the buying and selling side of the upstream market for gas would have been more equal (there being about eight or so moderately large suppliers operating in the North Sea, as well as three countries—Norway, the Netherlands, and Russia). In most natural resource markets, this number of buyers and sellers might normally be expected to lead to intense competition, but it is important to realize that gas is fundamentally different from oil, for the potential market area accessible to a given buyer or seller is severely limited by the pipe-line infrastructure and distance. On the most favourable scenario, in which the UK builds links to the continental gas grid and hence potentially allows buyers and sellers access to the widest possible market, it is unlikely that more than a few agents would be well enough placed to compete on each side of any transaction. Nevertheless, it can be argued that the increase in competitive pressure exerted on the firms when going from a single firm (like the BGC) to two competing buyers (two Area Boards, an Area Board and the CEGB, or possibly an Area Board and a foreign buyer) is as large as or larger than going from a small number of participants to a large number. The main benefits of such a move would be to improve the efficiency with which gas exploration and exploitation proceeds, as argued more carefully in Newbery (1986).

If the BGC is to be sold intact, the first question to ask, therefore, is whether there is any other way of exposing it to more competition in its dealings with gas suppliers. Two potential solutions have emerged—to allow gas suppliers to sell direct to UK customers, or to allow them to export gas, rather than being forced to sell to the BGC on its terms. Since the Oil and Gas (Enterprise) Act of 1982, UK suppliers have been free to negotiate direct sales to industrial consumers, though in most cases the supplier would still have to negotiate with the BGC for the use of the pipe-line system to deliver the gas to the contracting customer. In practice no use has been made so far of this provision. Part of the problem is that although the Act requires the BGC to make the pipe-line available to interested parties, the BGC has perhaps understandably not been in a great hurry to quote terms on which gas shipments may be made.[1] Here the two obvious steps which might improve matters are to allow gas to be used for electricity generation (this will increase the number of potential buyers large enough to make direct sales commercially viable) and to specify more carefully the terms on which pipe-lines are to be made available. It will also be necessary to prevent predatory pricing by the BGC designed to undercut the market for private sellers.

The second option, of allowing gas suppliers to export, appears to have been agreed in principle, though it will presumably require the supplier to provide his own pipe-line link to the foreign market. Whilst this may have some effect for a few fields in the southern North Sea, which is close to the Dutch gas-gathering system, its effect is likely to be rather small. Again, to widen the market, more needs to be done and, as mentioned above, the best remaining prospect is likely to be a link or links between the national transmission system and the continental grid, for this would allow a larger number of fields potential access to export markets. (If such a link were built, it might be attractive for Norway to

---

[1] There appears to be little change in the rules governing the obligation of the BGC to ship gas through its pipe-lines, and no obligation for it to publish tariffs for gas shipment, though the BGC is to be required to publish illustrative tariffs which would indicate to producing companies the tariffs to expect. Clause 19 'empowers the Director [of the Gas Users' Council] on application of a potential pipeline user to give the public gas supplier directions securing to the applicant the right to use a pipeline owned by the public gas supplier subject to such payments as may be specified.' Whether this would be enough to force the BGC actually to quote terms within a reasonable time-frame is unclear. There is apparently no mention of any change in the right of producers to export or import gas—this remains subject to the approval of Her Majesty's Government.

use Britain as a land bridge to the Continent for its new large offshore gas fields, and as a result, Britain would gain access to an additional large source of supply, as discussed in Newbery (1986). Whether this is still a viable prospect depends sensitively on how far future gas prices are expected to fall in sympathy with the current fall in oil prices.)

None of these proposals, even breaking up the BGC into competing Area Boards, would have much direct effect on the efficiency with which the Area Boards transmit the gas from the beach-head to the final consumer, for each Board remains a local monopolist for all except the largest consumers (who may be able to buy direct). Nevertheless, a significant opportunity was lost in selling the BGC intact, for it would have been much easier to regulate gas prices if the Area Boards had been set up as autonomous accounting units. The best way to regulate each Board would involve setting a price which is independent of the Board's costs, as this preserves incentives for efficiency. The problem is how to set the price, but with twelve competing regional companies there is no difficulty—the prices to consumers in any one region can be set as a mark-up on some weighted average of the remaining regional companies' costs, the weights allowing for regional similarities, as well as perhaps placing greater weight on lower-cost regions. Since each regional company has an incentive to minimize costs, it would be difficult for them to collude and defeat the intent of the regulation. (See also the discussion in Hammond, Helm, and Thompson (1985).) Whether this effect can still be achieved is considered in Section 6, on regulation.

Electricity

The 1983 Energy Act aimed to liberalize energy supply in the UK, by abolishing the statutory monopoly for the supply and distribution of electricity. It also required the Area Boards to publish tariffs at which they would be required, subject to technical feasibility, to purchase electricity offered by private producers. The Act also permitted private producers to make use of the transmission and distribution system at published rates so that they could supply final consumers directly. On the face of it, this appears to provide just the threat of competitive entry required to keep a natural monopoly operating at least cost, but, as Hammond, Helm, and Thompson (Chapter 8, this volume) argue, entry conditions into the electricity supply industry involve large sunk costs and a long-term commitment which greatly reduces the effectiveness of this entry threat. Suppose it were the case that a privately constructed coal-fired power-station located on a deep-water port using imported coal would be the least-cost method of generating power conventionally, and that a private contractor would be able to build such a power-station at lower cost than if under contract to the CEGB. It would then seem desirable that when the time came for the next conventional power-station to be ordered, it should be privately constructed and operated, and should sell electricity to the Area Boards as envisaged by the Energy Act. Does the Energy Act as presently drawn up provide the right incentives for this outcome to be likely? In particular, does it provide sufficient assurance about the future terms on which the electricity output would be bought so that the private supplier can predict the profitability of the investment with reasonable confidence?

The Energy Act requires that the private purchase tariff (PPT) be based on the Area Boards' avoidable costs in purchasing from the private supplier rather than from the CEGB, which in turn is specified by the bulk supply tariff (BST). If all operating decisions were based on the same principle (i.e. to produce electricity if the relevant part of the BST were above short-run marginal operating costs) then the CEGB would have a powerful incentive to set the BST at the efficient level, and a private producer who was in fact more efficient than existing public suppliers would be able to make a profit selling at these prices, and so would be encouraged to enter. But operating decisions are centrally made via the merit order, and although they are made to minimize system operating costs, there is no strong compulsion to align the BST to the relevant marginal operating costs, since this is not used to signal operating decisions. Moreover, the CEGB has been under repeated political pressure to change its pricing structure to meet a variety of changing objectives, and there is no guarantee that this process will not continue in the future. In short, the future course of the BST and with it of the PPT is hard to predict, and this uncertainty might deter potential entrants. Nevertheless, unsatisfactory though the situation appears, it could be argued that it will not deter *efficient* entry, for the following reason. If the entrant can produce electricity at lower cost than existing power-stations, and if the CEGB has to earn a required rate of return comparable to the private sector rate of return, then it must set the BST at a high enough level that the private power-station would earn a higher rate of return. Provided the CEGB is not able to place inappropriate costs into the category of unavoidable system costs, which Area Boards must pay regardless of whether they buy from the CEGB or private suppliers, then it would appear that the Energy Act offers the prospect of genuine competition.

Are there any flaws in this argument? What happens if the CEGB has excess capacity, and the government accedes to the compelling argument for setting the BST at short-run marginal cost (SRMC), at which level a new power-station is unlikely to make a profit? If the CEGB does have excess capacity then entry (i.e. additional capacity) is presumably undesirable, and the BST will give the correct signal. (Indeed, if the BST is set at long-run marginal cost (LRMC), as the CEGB claims it should be, there is a danger that entry will occur when it should not, and to that extent it is desirable that the BST be adjusted towards SRMC during periods of excess capacity.) But what if the CEGB has a built-in incentive to over-invest so that it almost always has excess capacity, and is hence protected from competitive pressures? Indeed, it has been argued that one of the main limitations of the pricing principles advocated by welfare economists and set out above is that they provide an incentive for 'appraisal optimism' when selecting new investments. The argument runs as follows. Suppose that the managers and staff of the CEGB derive utility from the size and rate of growth of the industry (because their promotion and pay prospects improve, as well as their prestige). Then there is an advantage in overestimating demand at LRMC prices and installing excessive capacity, for then efficient pricing dictates that prices be set at SRMC, below LRMC, and at these lower prices demand and the volume of output will indeed be higher. This incentive towards 'appraisal optimism' will be further enhanced if it is believed that it deters entry and hence preserves the position and size of the CEGB.

The solution to this problem is to require the CEGB to earn a required rate of return on its investment to offset this temptation, and if this is done (and is believed by potential entrants to be a permanent feature of electricity pricing and investment decisions) then entry should occur when efficient. The fact that such entry has not yet occurred may just be evidence that the CEGB is suffering from excess capacity, and will continue to do so in the near future. It may also be because no private constructer is willing to take on the massed forces of the British coal-miners and port-workers.

The other intriguing prospect is that French nuclear power will continue to undercut the cheapest British electricity, in which case the least-cost option for the UK would be to expand the number of cross-Channel cable links to France. If the CEGB is not willing to do this directly, then it is presumably open to Electricité de France to find a British company (perhaps a cable company) through which to sell electricity. In short, the Energy Act would seem to force the CEGB to consider this option very carefully indeed.

It might also be worth speculating whether it would be feasible to sell off the individual power-stations to private operating companies whilst retaining the grid as a common carrier. On the face of it, this proposal looks quite attractive, as there were about one hundred power-stations in England and Wales at the end of March 1983. However, about half the total electricity was produced by only ten coal-fired power-stations, and there were nine nuclear stations supplying just under 30 per cent of the total (Bending and Eden (1984)). Thus even at the level of the whole country, the degree of concentration would be quite high, whilst in any effective market area it would be very high indeed. Indeed, Schmalensee and Golub (1984) found that even in the US, where conditions appear far more favourable to deregulating the wholesale electricity market, some market areas exhibited high estimated effective concentration. This does not mean that competition is impossible, but it does mean that additional regulation would be required.

Again, as with the Oil and Gas (Enterprise) Act, the main weakness of the liberalization proposals is that they have little effect on the efficiency with which the Area Boards deliver electricity to final consumers. Given the local natural monopoly element of the distribution system, it is hard to see what further competitive (as opposed to regulatory) measures are available to improve this situation.

Coal

The most obvious measure to increase competitive pressure would be to allow customers to import coal freely. In effect, British Coal has already responded to this threat by offering to sell coal to the CEGB at import parity prices. The real issue arises when planning the location of new coal-fired power-stations (if and when they are needed). Provided the CEGB is free to choose their location and can negotiate long-term supply contracts with British Coal suitably indexed to competing fuels (imported coal, fuel oil, etc.) then the threat to locate at a deep-water port and buy imported coal should compel British Coal to price competitively (at least, on new contracts). The main problem is that British Coal may continue to cross-subsidize, in which case there would be no guarantee that power-stations were in fact located at least-cost locations using the least-cost fuel. Again, the logical next step is to remove British Coal's monopoly over the production of coal, so that new pits could be owned and operated by competing producers. Arguably, privatizing British Coal by selling pits to a variety of companies would seem the logical way of increasing the efficiency of the industry. Of all the nationalized industries, coal is arguably one of the best-placed on purely efficiency grounds to benefit from competitive privatization, as each mine operates under diminishing returns and has rather limited local monopoly power, greatly mitigated by the bilateral nature of most coal sales (to the CEGB). Whether it would adequately address the social problems is, of course, another matter.

6. The Regulation of the Privatized Fuel Industries

There would seem to be no need to regulate the prices of petroleum products, since there is an active spot market in Rotterdam and entry is reasonably easy for independents. For the immediately foreseeable future, that just leaves the regulation of gas, though it might be worth speculating on the need for regulation of electricity, were it to be privatized.

There is no doubt that gas prices will need regulation, for even if it had been the intention to break up the BGC into its separate Area Boards, each of these would still have had a total local monopoly in the supply of gas to most consumers. The evidence from Germany is that left to their own devices, gas boards set the price of gas about 10 to 20 per cent above the price of domestic heating oil, calculating that the additional advantages of gas in central heating allow them to set this price without losing too many customers to the competing fuel. At least until 1986, this would have resulted in too high prices for domestic consumers, too few consumers, and monopoly profits to the gas company.

Part II of the Gas Bill proposes that the BGC should be sold intact in its present form, and is to be subject to what appears to be rather weak regulation by a Director of an Office of Gas Supply (OFGAS), advised by a Gas Users' Council (with half the funding presently spent on the various Gas Consumers Councils). Consumers taking less than 25,000 therms per annum would be supplied at an announced tariff, which from the Draft Licence (published by the Department of Energy on 9 December 1985) is supposed to be no higher than a 'maximum average price'. This price is calculated according to the following formula:

$$M_t = (1 + i_t - X)P_{t-1} + Y_t - K_t,$$

where M_t is the maximum average price per therm in year t;
i_t is the fractional change in the RPI;
X is a number to be determined (perhaps 0.02);
P_{t-1} was the allowed gas price in year $t-1$, with P_0 specified;
Y_t is the allowable gas cost per therm in year t;
Q_t is the quantity sold to regulated customers in year t;
r_t is the Treasury bill rate in year t;
R_t is tariff revenue from tariff quantity in year t;
$K_t = (R_{t-1} - Q_{t-1}M_{t-1})(1 + r_t)/Q_t$; and
$P_{t-1} = P_{t-2}(1 + i_{t-1} - X)$.

The maximum price is thus made up of the allowable gas cost (roughly speaking, the cost of purchasing gas, though the definition takes nearly four pages to spell out) and non-gas costs, which are linked to the cost of living with some allowance (the X factor) for presumed increasing efficiency (as with telephone rates, which escalate at a rate equal to RPI – 3%). Larger customers would not be so protected and would have to negotiate tariffs, subject to a published maximum. There does not appear to be any requirement that larger customers should pay the same for the same type of contract, nor anything to prevent cross-subsidization, which, given the substantial rents earned on the earlier gas supply contracts, would be easy to finance, and would constitute a powerful barrier to the entry of suppliers attempting to make direct sales. The regulation of the gas industry has been scrutinized and strongly criticized in the Report of the Energy Committee (1986a).

The obvious questions to ask are whether this system of price determination encourages efficiency in gas use by consumers and in gas supply by the privatized British Gas Corporation or British Gas PLC (BGPLC). Consider first the question of efficiency in gas use for large consumers who must negotiate contracts. Most gas would be used for raising heat, in which use it is competitive with oil and coal. In the case where oil is the logical alternative, one could argue that bargaining between the purchaser and BGPLC will lead to efficient fuel choice, as it is probably cost-effective for users to install dual-fired burners which would permit them to burn oil or gas, depending on their relative prices. (To take a concrete example, my college signed a contract to run from February

1986 to January 1987, to take between 100,000 and 150,000 therms of gas. In February, gas was cheaper than fuel oil, but in April fuel oil was about 15 pence per gallon cheaper than gas equivalent, and the college switched to burning fuel oil. Next year we shall no doubt either bargain for a lower gas price or a lower minimum required gas take.) Faced with an essentially elastic demand for gas at the fuel oil price, BGPLC will undercut oil provided the opportunity cost of new gas (i.e. marginal cost plus any rent in the supply price) is below the price of oil.

In the case of customers for whom coal would be the cost-effective choice, matters are more complicated, as it is substantially more costly to retain a dual-firing capability. Here the danger must be that BGPLC would attempt to offer an initial contract which made gas attractive relative to coal, but once the investment decision had been made in favour of gas, it would then replace the contract by one less favourable. A shrewd purchaser would be well advised to insist on a long-term contract linked to the price of coal, preferably with a low minimum take, to allow subsequent switches to oil if that proved cheaper. Provided larger customers were reasonably intelligent in negotiating contracts, the system would seem to encourage efficiency in gas use, though there is a case for the Office of Gas Supply (OFGAS) to collect and publish the terms of existing contracts to increase the transparency of the market and to correct any imbalance in bargaining power between relatively smaller buyers and the monopoly gas supplier.

In the case of the regulated market, prices are related to the average cost of purchasing gas, and when examining whether this will lead to efficient gas use, four questions are important: first, whether prices should be based on the *average* cost of gas; second, whether the formula allows predatory pricing against alternative fuels; third, whether BGPLC will have adequate incentives to hold down the cost of acquisition; and fourth, whether the formula gives the regulated market adequate protection against 'unreasonable' price increases. The answer to the first question is that, theoretically, gas prices should be related to the opportunity cost of gas use, and there is little reason to believe that the present average cost of gas approximates this opportunity cost at all closely. A substantial fraction of currently supplied gas was purchased in the past under long-term contracts at favourable prices, well below the cost of replacement gas, and hence well below its opportunity cost. The Gas Levy was an attempt to increase the cost of gas delivered to the BGC somewhat towards the correct price, but it is a very blunt instrument for the purpose. Probably the best solution would be to change the basis on which the Levy were calculated, relating it to the difference between the past contract price for 'old' gas and some formula designed to measure the opportunity cost of the gas. Ideally, one would search for an observable market price for new gas contracts sold to buyers other than BGPLC, but the gas market is notoriously opaque. An alternative would be to take a weighted average of the price of new contracts signed by the BGC, and the index of fuels used in typical contract escalation clauses.

The second question is whether the formula allows predatory pricing against other fuels, and here the answer appears to be a qualified yes. Suppose that other fuel prices were suddenly to fall, making electricity or heating oil the efficient choice for customers installing central heating. It is open to BGPLC to match these price cuts for several years, and then to recoup the lost revenue by raising prices sharply later. The formula given above has the property that if BGPLC makes a short-fall in its revenue in one year, it can recoup it *plus interest* in subsequent years. BGPLC is thus indifferent to a lower revenue this year, because the present value of the extra revenue allowed in the future is equal to the loss now. The advantage of matching price cuts is that some consumers will make the incorrect choice (of gas), and once they are locked in, they can be squeezed, and will regret having made the choice. The regulations carry the provision that if BGPLC underprices gas by more than 90 per cent for two years running, then it may not be able to recoup the loss, and this will dampen the incentive to predatory pricing. The obvious solution to this problem is for OFGAS to publish forecasts of future gas prices, so that consumers can take informed long-run decisions when buying fuel-using equipment.

The third question was whether the ability to pass on increases in gas costs to the regulated consumers unduly weakens the incentive of BGPLC to negotiate an efficient contract with the gas suppliers. In the past the BGC has been criticized for, if anything, exercising too much monopsony power in bargaining with gas suppliers, and as a result discouraging exploration and efficient depletion, as well as transferring rent from the Treasury (as PRT and other taxes) to the BGC, where it is taxed less heavily, or to consumers, where it is taxed not at all. Whilst the earlier system can be criticized, the proposed alternative appears to go to the other extreme. One possible solution is to allow BGPLC to recover 90 per cent of the amount by which the allowable gas cost exceeds the initial, indexed allowable gas cost. (This would introduce a symmetry in the effective tax rate of profits/rents accruing to the gas producers and BGPLC.) In defence of the present system, it might be argued that BGPLC has an incentive to minimize the cost of new contracts because it retains the difference between its gas costs and the prices charged to the unregulated consumers, and it may be that this provides sufficient incentive to keep gas costs down, without the need for the 90 per cent cost recovery factor.

There are two additional aspects to setting the beach-head price of gas which are worrying. The first is that there is an incentive for BGPLC to shift non-gas costs onto the gas producers, since these can then be recovered in full. The costs most likely to be shifted are those of storage and managing the seasonality in demand, which might require BGPLC to invest in expensive facilities that could not be directly recovered through the pricing formula. If the gas supplier provides them and charges for them in the contract price then they can be recovered, but this might be a much less economic solution from the national viewpoint. The other point is that there may be an incentive for BGPLC to underprice gas from its own fields, to avoid the heavy rent taxes. Although it would not be able to recover these lost rents on sales to regulated customers, it could on sales to the non-regulated sector, and since the fraction of non-regulated sales is substantially larger than the fraction of rent retained after PRT, this form of transfer pricing will be attractive. Naturally the Treasury is concerned that gas sold by subsidiaries of BGPLC to BGPLC be fairly priced, but given the complexity of valuing gas contracts, this might be difficult to police.

The fourth question was whether the pricing formula allowed BGPLC to raise prices to the regulated market 'unreasonably', to which the answer seems to be no. In the case of British Telecom (BT), the tariffs cannot on average exceed the formula amount, but if the charges to large customers are cut in response to competition from Mercury (as appears to be happening) then BT is free to raise the charges to private subscribers. As I understand the definition of the maximum average price, it refers to quantities and revenues sold and earned in the regulated market alone, so that price cuts in the non-regulated market would not justify price increases in the regulated market. It is obviously important to check that this interpretation is correct, for if the average refers to total sales in both the regulated and unregulated markets, then if BGPLC cuts its price to the non-regulated market in response to competitive pressure, it would be able to increase its price to the regulated market, with little fear of losing many customers, most of whom are locked in by their past investment decisions. Whilst this may not lead to inefficiency, it will certainly be resented, and would reflect a presumably inequitable transfer from consumers to the equity shareholders of BGPLC.

The final issue to address is whether the regulatory system provides adequate incentives for efficiency in gas supply, and here the proposal is to adopt a scheme rather like the British Telecom pricing formula, where the maximum price assumes a predetermined rate of cost reduction in real terms, as yet to be decided. In principle, this is superior to a cost-based system, since BGPLC derives the full benefit of faster cost reductions. The difficulty lies in determining the rate of cost decline (the value for X in the pricing formula). Not too much should be made of this, however, for it will take many years before the discrepancy between estimated and minimum costs becomes large, and small discrepancies have a negligible efficiency cost and consist largely of transfers between gas consumers and gas shareholders. If the discrepancy becomes embarrassing, the Act contains provisions for revising the formula, and this should provide an adequate safeguard. The main implication is that the government should ensure that there is competitive tendering for the shares so that any underestimate in the value of X that makes the profitability of BGPLC higher is recouped in higher receipts from the sale of shares.

7. Pricing and Investment Policies for the Nationalized Fuel Industries

Once gas is privatized and allowed to price essentially as it pleases in the unregulated market, should the other fuel industries be given the same powers?

In the case of coal, this is presumably already the case, so the issue only affects electricity, where at the moment it is required not to discriminate between essentially similarly placed purchasers. The argument for not worrying unduly about the ability of the BGPLC to price-discriminate was that gas faces close competition from coal and oil as an under-boiler fuel, and sensible purchasers will use that fact in negotiating efficient contracts with the BGPLC. This argument applies much less strongly for electricity, which for many applications has no close substitutes, and hence represents a captive market with high entry barriers. There would thus appear to be good grounds for continuing to insist on non-discriminatory pricing for electricity.

More radically, should the Treasury abandon its required rate of return, which is a profits-based system of price regulation, and replace it by a maximum allowable price as for gas and telecommunications? If so, should it allow managers to receive some fraction of their salary as a share of the resulting estimated profits? On the face of it, the idea has obvious attractions, and might go some way to avoiding the excessively capital-intensive nature of electricity supply in the UK. It would also seem appropriate then to allow the electricity supply industry direct access to capital markets, perhaps at the same time removing the Treasury underwriting, so that the 'discipline of the capital markets', argued to be so good for the other privatized industries, could also be applied to the electricity industry.

Another issue to consider is whether the CEGB should be free to use gas in power generation, and again there seems no good reason for not allowing this. Indeed, now that private companies can supply electricity, one might well expect that companies with access to cheap gas (or gas with a low opportunity cost) might use it to generate power, and it would be illogical to prevent the CEGB doing likewise. There is the additional argument that it might be attractive for the CEGB to operate gas-fired small-scale combined heat and power systems for large consumers. The economics of these schemes looks attractive at the rates of discount used by the CEGB, but much less attractive at the rates of discount that firms apparently use when making energy conservation decisions. It has therefore been suggested that the CEGB build and operate the systems, selling heat and power to the firms (or, equivalently, leasing them the plant). The potential market amounts to about 4 per cent of the current generating capacity of the CEGB, with the attraction of a much shorter lead time in construction.

8. The Taxation of Energy

Two key reforms are long overdue in the energy sector. First, the taxation of rent is, outside the oil industry, in an unsatisfactory state, and second, the taxation of energy consumption has serious shortcomings. The issue of rent taxation has already been touched on, and primarily involves the taxation of the rent on old gas contracts which accrues to consumers, and not to the Treasury. A modified Gas Levy, as described above, would solve this problem.

Alternatively, the Treasury could auction the rights to the existing contracts to oil companies, which would then be free to renegotiate new contracts with the BGPLC. There is also the anomaly that British Coal collects the royalties on privately operated coal-mines—primarily the opencast pits. Logically the Treasury should set and collect such taxes, and again, logically, they should be rent taxes rather than royalties.

The taxation of energy consumption is one of the main instances where the British government appears not to have heeded the advice of Diamond and Mirrlees (1971). At the moment some intermediate goods (heavy fuel oil) are subject to an excise tax when they should be subject only to VAT, whilst other final consumption goods, notably gas and electricity, are zero rated. The fuel excise duty should be abolished, and VAT at the standard rate should be imposed, with compensating adjustments to supplementary benefits and the level of tax thresholds to offset any adverse distributional impacts. (Davis and Kay (1985) demonstrate what consequential changes would be needed to achieve this purpose.) The whole issue of protecting vulnerable energy consumers from hypothermia is best addressed through the system of benefits, not through concessionary pricing.

It might be argued that it would be political suicide to impose VAT on fuels, but it is hard to see the force of this argument. At the moment, the government effectively forces the electricity supply industry (ESI) in particular, and to some extent the BGC, to act as covert tax collection agencies, for the external financial limits and the financial targets effectively force these industries to raise prices in order to meet the required targets, even though on efficiency grounds they may feel that there is no case for so doing. The political attraction of this may be that the tax is not perceived as such, but merely evidence of the unsatisfactory nature of nationalized industry performance, but the economic cost is that whilst it is logical to raise prices to final consumers, it is damaging to raise them to producers. Consequently, the ESI is under pressure from industrialists to offer concessional tariffs to enable them to compete effectively in world markets, given that their competitors in Europe face lower energy prices. This could be done by abandoning the rather important principle of non-discriminatory pricing, but it would be far better to address one of the sources of the problem, namely the unsatisfactory system of taxing energy consumption. If the government were to introduce VAT either at a time of falling fuel prices or in exchange for relaxing the external financial limit, then domestic prices need not rise, whilst producer prices would, and everyone would be happy.

The only exception to the rule of exempting fuels from excise duty applies to motor fuel, where the excise duty is properly seen as a road user charge.

9. The Co-ordination of Energy Policy

The old view of energy policy attached high priority to the need for co-ordinating the decisions of the different fuel industries. Is this still an important objective, and, if so, how is it to be achieved when a growing fraction of energy is under private control? At the moment the main need for co-ordination is between the CEGB, British Coal, and the nuclear power industry, all of which remain under public control. The likely scale of gas use in electric power generation in the foreseeable future is sufficiently small that it is unlikely to require much sophisticated co-ordination, over and above the natural co-ordination that would arise in drawing up long-term contracts for gas supply to the CEGB. This leaves the main problem as the old one of forecasting the demand for British coal by the CEGB, and, given the lead times in power plant construction, it is as easy for British Coal to make this forecast as for the CEGB—the uncertainties on future relative fuel prices (which determine the merit order and demand for coal) are as difficult to resolve for either party. It is open to both to negotiate long-term contracts which share these risks. The real source of the difficulty lies in managing the coal industry, and dealing with the social problems and market failures in the labour market—a problem that has less to do with co-ordination than control.

An interesting test is provided by the dramatic fall in the price of oil from nearly $30 per barrel in 1985 to about $12 per barrel in April 1986. If coal is to be competitive in power-stations at this price, it will have to be delivered at a price of about £38 per tonne. The current average delivery price to the CEGB is about £45 per tonne, and even at this price a number of pits are still unprofitable (at market prices). At £35 per tonne, the London Business School calculates that the industry break-even output would be only 68 million tonnes per annum, employing only 110,000 miners (*The Times*, 28 April 1986, p. 17). What in these circumstances should happen to the price of coal delivered to the CEGB, the quantity of British coal used in electricity generation, and the price of electricity? The answer to the last question is in principle easy and given by the standard welfare economics arguments outlined above. The efficient price of electricity is the short-run marginal cost of producing the level demanded at the prevailing world prices for the marginal, least-cost fuel. Whether the marginal fuel is oil or coal is not immediately obvious—presumably there is not enough oil-fired and nuclear capacity to meet peak demand, and so it may be coal, assuming that the import price of coal is above that of oil on a thermal equivalent basis. The world price of coal is likely to fall partially in response to the fall in world oil prices, and so this is not much comfort to British Coal.

The efficient price for British coal will continue to be equal to the relevant world price of coal, and if, as seems likely, the oil burn of the CEGB increases, then coal is likely to be in export surplus and will have to be exported, perhaps at very disadvantageous prices. Consequently the efficient price of British coal will be the (low) export price level, and at this level the CEGB will presumably wish to continue to buy a large fraction of its previous purchases. The price actually paid for the coal is in a sense a relatively minor issue, as it is a straight transfer from one nationalized entity to another; the questions will be, who should show the losses arising, and how should they be financed? Provided the actual price paid is seen in this conventional way and has no consequential effects on future production decisions, then little is at stake.

The final issue is the effect the fall in oil prices has on the future scale of the coal industry, and here what is needed is a reappraisal of the likely future price of oil (and of the probability distribution around the forecast), together with a recalculation of the rate at which to reduce employment and output in the coal industry. The interesting question will be how to appraise new pits which offer potentially lower production costs. On the one hand, low current oil and coal prices must increase the prospect of higher future oil (and coal) prices, and hence make deferring development attractive, whilst on the other, the opportunity cost of miners available for transfer to new pits now might be well below their future opportunity cost, arguing for maintaining greater continuity in developing new pits.

10. Conclusions

The indirect aim of this paper was to argue that the old principles of pricing and investment continue to apply, though they need to be supplemented by explicit incentive mechanisms to ensure management efficiency. These basic principles indicate that the system of taxation still needs adjustment, and that further improvements can probably be made to increase the competitive pressures acting on the energy industries.

3 Energy Policy, Merit Goods, and Social Security*

ANDREW DILNOT AND DIETER HELM

1. Introduction

Energy, together with food, clothing, education, health, and home-ownership, is often thought of as a good that has special distributional merit in consumption. This is reflected in a variety of welfare policies—hardship payments in bad weather, allowances (either explicit or implicit) in the calculation of social security benefits, and special provisions against cut-off when difficulties arise in paying bills. In the UK, but not all European countries, energy in the form of electricity and gas (but not petrol) is also given special tax status, being zero-rated for value added tax. This status is now the subject of considerable political controversy.

Critics of these provisions point to the paternalism involved in the selection of priority for consumption of these types of goods by government rather than individuals. Economists typically rely on the price mechanism to allocate resources, where these prices are related to costs. Consumer preferences are treated as exogenous and sovereign. Distribution issues are separated from those of efficiency in resource allocation, and hence there is a general preference for providing the poor with additional income, rather than deliberately distorting prices to benefit the poor and providing quantities of selected goods that the poor may not have chosen had they had the income instead.

These criticisms are reinforced when the apparent inconsistency of current policy is examined. Food is not directly provided for the poor, though health and education are. Libraries and museums are made universally available, whilst theatres and footwear are not. Housing receives a special allowance, whilst food and energy do not.

In this article, the arguments for considering a select group of goods as distributionally 'special' or worthy of 'merit' are re-examined. The distinct claim that these should be provided directly by special benefits rather than through general income support is also examined. It will be argued that such

---

* This paper was first published in *Fiscal Studies*, August 1987.

Andrew Dilnot is a Programme Director at the Institute for Fiscal Studies. Dieter Helm is a Fellow in Economics at Lady Margaret Hall, Oxford and a Research Associate at the Institute for Fiscal Studies.

The paper is based on research conducted by the Institute for Fiscal Studies as part of a programme on energy financed by the Economic and Social Research Council. The authors are grateful for comments from David Thompson and Ian Walker, and for research assistance from Francis McGowan. The usual disclaimer applies.

'merit goods' do exist and, separately, that there are cases where these are better provided directly. These arguments concerning the theory of merit goods will form Section 2 of the article. We then go on in Section 3 to consider the empirical characteristics of energy demand, examining expenditure patterns of the poor. This section demonstrates that energy falls into the merit good category. Section 4 reports the current treatment of energy in the tax and benefit system, and the Codes of Conduct in terminating supply. Section 5 considers a number of alternative means of providing energy to the poor. Finally, Section 6 presents a number of conclusions which emerge from the analysis.

2. What is a Merit Good?

The notion that some goods might have a special status derives in the literature from a number of disparate sources. One major source of confusion is engendered by conflating arguments relating to fundamental welfare objectives and those relating to instrumental efficiency. Whether an item qualifies as a merit good depends on its distributional impact. It is quite a separate, *derivative*, and hence instrumental, question as to whether these goods should be provided directly in kind or through general income support. We therefore begin by identifying the rationale of merit goods at the general level, before moving on to consider the instrumental efficiency of alternative methods of provision.

Distributional Arguments and Merit Goods

The case for State intervention in the provision of merit goods rests ultimately on distribution. Though there are well-known efficiency arguments about the demand elasticities of particular goods and services, about the distortions caused by revenue raising and subsidy, and about price distortions which support intervention in energy, these are secondary in so far as they relate to the design of distributional policy, but not its rationale.[1]

Within the class of distributional arguments, we shall be exclusively concerned with poverty rather than inequality as such. Such a concern links directly with the principles behind the Welfare State. This has been based on the notion that an individual should have access to basic minimum social or merit goods to provide a minimum standard of living. The consumption of these goods provides the basic capabilities to enjoy a decent standard of living. This is a positive right or freedom.[2]

---

[1] The classic discussion of Ramsey pricing is relevant here, but in itself has nothing to do with the 'merit goods' arguments advanced here. See Atkinson and Stiglitz (1980) for an exposition of Ramsey rules.

[2] See Sen (1982) for an exposition of the capabilities approach. An accessible survey is provided by Dasgupta (1986).

Equality in provision may not, under this scenario, be desirable. The handicapped person needs far more resources to be capable of enjoying this minimum standard of living. Thus, on this view, goods are provided according to their ability to enhance capabilities. Some capabilities are, however, more important than others—the capability to read is much more basic or primary than the capability to enjoy beer. The demarcation principle is between more and less fundamental capabilities.

This demarcation can be made more precise. The most basic capability is that of survival, and a minimum provision of food, clothing, heating, health, and shelter are thus required. These *absolute* requirements for survival will change little over time. A second group of goods provide for a minimum *participation* in society—education, employment, and mobility may fall into this category. These groups are not entirely distinct, but they enable a separation between *absolute* and *relative* provisions in considering poverty. To be poor in this capabilities approach is to be *absolutely* deprived of requirements for survival and *relatively* deprived in participation at a minimum level in society. For example, a rural inhabitant is absolutely poor if he or she has insufficient food and shelter, and relatively poor if he or she is unable to participate as a result of the lack of a bus service.[3]

Thus some goods are more important than others in sustaining the capability to enjoy a decent standard of living, and which goods fulfil this role will depend upon whether it is absolute or participation capabilities with which we are concerned. Absolute ones will typically have priority, since these must be met before participation can be considered, but not vice versa. The fundamental question that arises in the context of this paper is whether energy is an absolute or participation merit good. We decide this question in Section 3 below.

Efficiency, Market Failure, and the Provision of Merit Goods

Because some goods have greater significance in yielding the capacity to enjoy a decent standard of living, it does not follow that these should necessarily be provided directly. Individuals may be better able to select amongst goods than the State, in deciding how best to meet capability requirements with a given income, since they will typically have a greater knowledge of their own peculiar requirements.

There may, however, be efficiency reasons for providing goods rather than money in some circumstances, and it is to these we now turn. To see how the provision of goods may be more efficient, we need to see how the actual economy may fail to meet the requirements of efficiency unaided by the guiding hand of the State. For the economy to allocate resources efficiently, individuals need to be able to make judgements about the desirability of all alternative

---

[3] Absolute and relative are employed here in a slightly different way from that traditionally used. See Sen (1984).

choices available to them. This can only be achieved if they are perfectly informed about the characteristics of goods and services available to them and about future changes and developments, and if they behave rationally. The goods themselves must be supplied competitively. In such an economy the direct provision of goods could not be superior to individual choice. Where, however, the conditions of perfect competition are not met, 'market failure' is deemed to occur, and we have a prima-facie case for government intervention. Violation of the conditions is, however, only necessary, not sufficient, for intervention. For against these problems of market failure must be set a further requirement that the government failure arising from intervention is not worse than the original market one. Clearly we must allow for the possibility that even if there were a market failure in merit good allocation, government officials may on occasions make a greater mess than the status quo ante. The decision to intervene is thus inevitably a pragmatic balancing one (see Helm (1986)).

Let us therefore begin by examining the types of market failure in the energy sector. They occur on both the demand and supply sides. Supply-side failures are substantial in the electricity and gas industries, which both exhibit considerable monopoly elements.[4] But it is on the demand side that the case for direct provision rests. The failures relate to preference and choice behaviour. Individuals may lack the requisite amount of information to make relevant choices. They may, for example, be unaware of the cost of energy or, more importantly, be unable to predict future relative fuel costs in non-repeated choices, such as in selecting an appropriate heating system. They may also behave inconsistently, or indeed be unable to form a preference.[5]

These energy market failures are reinforced by failures in the capital market. Seasonal fluctuations in demand may require access to credit markets for those on fixed incomes. For these groups, the imperfect market in bank lending may substantially bias these expenditure decisions.

Where market failures are substantial, the case for intervention is stronger. The State may provide the relevant information (in energy, through metering and other information services) or alternatively, where this is costly, intervene to replace or augment individual choice. Direct intervention requires that further conditions be met. In particular, the State needs: (a) to be better informed; (b) to know the preferences; (c) to have the appropriate incentives to maximize those preferences; and finally (d) the intervention costs must be less than the value of the distortion that would result if the individual made the choice unaided.

Thus the instrumental case for direct provision rests on the extent of these preference failures on the one hand, and the costs and effectiveness of alternative policies on the other. There is mounting evidence that casts doubt on

---

[4] See Hammond, Helm, and Thompson (1985 and Chapter 8, this volume) on gas and electricity respectively.
[5] It should be noted that even if the failure relates to lack of information, it does *not* follow that provision of information is *necessarily* superior to provision of the good directly.

Energy Policy, Merit Goods, and Social Security

the traditional assumptions concerning preferences, and hence utility functions (especially the requirements of transitivity, reflexivity, and completeness); see Kahreman, Slovic, and Tversky (1983).

Energy, Market Failure, and Welfare

So far we have argued that a merit good is one that relates directly to a basic or participatory capability. Whether it should be provided directly or indirectly depends on the extent of market failure. Delegating choice is more important, the greater the preference and informational failures. These in turn may be reinforced by monopoly elements.

That energy is a merit good should follow from its being necessary in order to fulfil the basic capability of living. Whether to provide it directly depends on the extent of preference and informational failure. The former can be empirically adjudicated by examining the importance of its role in the budgets of the poor. The latter can be addressed by considering the information available to individuals in purchasing energy and complementary durable goods. The first step is therefore to establish how important energy is in total expenditure by the poor. We then need to consider the extent to which the State's current methods of provision meet these requirements, and whether they could be bettered by alternatives.

3. Energy Consumption Characteristics

In this section we look in some detail at the pattern of household energy consumption. We begin by examining cross-section data, which provide us with information about differences in consumption at different income levels and for different household types, and time series data on the share of energy in aggregate consumption. We then move on to look at methods of heating for different groups.

Household Budgets

The most convenient source of information about the energy expenditures of the poor is the Family Expenditure Survey (FES). The FES is an annual government survey of around 7,000 randomly selected UK households which provides detailed information about expenditures, incomes, and demographic characteristics of the respondents.[6] Figure 3.1 illustrates the pattern of energy expenditures at different income levels in the 1984 FES. We can see that as income rises there is some increase in expenditure on energy, but this is only relatively small. It is also the case that much of this increase is related to the

---

[6] See Kemsley, Redpath, and Holmes (1980).

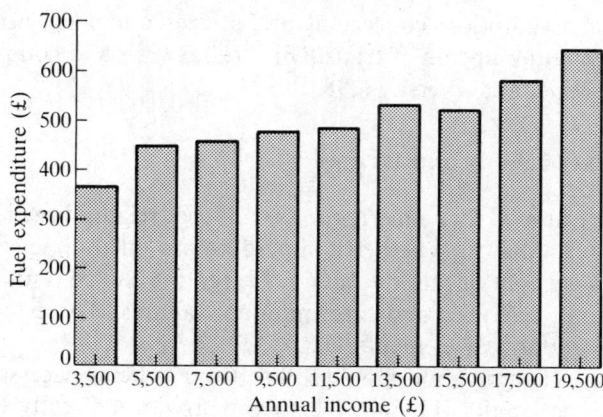

Figure 3.1 Annual household energy expenditures at different income levels (FES, 1984)

tendency for larger households to have higher incomes. Figure 3.2 illustrates the point that if we confine ourselves to looking at the relationship between income and energy expenditures for particular household types, we see almost no growth as we move from low incomes to high.

This pattern strongly suggests that we could think of energy as being a merit good, the consumption of which forms a vital part of life. Furthermore, the relatively constant level of expenditure tends to suggest that energy represents an absolute or basic capability. But such a conclusion may be premature, and for two reasons. First, we need to know how large a proportion of the expenditure of the poor is absorbed by energy: expenditure on matches may well be very similar at all income levels, but because it is so small an absolute expenditure, it is of little importance. What of energy? Figure 3.3 charts the proportion of total expenditure absorbed by energy and that absorbed by food at different levels of household income. For the poorest group the proportion spent on energy exceeds 13 per cent—less than the 30 per cent of their expenditure that goes on food, but more than the 10 per cent on housing. While there is no explicit benefit paid to the poor to cover their food expenditure, housing benefit is a means-tested benefit paid to almost one-third of the households in the United Kingdom. The proportion of total expenditure absorbed by energy falls very rapidly, to around 7 per cent for those with incomes in excess of £17,500.

It is perhaps also worth noting that while energy expenditure forms on average between 5 and 7 per cent of the budgets of non-retired households, it forms 17 per cent of the budget of single retired households mainly dependent on the state pension, and 12 per cent for retired couples mainly dependent on the state pension. Fuel expenditure forms a large part of the total budget of the poor, and rises in absolute terms very slowly with income. We would thus seem

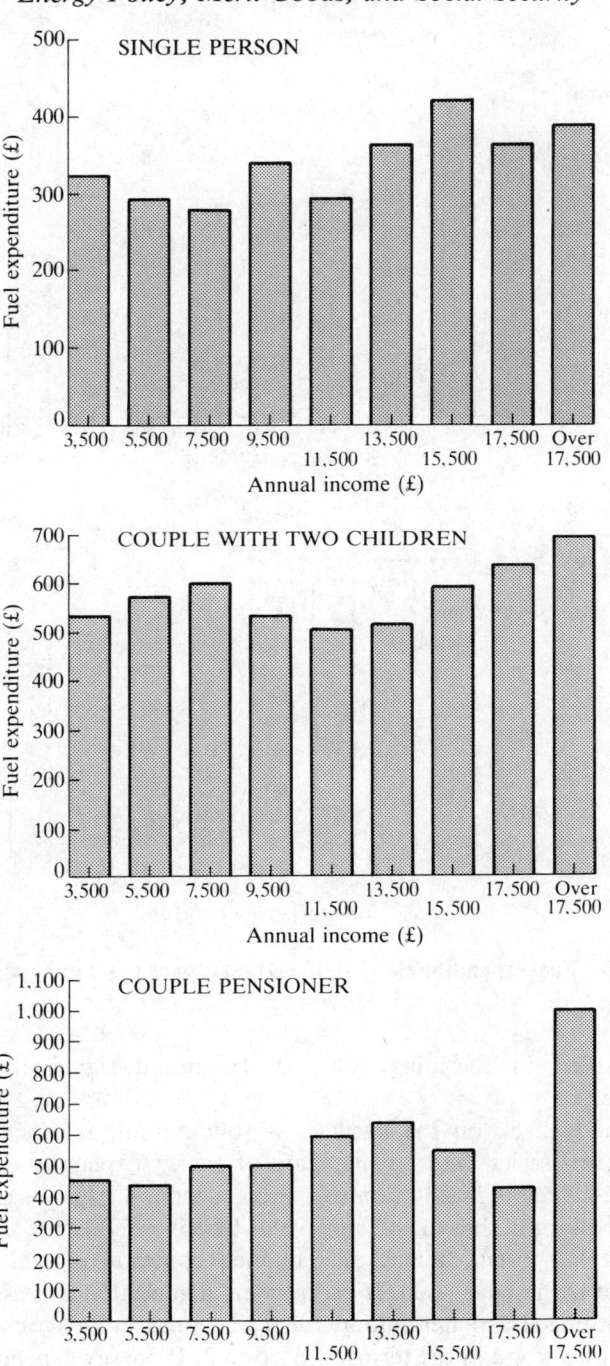

Figure 3.2 Annual household energy expenditures at different income levels

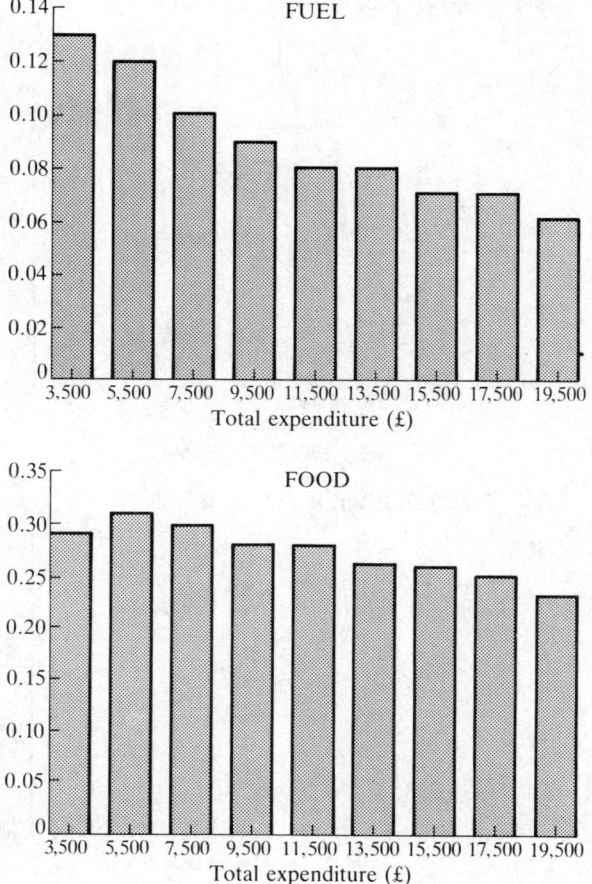

Figure 3.3 Fuel expenditure and food expenditure as share of total expenditure

to be justified in ascribing some distributional significance to energy expenditure.

The second check on the basic or absolute result relates to time series evidence. Time series data on the share of energy expenditure in household budgets are available in addition to the cross-section data we have examined above. The question that most concerns us here is whether the share of energy expenditure has risen, fallen, or remained constant as real incomes and standards of living have risen. If energy were a good that was essential to life, but not consumed in greater amounts as real incomes grow, we could think of it as an 'absolute' good in the terms of Section 2. If energy expenditure remains constant or grows as a share of total expenditure, this suggests that it has become a 'participation' good; that is, that increased amounts are required to

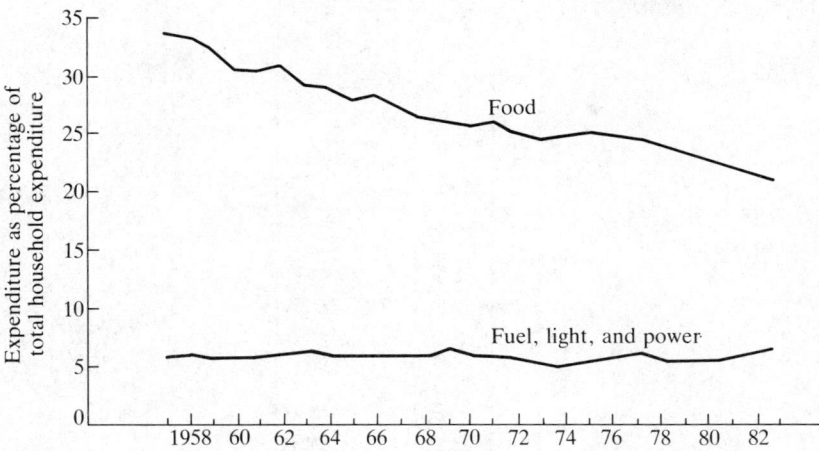

Figure 3.4 Changes in pattern of expenditure: all households (FES, 1957–83)

achieve the socially acceptable minimum standard of living. As Figure 3.4 shows, energy expenditure has remained constant as a share of the budget over a period when real incomes rose enormously. Food expenditure, by contrast, has almost halved its share of the budget. Despite the fact that, in a given year, energy expenditure seems relatively unrelated to income, as the general standard of living has risen in the UK, the required amount of heating and lighting has also risen, pointing energy out as an example of a 'participation' good. Thus it would seem that energy is both an absolute and a participation merit good.

Information and Market Failure for Energy

There are a number of further issues of fact which are relevant and can be assessed using FES data. As we noted in Section 2, one of the requirements for the market to be efficient in allocating resources is that individuals are free to make decisions about their own consumption. When considering freedom of choice and energy expenditure, we must be aware of the possible constraint imposed by the capital goods, especially central heating appliances, that households need to consume energy. Tables 3.1, 3.2, and 3.3 present an analysis of the type of central heating used by households at different income levels, in different types of housing, and of different composition. Table 3.1 demonstrates very clearly that as household income rises, the probability that some form of central heating will be available also rises, and that the type of central heating that is available also changes. While those with household incomes of less than £3,500 per annum are less than 3 times as likely to have gas central heating as electric, those with incomes in excess of £17,500 p.a. are 14

Table 3.1. Percentage of households with various types of central heating, at different income ranges

	Income range (£ per annum)								
	0–3,500	3,500–5,500	5,500–7,500	7,500–9,500	9,500–11,500	11,500–13,500	13,500–15,500	15,500–17,500	Over 17,500	Total
Electric	11	10	9	9	6	7	7	6	5	8
Gas	28	35	38	44	54	58	60	62	70	47
Other	8	12	11	13	12	11	11	12	14	12
None	52	44	42	34	28	23	22	20	11	33

Source: IFS analysis of 1984 Family Expenditure Survey data.

Table 3.2. Percentage of households with various types of central heating, by tenure type

	Tenure type			
	Council	Private rented	Mortgaged	Owned outright
Electric	12	7	5	9
Gas	34	20	65	44
Other	11	6	12	13
None	43	67	18	34
Percentage of population	30	8	38	22

Source: IFS analysis of 1984 Family Expenditure Survey data.

Table 3.3. Percentage of households with various types of central heating, by household type

	Single	Single-parent families	Childless couples	Couples with children	Multiple units	Single pensioners	Couple pensioners
Electric	10	8	9	5	6	15	9
Gas	40	42	49	58	50	30	40
Other	7	9	12	13	14	9	12
None	43	41	31	24	30	46	39

Source: IFS analysis of 1984 Family Expenditure Survey data.

times as likely to have gas as electric central heating. This pattern might suggest some constraint on the choice of the less well off over the form of central heating they should use. Table 3.2 may provide some partial explanation for this. Thirty per cent of households live in council accommodation, and a further 8 per cent in private rented accommodation. These households are typically not in a position to exercise any choice over the type of heating in their housing, are less likely to have any central heating at all, and are more likely than other groups to have electric, oil, or solid energy than gas if they do. In Table 3.3 we see that single pensioners are the group least likely to have any form of central heating, and couple pensioners the least likely to have gas if they have any form of central heating. While these tables present only circumstantial evidence, they appear to cast considerable doubt on the plausibility of a model where individual households choose freely between alternative forms of heating appliance. This qualification is particularly important when applied to those in rented accommodation, many of whom are among the poorer sections of the UK population. That a problem exists seems clear from the evidence of death from hypothermia. There were 578 hypothermia-related deaths in the first quarter of

1986, and over 6,000 more deaths than would have been expected during the five very cold weeks of February and early March of that year.

We conclude from the available empirical evidence that energy is distributionally sensitive, in the sense that it comprises a considerable proportion of the budgets of the poor. There is evidence that it has absolute characteristics. However, since time series data suggest that energy demand does not significantly decline with income, energy is also a participation merit good.

4. Energy and the Social Security System

In this section we examine the present social security treatment of energy, and describe briefly current disconnection policy. We identify the extent to which energy is in fact treated as a merit good, and examine the extent to which it is supplied directly to the poor.

The general principle underlying the treatment of energy expenditure in the social security system is that the standard levels of the main benefits are intended to cover energy expenditure without the need for special energy-related payments. Such payments will typically only be made in cases of special need, such as disability or inefficient/expensive heating systems.

The main benefit under which additional payments for energy are possible is supplementary benefit. The basic scale rates are supposed to cover energy costs, but weekly additions can be paid under a number of circumstances.[7] It is important to note that under the social security reforms which came into effect in April 1988, there are no heating additions within the income support system, which replaced supplementary benefit. The extent to which support is considered relevant only under exceptional circumstances can be seen from the list of entitled categories, which have built up into a complex maze.

(a) *Ill-health*. If the claimant, claimant's partner, or child need extra warmth, entitlement is increased by £2.20 per week. If the illness 'substantially reduces the claimant's capacity to lead a normal life', entitlement is increased by £2.55 p.w. Where more than one person satisfies the condition, only one addition is paid.

(b) *Accommodation*. An addition of £2.20 p.w. is available where a home is difficult to heat, particularly where the rooms are draughty, damp, or very large. Where the home is exceptionally difficult to heat, the addition is £5.55 p.w.

If a person or family satisfies the condition for a £2.20 p.w. addition on grounds of both ill-health and accommodation, an addition of £5.55 p.w. will be paid.

[7] See Child Poverty Action Group (1986) for a more detailed description of the treatment of energy costs in the supplementary benefit system.

(c) *Long-term sick or disabled householders* who are entitled to the long-term rate of supplementary benefit are entitled to an addition of £2.20 p.w. where their chances of finding work or their earning capacity are seriously limited.
(d) *Central heating*. An addition for central heating is now only available to those who were receiving supplementary benefit before 5 August 1985 and who had a central heating system before then.
(e) *Housing estate with high heating costs*. Where the Department of Health and Social Security (DHSS) designates an estate as having a heating system with 'disproportionately high' running costs, every claimant on the estate who is a householder should receive a weekly addition. This payment is £4.40 p.w. for a house with four rooms or fewer (excluding bathroom, lavatory, and hall) and £8.80 p.w. where there are five rooms or more.
(f) *Severe disablement*. An addition of £5.55 p.w. is automatically payable where any member of the family is receiving mobility or attendance allowance.
(g) *Age*. Where a member of the family is aged 65 or over, or under 5, an addition of £2.20 p.w. is available. Where a claimant or partner is 85 years or over, an addition of £5.55 is payable.

As well as these weekly additions, there is a possibility of receiving a one-off 'single payment' to help with energy debts. A grant will only be made if the claimant has no money to pay the bill or has put aside too little, and even then in only two situations:

(a) *Exceptionally severe weather*. Until 1986, there was no particular set of criteria for deciding whether or not the weather was exceptionally severe. One of the most used social security manuals commented 'Adjudication officers are expected to use their common sense and local knowledge when judging whether there has been a period of exceptionally severe weather'. Neither 'period' nor 'severe' is defined. Cold weather in the winter of 1985-6, and the realization that in some sense what mattered was how cold it was, rather than whether the temperature was exceptionally severe for a particular area, led to the introduction of the 'trigger temperature' system for single payments in exceptionally severe weather. Under the scheme, if the average temperature for a seven-day period, *beginning on a Monday*, was below -1.5 degrees Celsius, all supplementary benefit recipients in the area would receive a £5 payment for energy. The extremely cold weather in January 1987 saw John Major, the responsible DHSS Minister, announce on a Tuesday that the payment would be made for that week, before knowing what the average temperature would be over the week as a whole (although the weather forecast was fairly unambiguous, and largely correct). The same announcement was made in the following week, simultaneously with the decision to raise the 'trigger temperature' to 0 degrees Celsius from 26 January.

(b) *Lack of familiarity with a new heating system.*

An alternative arrangement to pay off accumulated energy debts is the Fuel Direct scheme. Under this arrangement, supplementary benefit recipients who are in debt can have part of their benefit withheld each week and paid direct by the DHSS to the fuel board. In return, the fuel board agrees not to disconnect. For the scheme to be available, the debt must be more than £30, there cannot be capital from which the debt could be met, there must be no non-dependants in the household who could reasonably be expected to clear the debt, and the benefit being paid must obviously be greater than the planned deduction. Where disconnection does occur, a single payment may be made for paraffin heaters, a cooker, and lighting equipment.

Each week some 1,800 households in England and Wales have their electricity supply cut off because they have not paid their bills, and some 900 households have their gas supply cut off.[8] Since 1976 the electricity and gas industries have voluntarily operated a Code of Practice on the payment of domestic energy bills (Electricity Council and British Gas (1982)). The Code covers the availability of easy payment schemes, the referral of debtors to the welfare agencies for help, and means by which the debt can be cleared. The Code was revised in 1978, 1980, and July 1982. As noted above, the social security system will help with energy debts in only very rare cases, so the Code of Practice largely determines ways in which debts can be repaid by the consumer, thus avoiding disconnection. The preferred route is for a payment plan to be agreed between consumer and industry; such a plan must then be adhered to. Prepayment slot meters can be fitted, and can be preset to recover the debt.

Table 3.4. Number of disconnections as a percentage of number of credit customers

Electricity (12 months to 31 March 1985)		Gas (12 months to 31 March 1985)	
London	0.85	North Thames	0.43
South Eastern	0.27	South Eastern	0.18
Southern	0.20	Southern	0.08
South Western	0.31	South Western	0.03
Eastern	0.60	Eastern	0.17
East Midlands	0.50	East Midlands	0.20
Midlands	0.46	West Midlands	0.41
South Wales	0.74	Wales	0.10
Merseyside & North Wales	0.73	Northern	0.28
Yorkshire	0.72		
North Eastern	0.48	North Eastern	0.18
North Western	0.37	North Western	0.24

Source: Electricity Consumers Council (1985) and National Gas Consumers Council (1985).

[8] See National Gas Consumers Council (1985) and Electricity Consumers Council (1985).

One of the most worrying features of disconnection policy is the enormous divergence of disconnection rates by region, as shown in Table 3.4. It is difficult to conceive of reasons for the disconnection rate for gas consumers to be 14 times as high in the North Thames region as in the south-west, particularly since the variation is so much smaller for the electricity industry.

In the year to 31 March 1985, some 90,000 households had their electricity supply cut off. In the year to 21 December 1986, some 45,000 households had their gas supply cut off. These households will typically be poor. For them, the system does not appear to work. An obvious solution to the problem of poor payment records is the installation of meters. However, neither the gas industry nor the electricity industry seems happy with the current technology, particularly because of security problems. Field trials are now under way on electronic meters which accept plastic cards bought from the fuel board rather than cash. Security is thus much improved. Meters not only solve problems of poor payment, but could provide far more information on the cost of energy as it was consumed. This information might help to prevent individuals from remaining cold through uncertainty as to the cost of heating their accommodation in very cold weather. Prepayment meters would give consumers a far more accurate picture of the price of heating and other services provided by energy. Lack of knowledge of the price of heating is a clear case of information failure, with an apparently obvious, if costly, solution. It will not solve problems of access to credit for the poor, but it will improve their ability to make use of whatever income and credit is available to them.

5. Alternative Methods of Provision

Given that we have identified certain sorts of energy consumption as 'merit goods', and given that we noted a number of arguments for direct provision, we now consider the distributional impact of some alternative means of providing further help with energy costs. Table 3.5 is derived from an analysis of energy expenditure in the 1984 Family Expenditure Survey. Four options, each costing some £1.5 billion, are examined:

(a) making 75 per cent of energy expenditure deductible for income taxation purposes (this policy is analogous to mortgage tax relief);
(b) cutting all energy prices by 15 per cent;
(c) cutting energy prices for single-parent families and pensioners by 60 per cent;
(d) paying all the energy costs of any household receiving supplementary benefit.

The table shows the average gains in pounds per annum that we estimate would accrue to different family types as a result of each of the four options. The least attractive in terms of concentrating help on those most in need is the tax

Table 3.5. Gains from alternative methods of reducing fuel costs for different household types (£ p.a.)

Type of family	75% tax-deductible	15% price cut	60% price cut for one-parent families and pensioners	Fuel on supplementary benefit
Single	44	47	—	55
Single-parent	28	74	297	265
Childless couple	82	70	—	32
Couple with children	96	83	—	56
Single pensioner	30	53	214	87
Couple pensioner	68	73	293	39

Source: IFS analysis of 1984 Family Expenditure Survey data.

deductibility option. This is obviously of no help to someone who pays no income tax. The poorest households tend not to be in work (they tend to be pensioners and the unemployed) and they gain nothing from this option; the group that does best is couples with children; single-parent families gain least. This is simply another possible reform which emphasizes the fact that the direct tax system is too blunt a tool to deal effectively with problems of poverty.

The second possibility is that of a general reduction of 15 per cent in the price of all energy. This is an analogous approach to the zero rating of energy for VAT which effectively provides a 15 per cent government subsidy. For the same reasons that zero rating for VAT is a poorly targeted expenditure, a 15 per cent cut in the price performs fairly badly. Nonetheless, such a price cut is probably better in this sense than tax deductibility, since at least some of the benefit goes to the very poor.

A more effective means of helping particular groups that can be identified as having low incomes is to cut prices only for those specific groups. At a similar cost to the other options, prices for retirement pensioners and single-parent families could be cut by 60 per cent. There are two obvious criticisms of such a proposal: one is that it gives nothing to poor households that do not fall into these categories; the second is that it provides assistance to richer members of the 'poor' groups. The obvious response to such criticism is to give help explicitly to those identified as being in need by the means-tested benefit system. The last option illustrated pays in full the energy bills of all households in receipt of supplementary benefit. The largest average gains go to single parents and pensioners, but poor members of the other groups also gain. The greatest objection to the use of the supplementary benefit system is its low take-up rate, which could lead to many of those most in need failing to receive their entitlements. These four alternatives illustrate a progression from highly untargeted to well-targeted payment, but none mirrors the UK system.

We noted at the beginning of Section 4 that the general principle underlying the treatment of energy expenditure in the social security system is that the standard levels of the main benefits are intended to cover energy expenditure without the need for special energy-related payments. There are a number of cases in which additions to the level of benefit are made in specific response to either particular needs for heating, as for the old or sick, or difficulty in achieving a given level of heating, as for accommodation that is classified as having high heating costs.

We have to ask why this form of half-way house has been adopted, in preference to a full expenditure-specific benefit like housing benefit, or the much more limited system of additions for specific dietary needs in the case of food. Slightly more than 7 million households in the UK receive means-tested housing benefit, more than one-third of the population. The main reason for the continued existence of this benefit is the continued existence of enormous regional variation in housing costs in the UK. A standard supplementary benefit level that included enough to pay for private rented accommodation in London would provide a household living in a low housing cost region of the north with a much higher standard of living. A standard supplementary benefit level that included only enough to pay for accommodation in a low housing cost region in the north would provide an unacceptably low standard of living in London. The variation in housing costs makes their inclusion in the standard benefit level almost impossible. The contrast with food is quite striking. While it is true that there is some regional variation in the cost of food, that variation is moderately small; thus including some notional standard amount in the benefit level is unlikely to lead to enormous inequities.

Figure 3.5 compares the distribution of housing expenditure and energy expenditure around their means in the 1984 FES. It is clear that the distribution of energy expenditure is much less variable than that of housing. This should come as no surprise to us, since the price of energy is very similar throughout the country. But before we jump straight ahead to saying that this relative stability of price should allow us to include a standard level of energy allowance in benefits, we must contrast the intrinsic value of energy with that of other goods. One characteristic of food bought in a supermarket is that little has to be done to it to produce that which is then eaten; most consumers have access to a cooker, the running costs of which are not large in relation to the cost of raw food, and the running costs do not vary enormously between different types of cooker.

Consumers do not desire energy as a final good, they desire warmth and power. Different forms of energy are more or less expensive, and more or less efficient at producing heat. Different forms of heating equipment are more or less efficient at transforming energy into heat. Different forms of housing are more or less efficient at retaining heat generated. Thus, there are a number of possible arguments for a social security system that includes specific payments for energy that are related to the circumstances of individual households. Such arguments are strengthened by the observations that many benefit recipients do

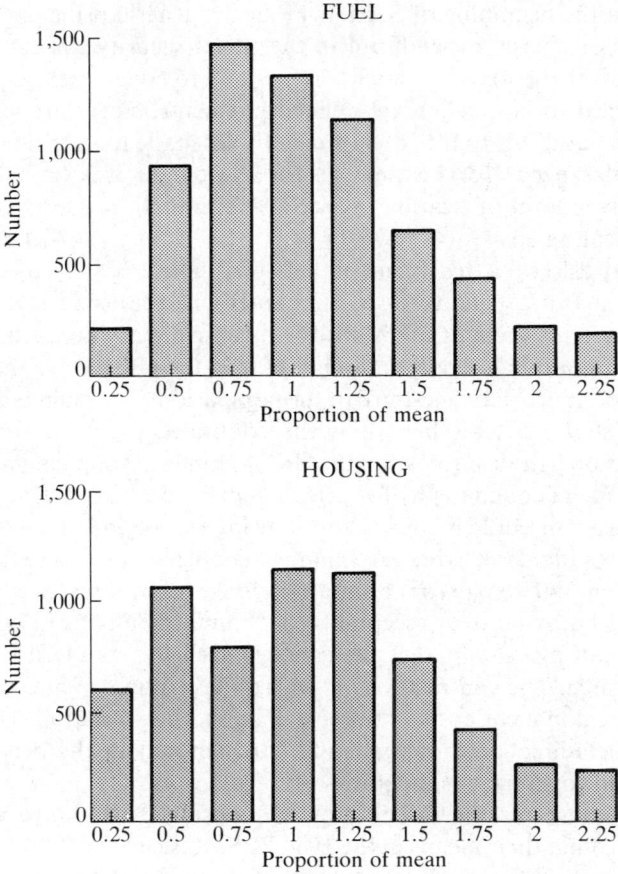

Figure 3.5 Distribution of expenditures relative to the mean

not choose their own form of heating system, since they live in rented accommodation, and that their accommodation may not be draught-proofed or insulated. Nonetheless, such additional payments complicate any system, and are costly to administer. One solution is to increase the general level of benefits to such an extent that all such problems are covered. There are two objections to this: such a change would be expensive, and it would be inequitable, providing substantial increases in the standard of living of some households while leaving others, with high heating costs or needs, at their initial level. If specific payments must be retained, it seems sensible to attempt to define the circumstances under which entitlement exists as clearly and objectively as possible, to simplify administration and to encourage high levels of take-up.

6. Conclusions

We have shown in this article that energy is a merit good. It has absolute characteristics, demonstrated by its role in the household budgets of the poor, and participatory ones, supported by time series data. There is thus a strong case for intervention on the basis of distributional arguments.

The present system only partly recognizes this, through the VAT treatment of gas and electricity, and the complex and specialized social security assistance. The desirability of existing policy can, however, only be judged in comparison with alternatives. From our examination of these, we conclude that general income tax deductibility and price cuts are amongst the least effective because they do not target the poor. A specific 'energy supplement' is much more desirable on both distributional and efficiency grounds—distributionally because it is of most benefit to the poor, on efficiency grounds because it can be designed to meet the informational and other market failures.

The comparison of policy options has an important corollary for the current debate regarding the imposition of VAT on electricity and gas. The VAT exemption is not targeted on poor households. It is not, therefore, a particularly good method of merit good provision; it is expensive; and it has undesirable efficiency characteristics. However, should VAT be imposed, it must be accompanied by appropriate adjustments to the social security system. An energy supplement would fulfil this requirement.

Part II

Energy in the Public Sector: Modelling and Performance

4 The Demand for Energy*

DEPARTMENT OF ENERGY

1. Introduction

This paper describes an econometric model of energy demand, disaggregated by sector and by type of fuel. It has been used by the Department of Energy in preparing projections of future energy demand, given projections of the exogenous variables, and in sensitivity analyses to assess the effect on demand of changes in the explanatory variables.

For the purpose of constructing this model of energy demand, the economy is disaggregated into seven sectors. Models for six of them are presented here: iron and steel, other industry, domestic, public administration, agricultural, and miscellaneous (the rest). A separate model not described in this paper is used for the transport sector.[1] For most sectors a basic energy demand model has been used in which a single equation is estimated for the total useful energy consumed (adjusted for efficiency in end-use) and a set of equations determine the fuel shares, given the total energy consumed; these are specified as depending on the paths of relative fuel prices.

The structure of the paper is as follows. In Section 2 the methodology adopted is outlined. Section 3 sets out results showing the elasticity of energy demand with respect to changes in output and with respect to changes in the overall level of real energy prices. Section 4 presents demand elasticities for individual fuels in response to changes in relative prices. These are based on model simulations rather than on coefficients in estimated equations. In Section 5 the robustness of the results is considered, in particular in relation to the various oil price shocks. Section 6 considers some other characteristics of the model.

2. The Model for Energy Demand

For most of the energy-consuming sectors, a common basic form of energy demand model has been adopted. A single equation is used to express the demand for total useful energy consumed in the sector. A further set of

---

* This chapter is based on a paper presented to the Sizewell B Inquiry by the Department of Energy in 1983 entitled 'Department of Energy Paper on Energy Projections Methodology', which is available from the Department on request. It is Crown copyright and is reproduced with the permission of the Controller of Her Majesty's Stationery Office.

[1] Details of the transport model are given in Department of Energy (1983b).

equations are employed to express the share of each fuel in the total demand derived from the first equation. Measures of useful energy consumption have been derived by applying average efficiency factors to data for energy supplied. These factors have been assessed for each fuel consumed in each sector over the data period and are treated as being constant. The main factors included as explanatory variables for energy demands are current and past values of the levels of economic activity and energy prices, together with a temperature variable.

The form of these equations is discussed below, using the model for the 'other industry' sector as an example.

Equation for Total Useful Energy

The equation for total useful energy demand in the 'other industry' sector is

$$\log E_t = 2.32 + 0.622(\log YE_t - 0.619 \log YE_{t-1}) \\ - 0.0871(\log T_t - 0.619 \log T_{t-1}) \\ - 0.0854 \log RPE_t + 0.619 \log E_{t-1} + e_t. \quad (1)$$

This equation expresses, in logarithmic terms, the total energy demand (E) in this sector in year t as depending on levels of economic activity (YE) and annual temperature (T) in year t and the previous year $t-1$, a real price of energy variable (RPE) in year t, the demand for total useful energy in the previous year, and an error term.

Estimation

Equation 1 may be rewritten as

$$\log E_t = a + b(\log YE_t - k^* \log YE_{t-1}) \\ + c(\log T_t - k^* \log T_{t-1}) \\ + d \log RPE_t + k \log E_{t-1} + e_t \quad (2)$$

This equation was estimated using ordinary least squares (OLS) by choosing a value for k^* and estimating a, b, c, d, and k, replacing k^* by the estimated value k and repeating until there was no difference between k and k^*.

Interpretation as a Partial Adjustment Model

By algebraic manipulation and setting $k^* = k$, equation 2 can be rewritten as

$$\log \bar{E}_t - \log \bar{E}_{t-1} = (1-k)(\log \bar{\bar{E}}_t - \log \bar{E}_{t-1}) + e_t \quad (3)$$

where

$$\log \bar{E}_t = \log E_t - b \log YE_t - c \log T_t, \quad (4)$$

$$\log \bar{E}_{t-1} = \log E_{t-1} - b \log \text{YE}_{t-1} - c \log T_{t-1}, \tag{5}$$

and

$$\log \bar{\bar{E}}_t = \frac{a}{1-k} + \frac{d}{1-k} \log \text{RPE}_t. \tag{6}$$

The interpretation of this reformulation is as follows.

Equations 4 and 5 express $\log \bar{E}$ as $\log E$ adjusted for the current levels of industrial output and temperature. Equation 6 expresses the 'desired' or 'equilibrium' level of energy demand in year t, after adjustment for output and temperature, in terms of the price variable in year t. Finally, equation 3 implies that the actual change, in logarithmic terms, in energy demand between years $t-1$ and t is a proportion $(1-k)$ of the equilibrium change from the actual level in year $t-1$.

The model has a number of implications. Firstly, in this form of model the level of energy demand in any year depends on the levels of industrial output and temperature in that year, and not in earlier years. Secondly, the level of energy demand in year t, once adjustments have been made for output and temperature levels, depends on the price in that year and the energy demand in the previous year. The last result is drawn on again in the next section when interpreting the nature of decision making implicit in this form of model.

Dynamic Properties of the Equation for Total Useful Energy

Equation 1 can be re-expressed in the following form (the derivation is set out in Department of Energy (1983b)):

$$\log E_t = \frac{a}{1-k} + b \log \text{YE}_t + \log T_t$$
$$+ d(\log \text{RPE}_t + k \log \text{RPE}_{t-1} + k^2 \log \text{RPE}_{t-2} + \ldots). \tag{7}$$

Equation 7 implies that changes in output YE_t and temperature T_t affect total energy demand E_t only in the year in question. However, E_t depends on current and all past values of real energy prices RPE_t, RPE_{t-1}, RPE_{t-2}, etc. Correspondingly any change in RPE_t will affect all later values of E, i.e. E_t, E_{t+1}, E_{t+2}, etc., but with decreasing weights as k^n decreases with n ($0 \leqslant k < 1$); specifically the elasticity of E_{t+n} with respect to RPE_t is $d.k^n$. For $k = 0.619$, the weights in the power series indicate the contribution to $\log E_{t+n}$ from a unit step change in $\log \text{RPE}_t$. These coefficients build up in the manner shown in Table 4.1. Thus 38 per cent of the effect of a change in the real price of energy on total useful energy demand will occur in the current year, 62 per cent by the first year after, 85 per cent by the third year, 97 per cent by the sixth year, and so on.

Although equation 7 implies that the past history of energy prices affects the current level of energy consumption in this sector, it is not suggested that

Table 4.1. Demand effect of an energy price change

Time period	Weight	Cumulative weight	Percentage
t	$d = .0854$	.0854	38
$t+1$	$dk = .0529$	.1383	62
$t+2$	$dk^2 = .0327$	.1710	76
$t+3$	$dk^3 = .0203$	.1913	85
$t+4$	$dk^4 = .0125$	.2038	91
$t+5$	$dk^5 = .0078$	.2116	94
$t+6$	$dk^6 = .0048$	.2164	97
$t+\infty$	$dk^\infty = 0$	.2241	100

industrialists consciously think back through the history of energy prices to determine their current level of energy consumption. Rather, the history of energy prices up to the previous year is embodied in the stock of energy-using equipment they had last year, and this determined their use of energy then (after taking account of the output level and the temperature in that year). The only decision industrialists make this year is to adjust their energy-using equipment from last year's position *taking account of this year's energy price only*, as indicated in equation 3.

This interpretation is preferred to an alternative view of equation 7 which views the term in brackets as an expectation of future energy prices based on past experience of energy prices. This interpretation would imply that industrialists could, within a given year, adjust their consumption of energy to the equilibrium level corresponding to their current price expectations. Since energy use is determined almost entirely by the stock of energy-using plant, equipment, and buildings, this is extremely unlikely to be the case. The interpretation given to the model is important because it affects the treatment of the error term when estimating the parameters of the model.

Equations for Fuel Shares

The model equations used to split the total demand for useful energy in the 'other industry' sector into the demands for individual fuels are set out in equations 8 to 11 below. The split is achieved in two stages. Electricity is used, in the main, for purposes other than bulk heating, so a single equation is used to determine the share of electricity demand in total energy demand. A further set of equations, 9 to 11, are then used to allocate the residual non-electricity energy demand between the three fuels coal, gas, and oil.

Equation 8 is similar in form to equation 1 above. In logarithmic terms, the share of electricity (including electricity consumed from own generation) in total energy demand (EL/E) is expressed in terms of the current price of electricity relative to the average price of energy (in useful energy terms)

(PEL/PAV), an electricity weighted output term (YEL), the share of electricity demand for the previous year, a dummy variable for the three-day week in 1974 (DUM), and an error term. Temperature was not found to be significant in this equation and has been omitted. The presence of the lagged term in the share of electricity demand implies, following the description given above for the total energy demand equation, that the relative price and output terms operate with a distributed lag on the share of electricity in total energy demand.

$$\log\left(\frac{EL}{E}\right) = -1.63 - 0.247 \log\left(\frac{PEL}{PAV}\right)_t + 0.244 \log YEL$$

$$+ 0.496 \log\left(\frac{EL}{E}\right)_{t-1} - 0.0509 \, DUM + e_t. \tag{8}$$

The model used to split the non-electricity energy demand into demands for coal, gas, and oil becomes more complex because of the need for the sum of the demands for the individual fuels to equal the corresponding total. A number of methods are available for this type of problem. The model chosen is a form of multinomial logit which is relatively easy to estimate and can incorporate the lagged effects of relative price movements.

This type of model has been used by a number of authors and is described by Pindyck (1979). Provided that the data available represent the results of energy-consuming decisions by a large number of individual consumers, ordinary least squares may be used for estimation. The model can be interpreted as defining a choice index for each fuel as a function of its attributes—in this case its price expressed in useful energy terms.

Equation 9 expresses, in logarithmic terms, the ratio of the share of oil to that of coal (SO/SC) in non-electricity energy demand as depending on the prices of oil (PIO) and coal (PIC) relative to the average price of all three fuels (PAM) in the current year, the shares of oil and coal in the previous year, and an error term. Equation 10 provides a similar relationship for gas and coal and equation 11 ensures that the shares add to unity. The dynamic properties of this system are complex, but they ensure that the mix of fuels does not respond in full immediately to a change in relative fuel prices, but rather that it responds over a period as capital equipment is adjusted to changing relative fuel prices.

$$\log\left(\frac{SO}{SC}\right)_t = 0.144 - 0.223 \log\left(\frac{PIO}{PAM}\right)_t + 0.585 \log\left(\frac{PIC}{PAM}\right)_t$$

$$+ 0.870 \log SO_{t-1} - 0.940 \log SC_{t-1} + e_t. \tag{9}$$

$$\log\left(\frac{SG}{SC}\right)_t = -0.213 - 0.432 \log\left(\frac{PIG}{PAM}\right)_t + 0.585 \log\left(\frac{PIC}{PAM}\right)_t$$

$$+ 0.724 \log SG_{t-1} - 0.940 \log SC_{t-1} + e_t. \tag{10}$$

$$SO + SG + SC = 1. \tag{11}$$

The estimated equations are reported elsewhere (see Department of Energy (1983b)). In the next two sections we discuss the main findings.

3. Sectoral Demand Elasticities

The main findings for each of the sectors modelled are set out in Table 4.2.

Table 4.2. Sectoral energy demand elasticities[a]

	Output or income elasticity		Average real price elasticity	
	Short run	Long run	Short run	Long run
Iron and steel	0.91	0.91	−0.11	−0.11
Other industry	0.62	0.62	−0.09	−0.22
Domestic:				
Space and water heating	0.53	0.89	−0.26	−0.43
Cooking	0.12	0.12	No price term	
Electricity (other)	1.9[b]	1.9[b]	No price term	
Public administration	0.43	1.16	−0.20	−0.54
Miscellaneous	0.80	0.80	−0.22	−0.22

[a]Excluding agriculture, for which there is no total energy equation, and transport, which is not covered in this paper.
[b]The equation is non-linear; the 1980 value of the elasticity is reported.

In the case of *iron and steel* a combination of a constant elasticity of energy demand with respect to production and a linear time trend was found to be the most satisfactory way of capturing the impact of technical change on energy demand. The equation implies that demand for useful energy varies with production with an elasticity of 0.91. At a constant output, demand for energy declines, according to the equation, at a proportional rate of 0.077 per cent per annum. The coefficient on the price term was not significantly different from zero at the 5 per cent level.

The output elasticity for *other industry* is lower than that estimated by Pindyck for the United Kingdom (0.78) and is at the lower end of the range of estimates for ten developed countries (0.62 to 0.86; see Pindyck (1979)). However, there are obvious problems in comparing estimates of elasticities from different studies. These estimated price elasticities for 'other industry' are generally lower than those quoted by other authors; for a survey of results see Hawdon and Tomlinson (1982).

The elasticity estimates for *domestic space and water heating* are within the very wide range of estimates quoted in Hawdon and Tomlinson (1982). The income term affects energy demand with a longer lag than the activity effect on the 'other industry' sector. The term in the real price feeds through rather faster than that in the 'other industry' sector. No significant price terms or lag structure were found in the case of *domestic cooking*. The equation for *other domestic* electricity consumption is non-linear, with an income elasticity that falls as income rises.

The major use of energy in *public administration* is for space and water heating. Lagged effects were found for both price and activity terms; the estimated elasticities are shown in the table.

The *agriculture* sector consists of two equations, one for oil and one for electricity. In the equation for oil demand, the elasticity estimates for price and output are of the correct sign but neither is statistically significant at the 5 per cent level. The equation for electricity has no price effect; the output elasticity is just significant. The two fuels are assumed to be non-substitutable.

The *miscellaneous* sector comprises mainly office and commercial activity in the private sector. Apart from specialist uses of electricity for lighting and electrical equipment, the main use of energy in this sector is for space heating. The output and price elasticities for total energy demand were estimated as 0.8 and -0.22 respectively.

4. Relative Price Elasticities of Fuel Shares

As before, the results for each sector of the economy are considered in turn. In the case of *iron and steel*, it is assumed that the scope for substitution between fuels is limited by technological considerations. The future paths of demand for the different fuels in this sector are therefore exogenously determined by changes in technology rather than by relative prices.

In the case of *other industry*, the price responses of the total demand model are complex because of the nature of the fuel mix equations. In this model the price elasticities for individual fuels are not constant but depend on the shares of each fuel in total useful energy demand. They thus vary as the fuel mix and total energy demand adjust to equilibrium following a change in fuel prices. Estimates of own- and cross-price elasticities for the four fuels electricity, coal, oil, and gas for this sector are shown in Table 4.3. They have been calculated by increasing each fuel price in turn and working out the impact on fuel use over time.

The estimated responses of fuel shares to relative price changes in the first year are considerably smaller than the longer-run responses, as expected. The own-price elasticities—the elasticity of each fuel share with respect to a change in its own price—are given in the diagonal elements of the tables and are all negative. The cross-price elasticities—the elasticity of each fuel share with respect to a change in the price of another fuel—are given by the off-diagonal elements in the tables. These are, in the main, positive and small.

Estimates of price elasticities of energy demand in the industrial sector vary

Table 4.3. Other industry: price elasticities

SHORT RUN (same year)

	Percentage change in demand for:			
	Electricity	Coal	Oil	Gas
1% increase in price of:				
Electricity	−0.2	0	0	0
Coal	0	−0.4	0	0.1
Oil	0.1	0.2	−0.2	0.1
Gas	0	0.2	0.1	−0.3

LONG RUN

	Percentage change in demand for:			
	Electricity	Coal	Oil	Gas
1% increase in price of:				
Electricity	−0.4	0.1	0.1	0.1
Coal	0.2	−2.4	2.9	1.4
Oil	0	1.2	−2.3	−0.1
Gas	0	0.8	−0.7	−1.5

considerably between studies; see Hawdon and Tomlinson (1982). Comparison is difficult because of the separation made here between the short run and the long run. However, the own-price elasticities obtained by Halvorsen (1976) for the USA in a static model are: electricity, −0.92; coal, −1.52; fuel oil, −2.82; and gas, −1.47. With the exception of electricity, these are not dissimilar to the estimated elasticities in Table 4.3.

The relative sizes of the long-run own-price elasticities have the following intuitive explanations. Electricity is used in the main for special purposes for which there are few substitutes, such as lighting and motive power; this suggests that its own-price elasticity should be lowest. Gas tends to be used as a premium fuel for heating where cleanliness and ease of control are important, so its own-price elasticity is likely to be less than those for coal and oil, which are the fuels used for bulk heating where substitution is more likely and where efficiencies in fuel use may be gained by investing in more modern equipment.

Table 4.4 sets out the price elasticities for *domestic* energy demand. These estimates are consistent with the rather wide range available from other studies; see Pindyck (1979).

The relative price elasticities for fuel shares in the *public administration* and *miscellaneous* sectors are shown in Tables 4.5 and 4.6. It has been assumed that there is no fuel substitution in the *agriculture* sector.

Table 4.4. Domestic energy demand: price elasticities

SHORT RUN (same year)

	Percentage change in demand for:			
	Electricity	Coal	Oil	Gas
1% increase in price of:				
Electricity	−0.1	0	0	0
Coal	0	−0.2	0	0
Oil	0	0	−0.2	0
Gas	0	0	0	−0.2

LONG RUN

	Percentage change in demand for:			
	Electricity	Coal	Oil	Gas
1% increase in price of:				
Electricity	−0.6	0.3	0.3	0.3
Coal	0.1	−2.1	0.3	0.3
Oil	0	0.1	−2.6	0.1
Gas	0.4	1.5	1.7	−1.1

Table 4.5. Public administration: price elasticities

SHORT RUN (same year)

	Percentage change in demand for:			
	Electricity	Coal	Oil	Gas
1% increase in price of:				
Electricity	−0.2	0	0	0
Coal	0	−0.3	0	0
Oil	0	0.1	−0.3	0.1
Gas	0	0.1	0.1	−0.3

LONG RUN

	Percentage change in demand for:			
	Electricity	Coal	Oil	Gas
1% increase in price of:				
Electricity	−0.2	−0.3	−0.3	−0.3
Coal	0	−3.0	0.6	0.6
Oil	0	0.7	−3.0	0.7
Gas	0	2.1	2.1	−1.9

Table 4.6. Miscellaneous sector: price elasticities

SHORT RUN (same year)

	Percentage change in demand for:			
	Electricity	Coal	Oil	Gas
1% increase in price of:				
Electricity	−0.1	−0.1	−0.1	−0.1
Coal	0	−0.6	0	0
Oil	0	0.3	−0.2	0
Gas	−0.1	0.2	0.1	−0.1

LONG RUN

	Percentage change in demand for:			
	Electricity	Coal	Oil	Gas
1% increase in price of:				
Electricity	−0.1	−0.2	−0.2	−0.2
Coal	0	−2.9	0.3	0.2
Oil	0	3.8	−0.9	0.3
Gas	0	3.1	0.5	−0.6

5. Robustness of the Results

Figures 4.1 and 4.2 provide statistical measures of the goodness of fit of each of the underlying equations, and illustrations of how these models track past movements in energy demands for the two main sectors—other industry and domestic.[2] If equations do not contain lagged values of variables, these figures simply plot both actual and calculated values of energy demands over time. Where lagged values of variables are present, the test is more stringent. The estimation procedure minimizes the sums of squares of residual errors over the data period using actual values of lagged variables. But the simulation exercise carried out uses the estimates of lagged energy demands from the previous year's calculations. In this way, the actual values of energy prices, industrial output, real income, and temperature (as appropriate) drive the dynamic models for energy demand.

In Figures 4.1 and 4.2 the solid line represents the actual levels of fuel demands and the crosses the calculated levels. In all sectors the models capture the principal features of changes in energy demand over the data period. The suggestion is frequently made that reactions to the oil price rise in 1973 and subsequent events have caused a breakdown in the relationship between energy

---

[2] Corresponding results for the other sectors are reported in Department of Energy (1983b).

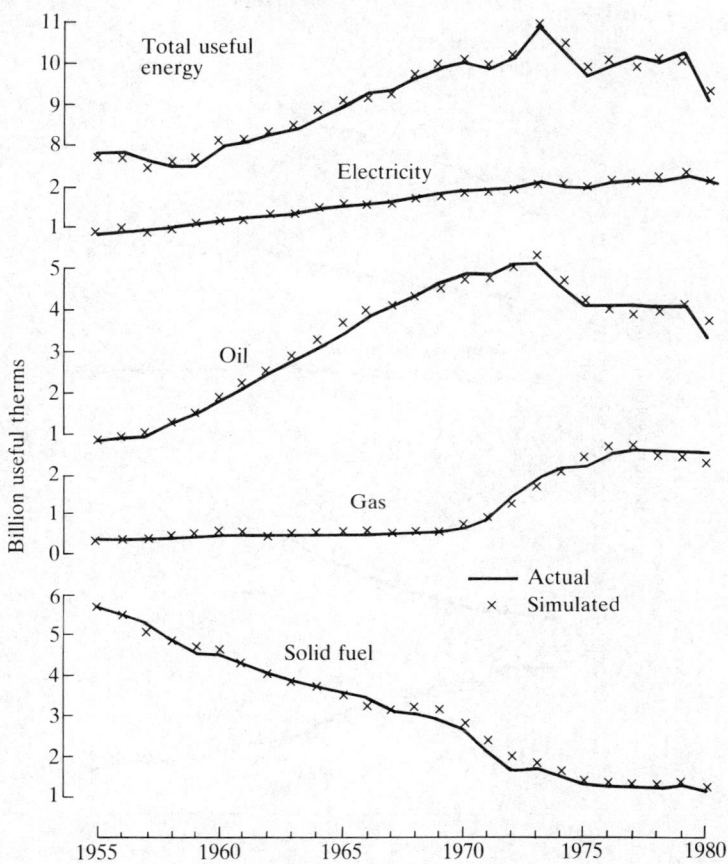

Figure 4.1 Other industry: simulated and actual fuel demands, 1955–80

demand and economic growth. As far as the relative movements of primary energy demand and gross domestic product are concerned, this seems to be the case in many countries, including the United Kingdom. However, it is possible that this relative movement results mainly from the changing economic and industrial structure within countries rather than from any major changes in the underlying relationships between energy demand, fuel prices, and economic activity for individual sectors within the economy.

The illustrations of goodness of fit in Figures 4.1 and 4.2 provide no indication of a fundamental breakdown in these relationships. In fact, these simple relationships with their lagged responses to changes in energy prices and economic activity continue to capture the movements in energy demand through both the pre- and the post-1973 periods. However, it should be noted that corresponding equations estimated *before* the first of the oil price shocks would

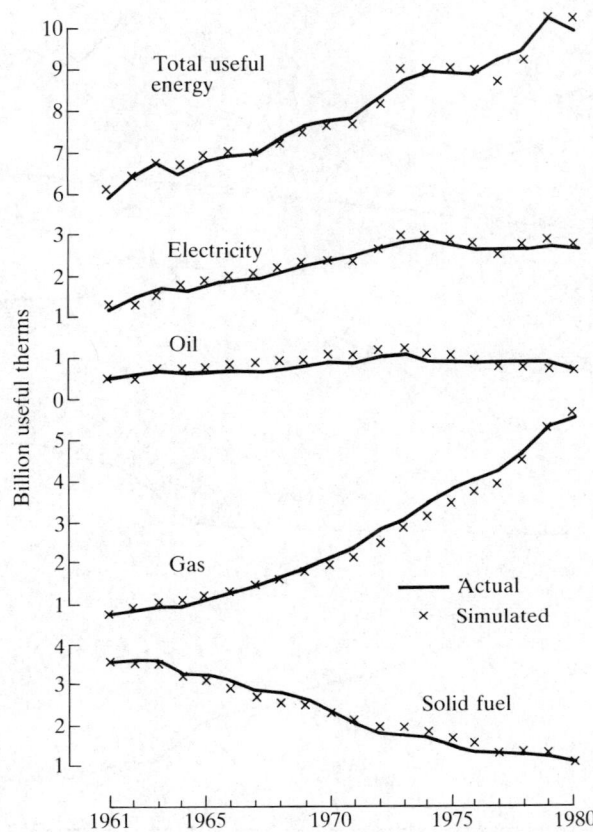

Figure 4.2 Domestic sector: simulated and actual fuel demands, 1961–80

have been significantly less successful; not surprisingly, the large relative price movements, and their responses in 1973 and 1979, are necessary to estimate reliable coefficients measuring the response to price changes.

Formal statistical tests for stability were carried out and results were presented in Department of Energy (1983b). The results provided support for the hypothesis that the output elasticities were stable, but were inconclusive regarding the stability of the price elasticities and the lag structure. Taken together with the elasticity values calculated for the models, the statistical measures of goodness of fit, and the dynamic properties, Figures 4.1 and 4.2 are reassuring in the context of the use of these models to provide long-run projections of energy demands.

6. Other Characteristics of the Model

Several properties of the model could be highlighted. For example, the estimated equations imply long lags in the adjustment of demand to price changes. Also, the treatment of conservation effects is worthy of consideration. These topics are discussed below.

Dynamic Properties

The dynamic properties of the model are illustrated in Figures 4.3 and 4.4, which show the way in which the models for other industry and domestic space and water heating adjust from the implied disequilibrium positions of 1980 to long-run equilibrium positions at 1980 price, output, and income levels.

In both cases there are long lags before energy demands settle down to their equilibrium levels. In the case of other industry, the levels of oil and gas demand reverse directions in response to earlier movements in prices.

It is important to stress that the measure of disequilibrium in 1980 and the dynamic properties illustrated in Figures 4.3 and 4.4 result from the

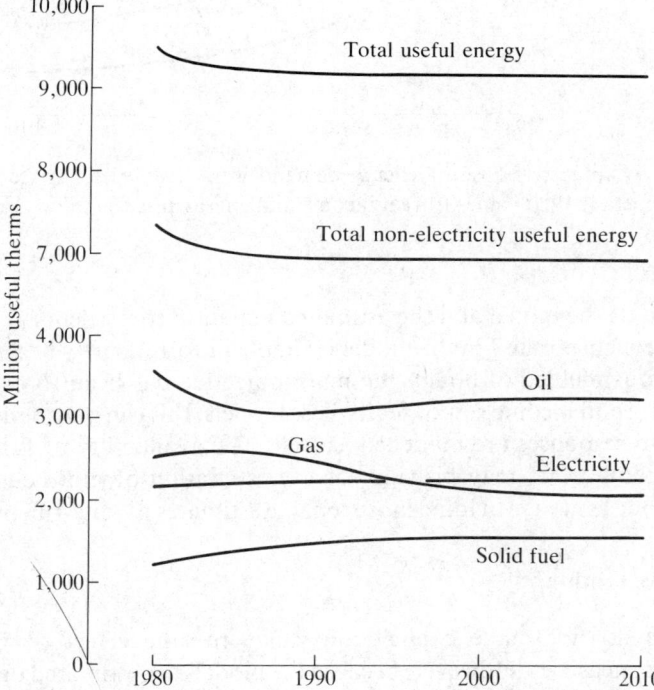

Figure 4.3 Dynamic response of energy demand model: other industry (at constant 1980 values of output level and energy prices)

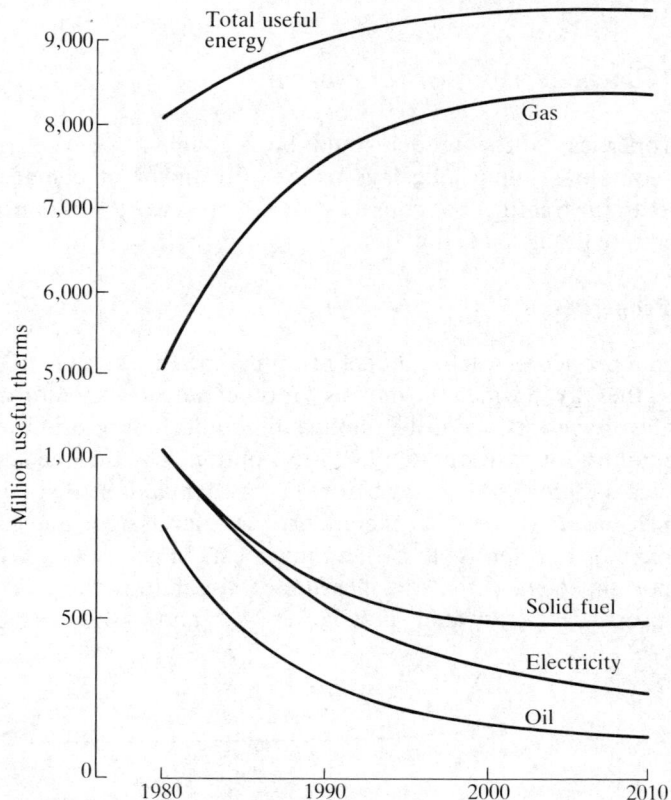

Figure 4.4 Dynamic response of energy demand model: domestic space and water heating (at constant 1980 values of real income and energy prices)

specification of the model and the estimated equation coefficients. The Koyck lag structure incorporated in the model is simple in form, partly because of the limited data available. Although the main objective has been to capture the long-run price and income responses in these models, this can only be achieved if the short-run responses are modelled as well. The availability of further data over the coming years may provide for the estimation of more complex lag structures and assist in providing more reliable estimates of long-run properties.

Energy Conservation

The models do not make explicit allowance for the effect of individual conservation measures. However, because the models are estimated on the basis of past trends in energy demand, in relation to output and prices, the model captures changes in the efficiency with which energy is used in two ways.

First, there is a non-price effect which arises mainly as a result of technical change. If fuel prices rise at the same rate as general price inflation (i.e. if real fuel prices are constant) then, for the aggregate of the sectors considered here, total demand for energy tends to rise less quickly than economic activity (gross domestic product). As a result, the model has the property that the use of energy per unit of economic activity falls faster, the faster economic activity increases. The second feature involves additional energy saving as fuel prices rise faster than the general rate of inflation. In this case, other inputs (such as capital for additional insulation or improved boiler controls) are substituted for energy. Consumers tend to use relatively less energy, to use it more efficiently, and to undertake more energy conservation measures, the faster energy prices rise in real terms. In this way, the 'price' and 'non-price' components of increasing efficiency in energy use are incorporated into the energy demand model. Quantitative estimates of the reduction in energy use arising from rising real energy prices are provided in Department of Energy (1982).

5 Modelling Public Enterprise Performance*

RAY REES

1. Introduction

Until quite recently, the supply side of the UK energy market was dominated by public enterprises, oil being the only major form of energy not supplied by a public corporation. The privatization of the British Gas Corporation has significantly altered the balance between public and private enterprise in the energy sector, and, looking ahead, it is reasonable to place quite a high probability on a state of the world in which electricity supply and much, if not all, of the coal industry are also privately owned. At the same time, the example of gas privatization suggests that private enterprise will be subject to regulation. It follows that in modelling energy supply we need a theory of the behaviour of public and regulated private enterprise. This paper outlines a theory of public enterprise that has been recently developed,[1] compares its predictions with some evidence on the decision-taking of the Central Electricity Generating Board (CEGB), and goes on to discuss how the theory might be extended to deal with regulated private enterprise.

The idea that a public enterprise is an organization with its own objectives, which it pursues subject to constraints arising out of the markets it faces and the control framework constructed by government, is an obvious but neglected one. The neglect arose because economists working on public enterprise economics were following an essentially normative agenda. It was tacitly assumed that ownership implied control, and that the government sought to maximize welfare, and so the problem was to work out how resources should optimally be allocated. Public enterprise economics therefore was, and largely still is, a branch of applied welfare economics, concerned with finding criteria for Pareto-efficient pricing and investment policies under 'second-best' conditions.

It was always clear that these pricing and investment 'rules' were not descriptions of how those running public enterprises actually behaved, and indeed it was often perceived that the policy objectives of governments, in so far as they could be spelt out, were not identical to those of the hypothetical 'central planner' whose job it was to solve the resource allocation problem. Those working on the 'normative agenda' would probably have rationalized their approach by the argument that whatever the problems of implementation, it is first of all necessary to have available a well worked out body of efficiency

---

* At the time of writing this paper, Ray Rees was Professor of Economics at University College, Cardiff; he is currently at the University of Guelth, Ontario.
The author is grateful to Dieter Helm for a number of helpful comments on this paper.
[1] See Rees (1984a, 1984b).

criteria which take account of the real characteristics of the economics and technology of the public enterprises, such as demand and cost uncertainty, peak-load problems, indivisibilities in capacity, the need to cover accounting costs under increasing returns to scale, the 'system' or 'network' technology of many enterprises, demand interdependence with monopolistic private enterprises, concern with the income distributional implications of prices and investment decisions, and so on. Some progress in implementing the criteria had, in fact, been made, often with active co-operation from some of the enterprises, particularly in methods of investment appraisal. There were, however, some strong sticking-points, particularly in introducing marginalism into pricing. A major problem also was that frequently government policy toward the industries was dominated by short-term macroeconomic concerns, and this tended to undermine all attempts to extend the scope of allocative efficiency.

Then came privatization. A critique of the normative agenda was put forward by Professor Stephen Littlechild,[2] who made the point, which had been made several times before,[3] that to prescribe rules on how prices should be related to marginal cost does nothing to ensure that the levels of cost are as low as they could be. More fundamentally, he emphasized the role of market forces in stimulating efficiency, innovation, and managerial dynamism, and argued that the particular set of incentives and opportunities created by the nature of public ownership served to stifle these. His arguments were important in directing attention away from the normative agenda and towards the question of institutional form—ownership structure, organizational structure, openness to market forces. They were also arguments whose time had come: they provided an academic reinforcement for the strong political drive toward privatization. It is impossible not to accept that this widening of the set of issues to which economists should address themselves was important, valuable, and overdue.

However, problems remain. As I have argued elsewhere,[4] in the case of the major enterprises that are or will be privatized—telecommunications, gas, water, electricity, airports—it is hard to believe that competition in output and capital markets will be strong enough to control the tendencies to market failure, as was indeed recognized in the creation of regulatory bodies such as the Office of Telecommunications (OFTEL) and the Office of Gas Supply (OFGAS), the imposition of 'RPI $-X$' pricing rules, and so on. Hence, privatization represents a change in the framework of control, from regulation by direct involvement with government department to regulation by indirect involvement with government department, *plus* quasi-governmental regulatory agency, *plus* private capital market. In some respects this has been a change for the better, but whether it will in the end represent an overall improvement in

---

[2] See Littlechild (1979).
[3] See, for example, the concluding section of Rees (1968).
[4] See Rees (1986b).

economic performance is still an entirely open question. The key point is that all the old issues remain: they are not simply wiped out by the change in the regulatory framework. We should still be concerned about the allocative efficiency of the pricing and investment policies of these large, essentially monopolistic enterprises. Just as importantly, we should be concerned about the levels of costs and productivity, given the incentives and constraints created by the new regulatory framework. Thus the normative agenda is still relevant, though it should be extended to include explicit analysis of managerial and technical efficiency in the context of new institutional structures.

This brings us back to the issue of a positive theory of public enterprise decisions. It seems to me to be impossible to address properly the question of the actual economic efficiency of these enterprises without an explicit analysis of their behaviour under whatever control framework prevails, whether 'public' or 'privatized'. It seems impossible to argue that, on the one hand, if publicly owned they will spontaneously choose policies in line with the prescriptions of the normative agenda, or that, on the other hand, conversion to private ownership allows market forces so to constrain the enterprises that only economically efficient policies are feasible. My aim in trying to develop a positive theory is then to allow formal analysis and prediction of enterprise decisions under alternative frameworks of control, and their assessment in terms of economic efficiency, both allocative and productive.

2. A Positive Theory of the Public Enterprise

Public enterprise decisions can be regarded as the outcome of a game played principally by four groups of players: senior managers, union officials, government Ministers, and Civil Servants. Each group has its objectives (which may well themselves be the outcome of some kind of coalitional compromise). The particular institutional framework, the formal and informal rules and processes governing how prices are set and changed, how wages, conditions of employment, and employment levels are determined, investment programmes formulated, and funding made available, determine the strategies open to each group and the way these interact to produce an outcome. One of the difficulties in studying this game (and indeed in playing it) is that the rules themselves are often quite vaguely defined, and external factors, the power of personality, and even random events can cause changes in them.

A way into the analysis of this complex game is to regard respectively unions, management, and government/Civil Servants each as if they were one player with well-defined objectives. The game could then be seen as being played in two stages. At the first stage, we take it that 'government' has set two constraints: a constraint on financial outcomes, such as a financial target or external financing limit (EFL), which determines the revenue and cost levels feasible for the enterprise; and a constraint on total investment, which puts a ceiling on capital stock. We then analyse the enterprise decisions relative to a given set of constraints. However, in reality the constraints are not arbitrarily imposed by government, but are the outcome of bargaining and negotiation between government and management with, quite probably, the union using whatever influence it can exert (and this will vary with the party in power) to slacken the constraints (especially the financial constraint) as much as possible. Thus the second stage of the analysis is to analyse how the constraints are determined. Both stages are necessary. The first stage gives us a mapping from constraints to decisions—prices, outputs, wages, employment, investment—and hence to the values of the outcomes—efficiency, profit, social welfare, and utilities of government, management, and union. The second stage then uses this mapping to assess the implications of different control frameworks. It is at this stage that the implications of the change from public to regulated semi-private ownership can be analysed.

Before going on to look at the details and results of the analysis at each of these stages, I should point out some limitations of the approach. As with any attempt to model a complex system of strategic interactions, it simplifies and abstracts, and so does not describe the whole picture. In taking government's role as essentially the setting of constraints, it assumes that there is no direct intervention in specific decisions, for example on individual investment projects, capacity closures, or wage negotiations. This is reasonably accurate for 'non-problematic' industries such as electricity, gas, water, and airports, but is much less appropriate for the problem industries like coal, steel, and railways, where more direct intervention has in the past taken place. For the latter group, a model that allows government a more active role should perhaps be considered.[5]

A second limitation is the absence of any direct constraints on prices—they are influenced only through the constraints on financial performance. This again is not entirely descriptively realistic, since there have been periods of time over which prices have been held down or, indeed, pushed up. However, it is relatively straightforward to incorporate direct constraints on prices into the model,[6] and no change in its fundamental structure is required.

We can now consider the details of the analysis and results of models that try to deal with both the stages of the game—enterprise decisions for given constraints and determination of the constraints themselves.

For any given set of constraints fixed by government, the 'management' and the 'union' negotiate to determine wages and employment levels, and these will then influence management's choices of outputs and prices. The preference of management is for size, in a static context, or for growth (minimal decline) in a

---

[5] For example, a fruitful approach could be one in which the union bargains directly with government, so that 'management' as a group with its own interests and influence is suppressed. For some implications of this approach, see Gravelle (1984).

[6] In fact it becomes essential to do so when we wish to model the type of regulatory system currently applied to British Telecom and British Gas. See Section 5.

dynamic one. In terms of their basic motivation, managers therefore are seen as not being essentially different from private enterprise managers (as distinct from owners). For the union, the higher are wage-rates and the greater the level of employment, the better—again this is a standard characterization of union preferences. Given the financial constraint set by government, there is then necessarily a conflict between the preferences of management and workers, since higher wages imply higher costs, higher prices, and lower output. However, this conflict is partly ameliorated by the fact that higher output permits higher employment, so much depends on the trade-off the union is prepared to make between wages and employment.

The outcome of the bargaining between management and union, for a given set of constraints, depends on 'relative bargaining strength'. This is itself a notoriously difficult concept to analyse, and discussion of it is postponed to Section 4. In the analysis, I simply take it as given. The general nature of the outcome of the bargaining process between union and management is determined by the assumption that this outcome is *efficient*, in the sense that it could not be changed in a way which would make *both* parties better off—we could not achieve, say, more output *and* higher wages *and* higher employment, subject also to meeting the government's constraints. Any improvement in the result for one party would have to be at the cost of a deterioration in the result for the other. The assumption that the outcome of the bargaining is efficient defines a set of possible outcomes whose general characteristics can be quite precisely described, and relative bargaining strengths then determine where in this set the specific outcome will be: towards the extreme more favourable to the union, with (for given constraints) higher wages and employment, higher prices, and lower outputs; or towards that more favourable to management, with lower prices, higher output, lower wages, and somewhat less employment.

Since the assumed motivation of management and union in this model is much as it would be in a model of private enterprise, the question arises of the sense in which this is a model of *public* enterprise. The answer lies, of course, in the constraints. The essence of public ownership, at the level of generality of this model, is that it replaces the constraints arising out of the capital market with a (possibly quite slack) constraint on the financial outcome and a direct constraint on capital (possibly accompanied by a lower *cost* of capital), and that it allows these constraints to be a subject of negotiation between enterprise and government.[7] At the same time, the constraints arising out of output market demand conditions and technology which would face a private enterprise continue to exist here.

The model yields a number of firm predictions about the equilibrium decisions of the public enterprise, chief of which are the following:

---

[7] What makes the model positive rather than normative is the assumption that the enterprise seeks to act in the interests of those who work in it, as defined by the preferences of management and unions, rather than to maximize social welfare, as usually defined.

(a) The financial constraint is never overachieved—it is always a binding constraint on the enterprise decisions. This is because any likely surplus over and above the constraint would be used to finance some combination of lower prices (and so higher output), higher wages, and more employment.

(b) Costs will be higher, at any output, than the minimum level achievable. This implies that for a given financial constraint, prices will be higher, and outputs lower, than they would be under cost minimization. There are essentially three reasons for this:

(i) Production will be excessively labour-intensive, given the wage-rate being paid and the cost of capital. Where the technology permits substitution of labour for capital, this will be because production techniques are chosen that have higher ratios of labour to capital than would be consistent with cost minimization. The obvious pragmatic explanation of this is that unions will have resisted the introduction of cost-reducing labour-saving equipment because of its employment implications. Where the technology does not permit substitution of labour for capital, there will simply be surplus labour—employment could be reduced, with capacity held constant, and output could be maintained. The extent of this 'overmanning' will depend on the union's bargaining strength, and on the strength of its preference for employment relative to that for high wages, since, given the financial constraint, one is obtained at the expense of the other. I would suggest two hypotheses to explain this relative preference. First, the preference for employment will be stronger when the total membership of a union is a subset of the labour force of a public enterprise, than when the enterprise labour force is a subset of the total membership. This for example suggests that the unions in the coal and rail industries will have a stronger preference for employment relative to wages than that in the electricity supply industry. The idea here is simply that union officials care very much about the size of their union membership, for both financial and political reasons, and when the union is enterprise-specific (as many are) they have to defend employment in that enterprise. Unions that are able to seek members outside the enterprise do not need to be defensive, and indeed may use success in negotiating high wages as a means of attracting new members. The second hypothesis is that unions in enterprises experiencing significant declines in demand—coal, rail, steel—will have a stronger relative preference for employment than those in enterprises experiencing demand growth—gas, telecommunications, electricity.[8] In the latter industries, productivity schemes in exchange for higher wages can be painlessly accom-

---

[8] This could be framed in terms of the distinction between 'inside' and 'outside' workers. In expanding industries, productivity can be increased by reducing jobs for 'outside' workers, whereas in declining industries the jobs of 'inside' workers are threatened.

modated through 'natural wastage', early retirement, and so on, while in the former group the problem is to minimize the extent of the forced redundancies caused by declining labour demand.

(ii) Wage-rates will be higher than the market level for the relevant categories of worker. This is because the monopoly position of the enterprise generates rents, at least some of which can be captured by unions, in the form of higher wages, by use of their bargaining strength. The extent of this wage differential, for given constraints, will again depend on the union's bargaining strength and relative preference for wages as opposed to employment.

(iii) A binding constraint on capital will reinforce the tendency for production methods to be excessively labour-intensive and so provides an additional reason why costs will be higher than they need be. The constraint prevents the enterprise substituting capital for labour to the extent that is desirable given the wage-rate being paid and the cost of capital set by government. Given this effect of the capital constraint, the obvious question is why it should exist. This can, however, only be answered by the analysis at the second stage, that of negotiation between government and enterprise, and so will be discussed below.

(c) The structure of prices set by management will be such that mark-ups over cost will be higher on outputs with lower elasticities of demand—there will be 'charging what the market will bear' to an extent determined by the financial requirement. This is in order to minimize the impact of the financial constraint and the cost levels implied by wage and employment decisions on the aggregate output of the enterprise. Loading more of the profit requirement onto less price-sensitive outputs leads to lower reduction in output overall.

(d) Tightening the financial constraint—requiring a larger profit or lower loss—results in a combination of higher prices and lower outputs on the one hand, and lower wages and employment on the other. The precise extent to which prices rise or labour costs fall depends on relative bargaining strengths of union and management, but both will generally occur. The extent to which pressure on labour costs affects wages *vis-à-vis* employment depends on the union's preferences. What is clear is that the utilities of management and union certainly fall as the profit constraint tightens.

(e) Tightening the capital constraint raises costs and prices, and reduces wage-rates. The effects on employment are ambiguous, since on the one hand labour will be substituted for capital at any given output, but on the other hand reduced output implies less employment. Overall therefore the effect is likely to be small. The cost increase arises because of the increased labour intensity of production, which is already excessive. Again, therefore, tightening a binding capital constraint reduces the utilities of management and most probably the union; relaxing the constraint increases them.

(f) A reduction in demand, with an unchanged financial constraint, leads to higher prices, reduced outputs, and lower wages and employment, all of which are required to maintain profitability in the face of lower demand.

These predictions seem to fit quite well with the (casually) observed facts of public enterprise performance, and so the model suggested provides a formal explanation of these. In the particular case of the electricity supply industry, the relationship between the model's predictions and actual outcomes will be examined in the next section. First, however, we consider models of the process of negotiation between government and enterprise by which the constraints are determined.

The constraints are set by a process of negotiation between enterprise and government, which can be thought of as proceeding as follows. The enterprise forms a plan which has two elements: a set of estimates or forecasts of demands, costs, prices, and profitability; and a 'bid' for resources, in the form of an investment programme. Discussion between government and enterprise takes place in the context of this plan, and the outcome is an investment allocation and a profit target, which then determine the constraints within which the enterprise will take its price, output, and employment decisions. At the core of this process is the problem of incompleteness of information, which takes two forms. There is first *general uncertainty* about the future values of parameters on which the plan depends, for example demand, technology, the macroeconomic environment. Secondly, there is *informational asymmetry*, in that the enterprise always possesses information about the relevant parameters which in quantity and quality is far superior to that available to government. In a sense, the planning process is a process of exchange of information for resources.

If we adopt a normative approach to the problem of investment allocation, there is a well-developed body of economic theory to draw upon, in the form of the theory of the principal–agent relationship.[9] Here, the agent is supposed to take decisions in the best interests of the principal, but, because of uncertainty, information asymmetry, and monitoring costs, is able instead to take decisions in his own interests, which do not coincide with those of the principal. In the particular version of the problem known as the 'adverse selection' model, the principal has to design a mechanism that induces the agent to reveal truthfully the information he possesses, upon which an allocation of resources is then based. Though application of this and related models to the problems of incentives and control in public sector decision-taking does, I think, yield useful insights, it does not describe how the parties in the public enterprise investment planning process actually behave. Here I think a somewhat different approach is required.

A characteristic of the actual planning process (not only in the public sector) is that the existence of uncertainty is allowed for in only a peripheral way. Essentially, planning proceeds by taking a 'central' set of assumptions which

---

[9] For a general survey of this theory, see Rees (1985a). In Rees (1985b, 1986a) I have tried to apply this theory to some specific problems of public enterprise control.

define a particular 'scenario' or 'state of the world'. Forecasts, estimates, and bids are prepared for this 'central' state of the world, while uncertainty is taken account of by, at best, some degree of piecemeal 'sensitivity analysis' and the allowance of some kind of margin for contingencies. To all intents and purposes, however, the 'central scenario' is treated in the planning process *as if* it were certain. The key advantage enjoyed by the enterprise is that, because of the information asymmetry, it selects the state of the world on which to base the plan and associated bid for an investment allocation. *Plan optimism* then corresponds to the case where a 'central' state is chosen that is in fact much more favourable than the true central or most probable state. Forecasts of demand input productivities and output prices are relatively high, forecasts of costs are relatively low, and so the investment programme will have an optimistically high rate of return—and be larger in consequence. *Plan pessimism* is then the converse case, where the plan is based on a relatively unfavourable 'central' state which is worse than the most probable state. The aim of a positive model is then to predict the type of plan that will be chosen, given

(a) *the preferences of the enterprise*, which, from the discussion given earlier, take a form in which a lower profit constraint is preferred to a higher one and more capital is preferred to less (at least up to the point at which the capital constraint becomes non-binding);
(b) *the characteristics of the plan appraisal mechanism*, the chief of which are:
   (i) state responsiveness: the mechanism is state-responsive if choice of a more favourable state leads to a higher capital allocation;
   (ii) profit consistency: the mechanism is profit-consistent if the profit target that is set is precisely the profit that is forecast to result from the final capital allocation in that state of the world which is the basis of the plan—in other words, the enterprise is constrained actually to earn the rate of return it says it is going to earn when it makes its bid for a capital allocation;
(c) *the information possessed by government*, which essentially places (possibly quite wide) limits on the degree of optimism or pessimism that can be built into the enterprise plan, i.e. the extent to which its 'central state' can diverge from the most probable state of the world.[10]

In general, I would describe the actual public enterprise planning process as state-responsive, while the degree of information possessed by government is usually such that plan optimism or pessimism can exist. The key question is that of profit consistency. It can be shown that if the appraisal mechanism is profit-consistent, then the enterprise will submit the most pessimistic plan it thinks the government is willing to accept—it will understate the true rate of return achievable on its investment. The reason is that although it will thereby receive a lower capital allocation, this is more than compensated for by a lower profit

---

[10] See Rees (1984b) for details of the proofs.

target, so the enterprise ends up better off overall (in terms of its own objectives). On the other hand, if the appraisal mechanism is not profit-consistent, in that the profit target is independent of the capital allocation, then the enterprise will submit the most optimistic plan it thinks the government will accept. The reason is that this increases the capital allocation without carrying with it the penalty of an increased profit target. The enterprise can make an inflated bid for capital expenditure, secure in the knowledge that it will not actually have to generate the returns it is forecasting in its plan.

I would argue that the system of appraisal of public enterprise investment programmes as it exists in the UK is not in fact profit-consistent, essentially because of a difference in time horizons for investment planning on the one hand, and financial target setting on the other. Typically investments have very long gestation periods—this is particularly true in the energy sector—and so the flow of profits resulting from an investment programme currently under consideration will begin well into the future. Now, a financial target is set in the light of the immediate and short-term conditions confronting the enterprise, and no attempt is made to tie the enterprise to the profit projections it made when the investments currently coming on stream were being evaluated in the past. Thus, what matters for a profit target is the *actual* state of the world that currently faces decision-takers, *not* the state of the world that formed the basis for the past investment plan. It follows that an optimistic choice of state can be made for an investment plan, in the expectation that (very probably) the state of the world that prevails when the profit target will have to be set will be less favourable than that assumed in the plan, and so the investment will not in fact be constrained to earn the profit forecast for it at the planning stage.

This concludes the discussion of the general theoretical aspects of modelling public enterprise performance. We can now turn to the empirical evidence. In the following section, I examine the report of the Monopolies and Mergers Commission (MMC) on the Central Electricity Generating Board (CEGB), to see to what extent the performance of that public enterprise matches the predictions of the theory outlined in this section.

3. Predictions and Performance: The Case of the CEGB

The MMC report on the CEGB (1981) is a useful source of information on the economic performance of that enterprise for the period to which it relates (broadly, 1965–81). In comparing this performance with the theoretical predictions set out in the previous section, I shall consider in turn three main areas:

(a) wages, employment, and productivity;
(b) pricing policy;
(c) investment planning.

Wages, Employment, and Productivity

To put the discussion in context, we should first note that electricity generation and high voltage transmission are highly capital-intensive, with labour accounting only for about 10 per cent of total costs. Fuel on the other hand accounts for around 60 per cent of costs. Accordingly, the major issues in relation to input choices are ones of the scale and timing of investment and the design of new generating capacity—fuel use, size of generating sets, etc. Certainly labour–capital substitution is not a major issue—the technology can essentially be regarded as having a fixed labour requirement once the choice of power-station type has been made, predominantly on capital and fuel cost comparisons. Excessive labour intensity, if it exists, would then have to take the form of straightforward overmanning rather than choice of excessively labour-intensive (but technologically efficient) technique. The MMC report does not provide information on which to judge whether overmanning does in fact exist. What it does show is that wage levels of the two main groups of workers—manual workers (referred to as NJIC staff) and electrical/electronic engineers (referred to as NJB staff)—were each well above the national average levels for those groups of workers as a whole over the years 1971-81. This is shown in Figures 5.1 and 5.2, which are reproduced from the MMC report. The fairly dramatic change in relative earnings of manual workers that took place in 1970 was a result of the introduction of far-reaching productivity agreements in the late 1960s, which led to more efficient working patterns, work measurement, and productivity payment schemes. Over the period 1966–80, the

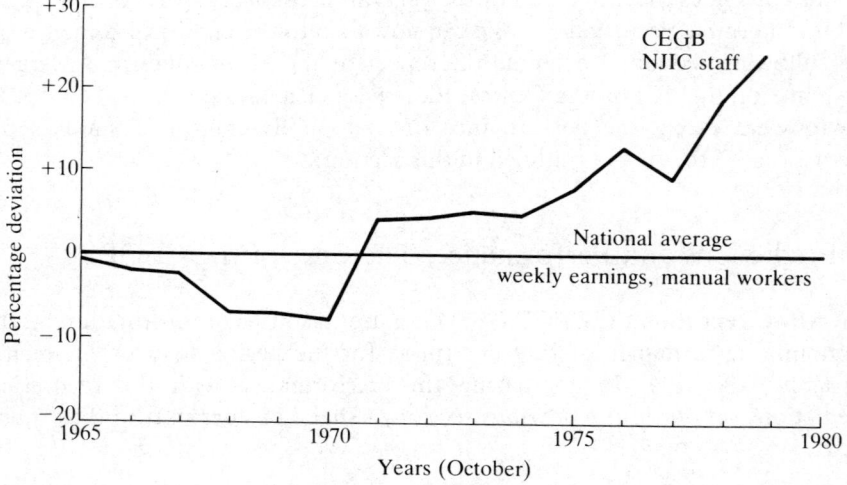

Sources: CEGB and Department of Employment.

Figure 5.1 Percentage deviation of the average weekly earnings of CEGB NJIC staff from the national average for manual workers

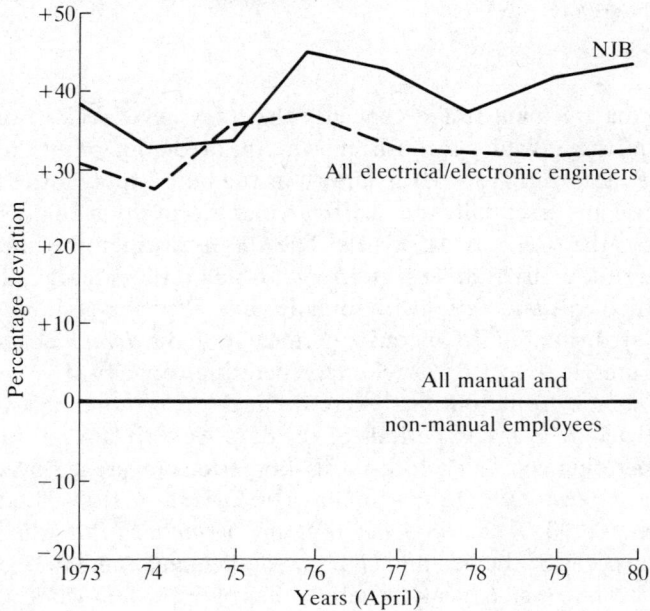

Sources: CEGB and New Earnings Survey.

Figure 5.2 Percentage deviation of gross average weekly earnings of CEGB NJB staff and electrical/electronic engineers from all manual and non-manual employees in all industries and services

overall labour force fell by some 20 per cent, with a 31 per cent fall in the number of manual workers. Since output continued to grow, there was a 75 per cent increase in average labour productivity over the same period. Underlying these changes was the major programme of investment in generating capacity over the period 1966-72, which was associated with a significant increase in the scale of generating sets and average size of power-stations. The required changes in size and composition of the labour force were achieved by natural wastage, redeployment, and voluntary redundancy, and appear to have presented few industrial relations difficulties.

I would conclude therefore that in this enterprise the bargaining power of the unions was used primarily to obtain higher wage-rates than the average for the relevant groups of workers, rather than maintaining an inefficiently large labour force, though overmanning, of the 'pure surplus labour' type, could still exist. It is worth noting that the labour force is represented by a number of unions, the majority of which have a large proportion of their membership *outside* the CEGB.

Pricing Policy

The CEGB management stated explicitly that they never seek to overachieve their financial target, which is consistent with the prediction of the theory. Their prices are set according to what is known as the bulk supply tariff (BST), so-called because it is essentially the tariff at which electricity is sold wholesale to the 'retailers', the twelve Area Boards. The CEGB distinguished between two components of the tariff: an *energy charge*, which reflects marginal operating (primarily fuel) cost and varies with total demand as power-stations of varying efficiencies are brought into operation to meet it; and a *demand charge*, which reflects the marginal cost of providing generating capacity. For purposes of the energy charge, four 'outputs' were defined, corresponding to electricity supplied in each of four sub-periods of the day. An estimate was made of the marginal operating cost in each of these sub-periods (averaged over an entire year) giving a *peak rate* (18.00–18.30), a *night rate* (01.00–08.00), a *night shoulder rate* (00.00–01.00, 08.00–09.00), and a *standard rate* (the rest of the day). For purposes of the demand charge, two demands were defined: that on the day of absolute system peak (generally the coldest winter working day), the *peak demand*; and that on average over the periods of the year when demand is at around 85 per cent of system peak, which is called *basic demand*. The demand charges differed between these periods, essentially because a (permanent) increase in peak demand would be met by an increment in gas turbine capacity, while an increase in basic demand would be met by an increment of 'base-load' capacity—coal-fired or nuclear generating plant. Hence marginal capacity costs differ between the two demands.

In the period to which the MMC report relates, the CEGB considered that setting these six charges at the relevant marginal costs would have generated more revenue than was necessary to meet its financial target, and so it adjusted charges downward according to the principle that 'if the Tariff rates have to differ from the marginal rates the brunt is borne by the *less elastic features of the tariff*, in particular the basic capacity charge and the standard kwh (energy) rate'. Thus, the CEGB appeared to implement a form of Ramsey pricing, as the model predicts, though its actual adjustments to 'marginal cost prices' were quantitatively quite small.

However, the CEGB's discussion of its own pricing policy is based upon a misinterpretation of 'marginal cost'. At the time, it had a quite considerable degree of excess generating capacity, enabling increases in demand at all times, including the peak, to be met not by capacity expansion, but simply by deferring closure of old plant. Hence *true* marginal capacity costs were considerably below the cost of new capacity: under a properly defined *marginal* cost based tariff, its demand charges would have been around £5–£10 per kilowatt (kW). Of course, such a pricing policy would have implied a failure to meet the financial target. Logically, therefore, we should interpret the actual charges of

£30 per kW basic capacity charge and £14 per kW peak capacity charge as the means adopted of meeting the financial target, rather than as a reflection of current marginal costs. This also explains an inconsistency in the CEGB's approach to the two demand charges. It argued that if the system had been at capacity with an optimal mix of generating plant, then its marginal cost of basic capacity would have been £31 per kW, while its marginal cost of peak capacity would have been £33 per kW. It reduced its peak capacity charge to £14 per kW in the light of the fact that the *actual* marginal capacity cost was only £5–£10 per kW, but did not do the same for the basic capacity charge. Thus, in effect, this 'less elastic feature of the tariff' carried the main burden of the financial target. This accords well with the predictions of the theory.

Investment Planning

Throughout the period covered by the MMC report, the investment planning of the CEGB can only be described as extremely optimistic in respect of the three main components of the plans: the forecasts of demand upon which estimates of required capacity were based; the estimates of the costs of new power-stations; and the estimates of the time that would be taken to complete construction of the power-stations, particularly of oil-fired and nuclear power-stations. Tables

Table 5.1. Forecasts and out-turn of maximum electricity demand[a] (GW) in England and Wales

Date when forecast made	Forecast for demand in	CEGB Planning Department	ESI-adopted[b]	Out-turn	% over forecast by CEGB planners	% over forecast of ESI-adopted
Mar. 1969	1974–5	54.1	53.2	41.9	29.1	27.0
Mar. 1970	1975–6	57.1	54.0	41.1	38.9	31.4
Mar. 1971	1976–7	58.7	54.0	42.0	39.8	28.6
Mar. 1972	1977–8	60.6	55.0	42.4	42.9	29.7
Mar. 1973	1978–9	56.8	56.5	43.8	29.7	29.0
July 1974	1979–80	58.2	56.5	44.1	32.0	28.1
Mar. 1975	1981–2	53.7	54.0			
Mar. 1976	1982–3	52.5	52.0			
Mar. 1977	1983–4	51.0	51.5			
Nov. 1977	1984–5	53.0	52.0			
Oct. 1978	1985–6	50.9	50.6			
Oct. 1979	1986–7	48.9	50.3			
Feb. 1980	1986–7	48.5	46.8			
Oct. 1980	1987–8	45.3	47.5			

[a] In 'average cold spell' conditions.
[b] The 'ESI-adopted' forecasts represent the 'official' forecasts adopted by the Electricity Council, the CEGB forecast being one of the inputs into this.
Source: CEGB.

Table 5.2. Investment decisions and commissioning lags

Scheme	base dates	Type	No. of units and design net capability (MWso)	Planned year of commission	Year of start on site to probable year of commissioning of last unit
Dinorwic	Dec. 1973	Hydro (pumped storage)	6 × 250	1979–81	1974–83
Drax completion	Mar. 1978	Coal	3 × 660	1984–6	1978–86
Grain	Dec. 1970	Oil	5 × 660[a]	1976–9	1971–83
Ince 'B'	Jan. 1972	Oil	2 × 500	1977	1972–82
Littlebrook 'D'	June 1973	Oil	3 × 660	1979–81	1974–83
Dungeness 'B'	Mar. 1965	AGR	2 × 600	1970–1	1966–81
Hartlepool	May 1967	AGR	2 × 660	1974	1968–82
Heysham I	Dec. 1969	AGR	2 × 660	1976	1970–82

[a]Three units only currently programmed for completion. No decision has been taken by the Board to complete units 4 and 5.

Table 5.3. Generation Development and Construction Division: estimated costs to completion (ECTC) of power-stations currently under construction (excluding Heysham II[a]) (£ million)

	Scheme base dates	Scheme base date prices						Price escalation to 31 Mar. 1980	ECTC including price escalation to 31 Mar. 1980
		Original sanction excluding tolerance/ risk margin	Changes in design or policy	Net under/(over) estimations	Estimated costs of delays[b]	Unallocated scheme tolerance/ risk margin	ECTC at scheme base date prices		
CONVENTIONAL STATIONS									
Dinorwic	Dec. 1973	116.0	21.8 (changes in accounting practices)[c]	14.6	25.4	5.1	182.9	219.1	402.0
Drax completion	Mar. 1978	606.3	3.7	—	—	78.8	688.8	201.8	890.6
Grain	Dec. 1970	209.9	8.7	14.2	41.4	4.5	278.7	295.2	573.9
Ince 'B'	Jan. 1972	109.5	1.4	1.6	33.0	0.9	146.4	125.6	272.0
Littlebrook 'D'	June 1973	183.2	22.4	5.7	29.3	2.4	243.0	252.2	495.2
NUCLEAR STATIONS (excluding nuclear fuel)									
Dungeness 'B'	Mar. 1965	88.5	24.8	63.0[d]	40.2	—	216.5	243.5	460.0
Hartlepool	May 1967	91.8	52.1	9.9	59.2	6.5	219.5	235.6	455.1[a]
Heysham I	Dec. 1969	142.3	33.9	10.1	35.5	8.0	229.8	221.0	450.8[a]

[a] Not yet under construction.
[b] Includes settled claims at prices ruling at date of settlement and outstanding claims at March 1980 prices.
[c] Excluding possible provision for buffer fuel storage, estimated at £12.5m per station.
[d] Comprises changes due to underestimation of original design, fundamental data during construction, and new commercial arrangements following collapse of original consortium.

5.1 to 5.3 present the evidence for this assertion. The MMC concluded that 'during the decade 1970–80 there would have been sufficient capacity to meet demands without completing any new stations'. It estimated the capital expenditure on these unnecessary power-stations as at least £2,500 million at 1980 prices. Table 5.2 suggests that on average this investment was undertaken roughly five years before it was required. This implies that, at a real interest cost of 5 per cent per annum, the cost of the premature investment was of the order of about £600 million, well above the cost of all but one of the power-stations in Table 5.3.

4. Privatization

There has been considerable debate about privatization and, explicitly or implicitly, many of the arguments rest on predictions about the economic performance of the privatized enterprises. However, there seems to have been virtually no formal modelling of the behaviour of these enterprises which could provide a rigorous basis for such predictions. In this section I adopt the theoretical approach set out in Section 2 to consider this question.[11]

Given the model of the public enterprise set out earlier, the effects of privatization can be considered under six headings:

(a) the effect on the profit constraint;
(b) the effect on the constraint on capital;
(c) the effects on long-run investment planning;
(d) possible change in the bargaining power of unions;
(e) possible change in the preferences of managers;
(f) the impact of price constraints of the 'RPI $- X$' type.

I shall take each of these in turn.

The Effect on the Profit Constraint

It seems reasonable to suppose that privatization would imply at the very least a significant tightening of the profit constraint on the enterprise. Thus, *everything else remaining unchanged*, the model would predict for the privatized enterprise higher prices, lower outputs, higher labour productivity, lower employment, and lower wage-rates, since these are the consequences in the model of tightening the profit constraint.

The Effect on the Capital Constraint

Since privatization implies unrestricted access to the capital market, the capital rationing constraint as such would be removed. If everything else remained

---

[11] For further discussion of the likely behaviour of the electricity industry under privatization, see Yarrow (Chapter 10, this volume).

unchanged, and the constraint had previously been binding, the result would be an increase in capital intensity and a fall in labour intensity of production, and a concomitant reduction in costs and prices. It could be argued that against this should be set the likelihood that the cost of capital to the enterprise will rise. Certainly, the enterprise cost of borrowing is likely to be higher than the rate at which it previously borrowed from government. On the other hand, it should previously have used as a cost of capital the so-called required rate of return (RRR) of 5 per cent in real terms, which was meant to be closer to the marginal rate of return on private sector investment. Moreover, the privatized enterprise will be able to finance itself with a mix of debt and equity capital, rather than the former exclusively. On balance, therefore, the effects of the change in cost of capital may not be great. Of much more probable significance is the effect on the degree of realism of long-run investment planning.

Investment Planning

Perhaps the single most important effect of privatization on economic performance is likely to be a greater degree of realism in long-run planning. Earlier it was argued that public enterprise investment plans are likely to be over-optimistic, because management are not constrained to earn the returns they predict for their investment programme. Although its effectiveness can be overstated, it is probable that the capital market is able to impose sufficient discipline on management to ensure more realistic plans, essentially because the results of over-optimism are brought to bear on management through effects on profitability and the share price. Perhaps the greatest single failure of the system of government control of public enterprise, at least up to the early 1980s, was its inability to find an equivalent mechanism for enforcing realism in investment planning.

Unions' Bargaining Power

It is unlikely that privatization alone will affect the major determinants of a union's bargaining power: its degree of membership coverage among the workforce, closed shop agreements, membership loyalty and solidarity, resources for funding activities including strikes, and the extent of the costs it can impose by withdrawing labour. However, one additional factor has in the past sometimes greatly increased unions' bargaining power: that is, when they have been able directly to call on government support and influence in negotiating with management. This of course tends to vary with the party in power: the party that supports privatization is also unlikely to lend its support in this way, while the party that would do so is likely to reverse the privatization policy in any case. Thus, if private ownership of the relevant group of enterprises continues, there is unlikely to be any real difference in the degree of union bargaining power as such (as opposed to the willingness of management to resist its exercise). Improvements in labour costs and productivity are therefore likely to be due essentially to the tightening of profitability constraints and the greater orientation toward cost reduction that this creates.

Preferences of Managers

Implicit in the discussion so far has been the assumption that the senior management of a privatized enterprise will have the same underlying objective—size—as they had under public ownership. Certainly their behaviour will become more profit-orientated, but this can be regarded as due to the tightening of the profit constraint and concern with stock market valuation of the company's shares. However, it could be argued that privatization will bring about a more fundamental change in the motivation of managers, possibly through change in personnel, and possibly through participation in incentive schemes such as stock options and profit-related bonuses. To assess the implications of this, a model which assumes that managers seek to maximize profits is analysed in the Appendix, and the results of this are reported in Section 5 below.

Price Constraints

Restrictions on the rate of increase of public enterprise prices have quite often been imposed in the past, sometimes as part of an explicit prices and incomes policy and sometimes by covert political intervention. The novelty in the price constraints imposed on some of the privatized enterprises is that they are explicit, set for a fixed time period (usually five years), and are aimed at curbing abuse of monopoly power rather than countering general inflation. In each of these respects there is an improvement over what has gone before. This is not to say that they, and the associated regulatory apparatus, are beyond criticism. However, the purpose of the present paper is to conduct a positive rather than normative analysis. The question therefore is, how are these constraints likely to influence the behaviour and performance of the privatized enterprises? A formal analysis of this question is carried out in the Appendix, and the results are reported in the next section.

5. The Implications of the Price Constraints

Let us suppose that the management of the privatized enterprise become entirely profit-motivated—in the absence of regulatory constraint they would seek to maximize profit rather than size. They will of course still be faced by a union with a positive preference for high wages and employment, and, given the union's bargaining strength, this will imply that costs will be higher than they would be under strict cost minimization. *Given* the cost levels which the union's bargaining strength allows it to impose, however, profit maximization would result in a pattern of prices and outputs that equalizes marginal revenues and marginal costs, and this would in general imply higher prices and lower outputs than in the case where management seek to maximize size. A regulatory constraint placed directly on the rate of price increase will, if it is binding,[12] modify this result.

Given an initial set of prices for the outputs concerned, application of the constraint on permissible price *increases* will imply for the next period a set of initial price levels, which can be taken as the constraint in a static model such as that considered here. The constraint can be expressed in the form: that an index consisting of a weighted average of a particular group of prices may not exceed some particular value. The question then arises of how the constraint will affect the structure of prices that a profit-maximizing management would seek to set. The answer is provided by the analysis in the Appendix, and can be expressed in the following way. Raising the price of a specific output yields an increment of profit, but also incurs a cost in the sense that it causes a marginal increase in the value of the index of prices that forms the basis for the price constraint. The firm therefore must trade off the increase in profit against the impact on the price index. The equilibrium condition for any one output[13] can be expressed as

$$(MR_j - MC_j) \frac{dy_j}{dp_j} = \rho \alpha_j$$

where MR_j and MC_j are marginal revenue and marginal cost respectively, dy_j/dp_j is the slope of the demand curve at the optimal point, ρ is a measure of the marginal cost of the price constraint to the enterprise, and α_j is the weight given to the price of the jth output in calculation of the price index. The left-hand side is the marginal profitability of a price increase, the right-hand side can be interpreted as the marginal cost of a price increase, and the optimal price then equates these. If we denote the left-hand side by $d\pi_j/dp_j$, we have for any two outputs i and j

$$\frac{d\pi_j/dp_j}{d\pi_i/dp_i} = \frac{\alpha_j}{\alpha_i}.$$

---

[12] Of course, if the constraint were not binding—lower price increases are chosen than would be permitted under the RPI − X constraint—this is bad news, since it implies that the enterprise is able to set profit-maximizing prices without violating the price constraint. The observation that the enterprise increases prices by the maximum possible is evidence that the constraint is at least effective, even if nothing can be said about how far away it is from the theoretical optimum.
[13] This condition assumes that the demand for each of the enterprise's outputs depends only on its own price. Note also that $MR_j < MC_j$ if the price constraint binds, while $dy_j/dp_j < 0$ in general. See the Appendix for the general derivation.

Thus, *relative* deviations of price from the profit-maximizing level depend only upon the relative weights in the price index. Finally, note that either $\alpha_j = 0$ (the output is not included in the constrained price index) or $\rho = 0$ (the price constraint is non-binding) implies the profit-maximizing condition $MR_j = MC_j$. These results suggest that there is no reason in general for the structure of prices of the relevant group of outputs to conform to an allocatively efficient structure. It shows that if we were interested in a normative application of these results, a key question would be the basis of the weights that are used in constructing the price index in terms of which the price constraint is expressed.

Note finally that this analysis, in taking the price constraint as given, has ignored what is in reality an important problem: the constraint on the rate of price increase is set for a specified period—usually five years. Given that the sum of the gestation period of the enterprise investments and their lifetimes typically extends well beyond this period, how is the enterprise to determine an investment programme when its future price constraints are unknown? Clearly, this adds an important element of uncertainty to investment decision-taking, but space precludes here a rigorous analysis of the implications of this.[14]

6. Conclusions

In modelling the supply side of the energy sector, it seems to me to be important to have available a reasonably rigorous analysis of the behaviour of the enterprises concerned. This requires a theory of the motivation of the enterprise, and a formulation of the constraints on its choices arising out of the particular regulatory structure to which it is subject, whether under public or private ownership. This paper has outlined an approach to this question which, in the light of some readily available empirical evidence, appears to be quite fruitful. However, much remains to be done.

Appendix: The Regulated Monopoly

The purpose of this Appendix is to analyse a model in which profit rather than size is the objective of management, and the constraints on profit and capital that exist for a public enterprise are replaced by one on price levels. Thus the model seeks to capture the main features of the regulated, privatized enterprise. Specifically, the enterprise seeks to maximize the objective function $\pi + \lambda v(w,L)$, where

$$\pi = R(y) - wL - rK \tag{1}$$

is profit, v is the union's utility function, w is the wage, L is total labour input into the enterprise, and λ is a constant which could be taken to represent union bargaining strength. We can think of y as in general a vector $[y_j]$, with y_j being

---

[14] This point is further discussed by Yarrow (Chapter 10, this volume).

output of the jth good, $j = 1, \ldots, n$. $R(.)$ is then a concave revenue function with $R_j \geq 0$, $\forall j$. K is total capital and r is the cost of capital. The enterprise has a transformation function

$$g(y,\ell,k) \leq 0 \tag{2}$$

where $\ell = [\ell_j]$ is a vector of labour inputs and $k = [k_j]$ a vector of capital inputs.

In practice, the price constraint takes the form of a constraint on the rate of growth of an index of enterprise prices. However, in a static model it suffices to represent this as a constraint on price levels, since any specific constraint on the *rate of growth* of prices between two periods (of the RPI-X form, for example) can obviously be expressed as a constraint on the *level* of prices at the subsequent period if the initial prices are given. Thus the price constraint takes the form

$$\sum_j \alpha_j p_j \leq I^0 \tag{3}$$

where the α_j are weights used in constructing the price index, and depend in part on the initial levels of prices.

The necessary conditions for a maximum of the objective function subject to the constraints (2) and (3), given also the (inverse) demand functions $p_j(y_1, \ldots, y_n)$, are

$$p_i + \sum_j y_j \frac{\partial p_j}{\partial y_i} - \frac{\mu \partial g}{\partial y_i} - \rho \sum_j \alpha_j \frac{\partial p_j}{\partial y_i} = 0 \quad i = 1, \ldots, n; \tag{4}$$

$$- w - \mu \frac{\partial g}{\partial \ell_i} + \lambda \frac{\partial v}{\partial L} = 0 \quad i = 1, \ldots, n; \tag{5}$$

$$- L + \frac{\partial v}{\partial w} = 0 \tag{6}$$

$$- r - \frac{\partial g}{\partial k_i} = 0 \quad i = 1, \ldots, n; \tag{7}$$

$$g(y,\ell,k) \leq 0 \quad \mu \geq 0 \quad \mu g(y,\ell,k) = 0; \tag{8}$$

$$\sum_j \alpha_j p_j - I^0 \leq 0 \quad \rho \geq 0 \quad \rho \left(\sum_j \alpha_j p_j - I^0 \right) = 0 \tag{9}$$

where μ and ρ are Lagrange multipliers.

Conditions 5–8 are essentially similar to those derived in Rees (1984) for the public enterprise. Essentially they show that the regulated private enterprise will be subject to excess labour intensity and wage-rates above the competitive level to an extent dependent on the union's bargaining power—this is the way in which labour appropriates some of the rents arising out of the monopoly position of the enterprise. If we note that $\mu \dfrac{\partial g}{\partial y_i}$ can be interpreted as the marginal cost of output y_i, then condition 4 says that the price of output y_i will depend on its own elasticity of demand, the relationships of complementarity or substitutability with other outputs of the enterprise, the pattern of weights in the price index, and its marginal cost. To simplify, suppose $\frac{\partial p_j}{\partial y_i} = 0$, $i, j = 1, \ldots, n$, $i \neq j$. Then the 'pricing rule' for each output becomes simply

$$p_i\left(1 - \frac{1}{e_i}\right) = \mathrm{MC}_i + \rho\alpha_i\frac{\mathrm{d}p_i}{\mathrm{d}y_i} \tag{10}$$

where e_i is price elasticity of demand.

6 Performance of the Public Sector Energy Utilities between 1968 and 1978*

RICHARD PRYKE

1. Introduction

The assessment in this paper of the performance of three nationalized energy industries—gas, electricity, and the coal-mines—is drawn from Pryke (1981). The assessment of each sector is in two parts. The first focuses upon productive efficiency and examines employment and labour productivity. The second part widens the examination of productive efficiency to consider the contribution of other factor inputs and to examine, on this basis, trends in total costs. It also assesses the efficiency of the energy utilities' pricing policies by examining the relationship between prices and costs and by looking at profitability.

The paper reaches a number of specific conclusions on the success, or otherwise, of the three energy industries in performing efficiently within the framework for the economic and financial regulation of nationalized industries which prevailed during this period.

The paper does not, however, consider in detail how far deficiencies in this framework of control may have contributed to the observed performance record. On this, the reader is referred to Pryke (1981).

2. The Gas Industry

Employment and Productivity

Between 1968 and 1978 the Gas Corporation's labour force was reduced from 122,000 to 100,000, although, as can be seen from Table 6.1, most of the decline had already taken place by 1974. The fall in employment was entirely explained by conversion to natural gas and the virtual elimination of staff engaged on production. As there was a substantial increase in output, productivity rose sharply. Over the period 1968–78 output per equivalent worker (OEW) increased by 125 per cent. OEW allows for the change in the number of hours worked by manual, but not salaried, employees and is to be distinguished from

---

* This article is based upon the relevant sections in Pryke (1981), originally published by Martin Robertson. For a discussion of developments since 1978, see Molyneux and Thompson (1987). Richard Pryke is Senior Lecturer in the Department of Economic and Business Studies, University of Liverpool.

Table 6.1. British Gas: output, productivity, and finances

	Gas sold (m therms)	Output (1968=100)	Employment (000)	Average weekly hours worked by manual workers, October week	Output per equivalent worker (1968-9=100)	Real unit operating costs[a,b] (1968-9=100)	Real unit staff costs[a] (1968-9=100)	Real staff costs per employee[a] (1968-9=100)	Real staff costs per employee in manufacturing (1968=100)	Relative weekly earnings of manual men, October week (1968=100)	Real cost of fuel per unit[a] (1968-9=100)	Real revenue per unit[a,c] (1968-9=100)	Revenue[a,c] (£m, 1978 prices)	Gross surplus[a,d] (£m, 1978 prices)	Gross surplus as % of revenue[a] (%)	Net surplus[a,e] (£m, 1978 prices)
1963	2,894	71.7	124.6	46.7	69.9	130.3[f]	130.4	90.4	87.2	98.3	162.3	119.4[f]	1,737	256	14.7	..
1968	4,416	100.0	122.1	46.3	100.0	100.0	100.0	100.0	100.0	100.0	100.0	100.0	1,812	350	19.3	..
1969	4,973	106.3	120.0	47.0	107.3	94.6	96.9	105.0	103.0	101.3	76.8	93.3	1,781	326	18.3	..
1970	5,750	108.9	118.7	46.3	112.1	88.1	93.4	112.4	107.4	102.5	61.4	84.2	1,742	285	16.4	..
1971	7,490	121.4	114.9	45.2	130.5	74.8	82.5	121.0	110.3	105.8	48.1	75.2	1,804	367	20.3	..
1972	9,757	135.5	109.4	45.2	152.9	64.3	69.7	127.9	114.0	103.4	41.2	68.3	1,943	486	25.0	..
1973	10,700	143.5	104.8	46.2	167.4	58.8	65.1	136.6	118.1	100.9	42.4	61.0	1,903	447	23.5	..
1974	12,634	152.2	102.3	46.3	181.7	53.3	60.9	142.8	118.5	106.3	38.9	56.7	1,958	484	24.7	..
1975	13,081	154.9	102.7	43.8	188.4	50.3	63.5	148.5	123.7	114.0	33.4	57.5	2,051	614	29.9	..
1976	13,969	155.1	101.2	43.1	192.5	47.3	61.6	148.0	126.0	111.6	32.0	62.0	2,261	875	38.7	496
1977	14,549	167.2	99.8	43.4	209.7	49.3	58.9	157.2	123.3	111.2	42.3	65.2	2,607	1,013	38.9	584
1978	15,281	181.8	100.4	44.0	225.7	54.3	58.1	163.2	128.9	118.1	57.9	64.8	2,761	889	32.2	432

Note: Output and employment figures were derived from financial year data. The output index consists of sub-indices (combined with 1963 and 1968 net output weights) for (i) gas distribution and appliance sales and (ii) activity at gasworks as reflected by gas made and reformed, etc. The distribution sub-index comprises 8 weighted indicators: gas sales by type of contract and consumer (using revenue weights) and number of appliances sold (net output weights for gas and appliances). Instead of treating gasworks as a separate activity, double deflation could have been used, in which case the cessation of production at works would have been reflected by increased purchases of natural gas. However, due to the very low price that was paid, an enormous increase in net output would have been shown. The unit series were estimated using gross output at 1968–9 (and 1963–4) prices.

[a] Financial year (April–March) beginning during year shown.
[b] Excludes replacement expenditure now charged to revenue. Also excludes central heating installation and oil production as represented by their revenue. Includes expenditure on conversion to natural gas. Also includes depreciation and interest of North Sea exploration and production subsidiary in order that its gas should appear at cost.
[c] Excludes central heating installation and oil.
[d] Replacement expenditure now charged to revenue not deducted. Conversion expenditure and depreciation and interest of North Sea subsidiary deducted. Before stock appreciation; but after interest of North Sea subsidiary.
[e] Excludes sale and installation of appliances.

output per man-year (OMY) which does not allow for variations in average hours.

Because gasworks have been treated as a separate activity in my output estimates, the growth in OEW is not directly attributable to conversion to natural gas; although conversion has led to a reduction in employment, this has been offset by the decline in production at gasworks. However, natural gas has indirectly been responsible for much of the rise in productivity because it has boosted sales and there is only a weak link between output and the number employed on distribution. About 30 per cent of all workers are employed on customer servicing, which includes the installation and maintenance of appliances and meters. Figures are available from 1972-3 onwards for the number of jobs performed and they show that there has been a small decline. In addition, 15 per cent of the labour force are engaged on customer accounting and work at showrooms. Their work-load largely depends on the number of customers, which increased by 10 per cent over the period 1968-78, and on the number of appliances sold, which fell by 10 per cent. The only other large block of workers, apart from administration and general services, is employed on transmission and distribution. There must have been an increase in activity here but it cannot have been anything like as large as the increase in sales. The mileage of mains in use increased by only 18 per cent, and the length laid and relaid fell by 16 per cent, although this work is largely undertaken by contractors.

Back in 1968 British Gas had ample scope for a more efficient use of labour. Only about 14 per cent of all manual employees were on incentive payment schemes based on work study (WSIP) and little progress had been made in measuring and then improving the productivity of salaried staff. Those workers who were on WSIP had a performance of around 90, where 100 represents the standard that should be readily attainable. In contrast, the performance of other manual workers only stood at around 50 and, after a thorough review of the situation, the Prices and Incomes Board concluded that there was 'considerable scope for improving labour productivity and securing cost reductions'. It also found that work measurement could be applied on an extensive scale to salaried staff and that where it had been used, performance had increased, on average, by some 27 per cent.[1] By 1978 about 75 per cent of the industry's manual employees were on WSIP and productivity measurement had been applied to 25 per cent of salaried jobs. According to British Gas, workers who are on WSIP now have a performance of over 90. However, this gives a misleading impression of the gains which have been made because there is a tendency for incentive schemes to get slacker as time goes by. The performance of workers engaged on customer servicing ranged from less than 44 in the worst region to about 80 in the best, as measured under a scheme for monitoring productivity which is based on work study times.[2] It is clear that the British Gas

[1] National Board for Prices and Incomes (1970, pp. 8, 11, 25).
[2] Price Commission (1979a, pp. 63, 65).

Corporation (BGC) still has considerable scope to improve the efficiency with which labour is used but that substantial, although not enormous, progress has already been made. Between 1972-3 and 1978-9 there was a rise of 10 per cent in the number of jobs per manual employee on customer servicing and there has been some decline in the number of hours worked.

Costs, Prices, and Profits

During 1978-9 the Gas Corporation's unit costs were about 45 per cent lower than they had been in 1968-9. They fell steadily until 1976-7 but then increased. This is largely explained by the movement of unit fuel costs. Between 1968-9 and 1976-7 there was a fall of two-thirds due to the decreasing use of coal and light oil and the increasing use of natural gas, which became progressively cheaper. However, after 1976-7, unit fuel costs shot up because of the use of gas from the Frigg field for which BGC was having to pay a market price. Nevertheless, during 1978-9 the cost of fuel was still around 40 per cent lower than it had been a decade earlier. Due to the huge rise in productivity, there has also been a spectacular decline in unit staff costs. By 1978-9 they were about 40 per cent lower than in 1968-9, and this despite increases of 18 per cent in the relative earnings of the industry's manual employees and of around 60 per cent in staff costs per employee. This compares with a rise of only 30 per cent in manufacturing.

In 1978-9 BGC's prices were 35 per cent lower than they had been 10 years earlier, although they were somewhat above the low point which had been reached in 1976-7. Because unit costs have declined more than unit revenue, the Corporation's gross margin has improved from 19 per cent in 1968-9 to 32 per cent in 1978-9. During 1978-9 BGC had a net profit of £430 million which represented a return of 6.3 per cent on capital employed at replacement cost. However, the financial position would look very different if BGC had paid the full market price for North Sea gas. If it had had to pay a price equal to that of Frigg gas, the Corporation would have incurred a net loss of approximately £830 million, which is another way of saying that its prices were too low.

3. Electricity

Employment and Productivity

Between 1968 and 1978 British Electricity Boards (BEB) cut its labour force from 238,000 to 177,000 which was a reduction of more than a quarter (Table 6.2). Over the period 1968-73, output per equivalent worker shot up by more than 50 per cent. Between 1973 and 1978, progress was less rapid because output scarcely increased and the labour force declined more slowly. Nevertheless, by 1978 OEW was two-thirds greater than it had been in 1968. The industry's

Table 6.2. British Electricity Boards: output, productivity, and finances

	Output (1968 = 100)	Electricity production as % of capacity[a] (%)	Employment (000)	Average weekly hours worked by manual workers, October week	Output per equivalent worker (1968 = 100)	Quantity of capital (1968 = 100)	Output per unit of labour and capital (1968 = 100)	Real unit operating costs[b] (1968 = 100)	Real unit staff costs (1968 = 100)	Real staff costs per employee (1968 = 100)	Relative weekly earnings of manual men, October week (1968 = 100)	Real cost of fossil fuel per unit (1968 = 100)	Real revenue per unit (1968 = 100)	Revenue (£m, 1978 prices)	Gross surplus[b] (£m, 1978 prices)	Gross surplus as % of revenue[b] (%)
1963	81.4	46.9	225.8	48.8	76.6	75.8	96.1	109.8	108.2	93.0	112.7	119.0	101.4	3,682	1,305	35.4
1968	100.0	44.5	237.8	40.9	100.0	100.0	100.0	100.0	100.0	100.0	100.0	100.0	100.0	4,597	1,857	40.4
1969	105.7	44.9	224.2	40.6	112.5	104.5	104.2	95.5	93.3	103.6	100.1	96.4	93.8	4,614	1,832	39.7
1970	109.4	44.5	213.1	40.8	122.2	110.8	104.7	98.3	91.3	109.8	102.2	102.5	88.2	4,544	1,560	34.3
1971	111.7	42.5	203.4	40.7	130.9	120.1	101.7	100.9	92.2	118.0	114.6	104.2	88.7	4,659	1,511	32.4
1972	115.2	40.3	194.0	40.4	142.3	127.5	101.2	101.1	91.7	126.4	114.9	101.1	87.1	4,817	1,551	32.2
1973	121.2	41.7	190.1	40.6	152.3	132.5	103.6	98.5	87.9	128.3	114.0	99.5	81.1	4,651	1,288	27.7
1974	116.3	39.2	189.0	41.0	146.1	135.5	97.7	119.6	94.5	133.0	112.5	138.3	85.4	4,923	1,002	20.3
1975	116.0	39.3	188.7	39.2	149.7	136.1	97.6	121.4	96.6	137.1	117.4	144.8	92.3	5,182	1,212	23.4
1976	117.2	39.9	183.0	40.3	153.4	134.0	100.1	122.9	93.6	136.9	119.0	144.5	97.2	5,460	1,402	25.7
1977	119.8	41.7	177.8	39.5	163.3	132.2	104.3	123.7	87.7	134.6	116.2	148.2	98.4	5,630	1,448	25.7
1978	122.8	42.4	176.6	39.8	167.9	132.2	106.9	127.6	91.2	143.0	124.2	152.6	99.6	5,894	1,471	25.0

Note: The output index includes 18 weighted indicators: quantity of electricity supplied distinguishing between ordinary and off-peak sales, number of appliances sold, and contracting and capital work (expenditure on materials deflated). Revenue weights for 1968 (and 1963) were used for electricity itself; but this, contracting, and capital work were combined with net output weights. The per unit series were estimated using gross output, apart from capital work, at 1968 (and 1963) prices. The financial figures, employment, and 3 output indicators were derived from financial year data. The CSO's figures for the gross capital stock were used as bench-mark estimates for the quantity of capital in 1963 and 1978 as the CEGB's capital work in progress represented roughly the same proportion at both dates. BEB's generating capacity was used to interpolate the figures for the other years. A 1963 capital weight of 63.1% was used.
[a]The plant load factor.
[b]Amount written off nuclear stations' initial fuel has been excluded throughout.

performance appears much less impressive when allowance is made for the use of capital. Between 1968 and 1978, the quantity of capital in use rose by approximately a third, due to the completion of stations which were begun during the period of optimistic growth forecasts and heavy plant ordering in the 1960s. As a result, output per unit of labour and capital increased by only 7 per cent over the period 1968–78, and the picture does not look any better if allowance is made for the amount of fuel that was used.

The rise in labour productivity has been partly due to the closure of old labour-intensive power-stations and the introduction of big units which, because there are large economies of scale in generation, require relatively little manpower. The growth in OEW has also been due to the contraction of activities that are especially labour-intensive, namely capital work and the sale and installation of appliances. Another factor has been the growth in sales of electricity, as a large part of the labour force in distribution is engaged on work that is related to the number of customers or the size of the transmission system, both of which grew more slowly than consumption. Despite this, it seems clear that there was a substantial improvement in the efficiency with which labour was used.

The Status Agreement of 1964–5 led to a large reduction in overtime through the introduction of staggered work patterns. The Agreement also contained a declaration that employees should co-operate in the adoption of the most efficient work practices and that flexible use could be made of labour, regardless of the old boundaries between jobs, provided that workers received the rate for the job and were not temporarily upgraded to craft duties. At a later stage, the unions agreed to support the use of work study, which had hitherto been little used. As a result, some progress was made towards the adoption of better working methods, but it was sporadic and tailed off. In September 1967 a new agreement was concluded. This affirmed management's right to make any changes in working arrangements that it considered necessary and spelt out the ways in which labour could be redeployed. The agreement also provided for the establishment of a central register of the best labour practices and for the introduction of manpower utilization yardsticks by which progress would be jointly monitored. This concordat led to a dramatic improvement in efficiency. The size of work teams was reduced, craftsmen started to do work that had previously been undertaken by other craft grades or by those with less skill, and there was also greater flexibility in the use of semi-skilled and unskilled workers. Jointing teams, for instance, were cut from three men to two, and craftsmen began driving themselves around and doing carpentry and other work where this was part of the job.[3] During 1968 a further and even more important agreement was concluded. It provided for the introduction of incentive payments based on work study. Bonus payments would commence when workers' performance reached 65 per cent of the standard level, and their size would increase until it

---

[3] Edwards and Roberts (1971, pp. 109, 110, 137, 141, 150, 172, 173, 274, 295–9).

reached 100 per cent, at which point a maximum addition equivalent to one-third of basic pay would be received. As the general level of performance was still very low, at something under 65 per cent, there was ample scope for improvement.[4] Relatively little progress was made with the introduction of the incentive payments until 1970, but after that, progress was rapid, and by 1973 the great bulk of all manual workers had been covered.

The Electricity Council has made estimates, based on activity yardsticks and on work study data, of the increase in the productivity of manual workers in distribution. These suggest that there was a rise of over 60 per cent during the period 1968-77, and it appears from a survey that capital expenditure was responsible for only about 10 per cent (6 percentage points) of this increase. The Central Electricity Generating Board (CEGB) claims that even larger gains have been made at generating stations.[5] It is difficult to believe that the increase has been so large, but there is no doubt that there has been a marked improvement in efficiency, for which the industry deserves considerable credit. Many British industries use labour inefficiently, but few have made a sustained effort to improve the situation, and even fewer have had much success. That the electricity industry has succeeded has been due partly to the vision and persistence of its management and the good labour relations that had been fostered through joint consultation. However, the *sine qua non* of success was the co-operative attitude of the electricity unions, which were strongly committed to the cause of higher productivity and mounted a determined campaign to persuade their reluctant rank and file to accept the agreements that had been negotiated.

International comparisons of the growth in labour productivity confirm that BEB has been fairly successful at improving the efficiency with which it uses its manpower. Between 1968 and 1978, there was a growth in electricity sales per man-hour of 77 per cent at BEB, 37 per cent at the American investor-owned utilities, 83 per cent in Germany (1968-77), 104 per cent at Electricité de France, and 112 per cent in Belgium (1968-77).

However, BEB continues to use labour less efficiently than the American private utilities. During 1978, the quantity of electricity sold per worker engaged on its production and distribution was nearly three times greater in the US than in Britain.[6] This, however, gives an exaggerated impression of American superiority, since there is a close association between sales per employee and sales per customer, and the level of consumption is far greater in the US.

But high sales per customer are only an advantage for the distribution side of the industry. In production the quantity of electricity generated—or, what is preferable, capacity—can fairly be related to employment. If such a comparison is made, it appears that during 1977-8 the number of workers per unit of

---

[4] National Board for Prices and Incomes (1968, pp. 2, 4).
[5] Select Committee on Nationalised Industries (1977-8, pp. 106, 107); Edwards and Roberts (1971, p. 274).
[6] Edison Electric Institute (1978, pp. 31, 49).

capacity at power-stations owned by the American investor-owned utilities was about 70 per cent smaller than in Britain, and that in France it was 50 per cent lower.[7] These comparisons do not allow for any variation between the countries in average hours worked, the use of contract labour, or other complications. However, a study of a Scottish station (Cockenzie) and a similar coal-fired generating plant in America (Will County), in which such factors were taken into account, showed that the number of equivalent workers per unit of capacity was about 45 per cent lower in the US. The contrast was less great for two nuclear stations as the number of equivalent workers per unit was only 20 per cent lower at the American plant (Dresden) than at the Scottish (Hunterston). Labour is also used somewhat more efficiently away from generating stations in America. In the investor-owned utilities, there were 18 per cent more customers per worker than in BEB, ignoring those employed at power-stations, on capital work, and (in Britain) on retailing and contracting (which are not undertaken by the American utilities).[8] However, this almost certainly understates the Americans' lead and it is reported by a high-ranking BEB official who has made an on-the-spot investigation that, where comparisons can be made, distribution manning levels are 30 to 40 per cent higher in Britain. There are a number of reasons why American productivity is higher. At power-stations there are fewer men watching machinery and dials because it is assumed that plant will function properly, there is greater flexibility of labour, and it is reported that American maintenance staff get down to work more promptly. In distribution, too, less time is wasted because materials and parts are assembled overnight and are ready to be picked up by working parties. Moreover, some activities, such as overhead line work, are very efficiently organized and workers seem to be more highly motivated.

Costs, Prices, and Profits

Between 1968 and 1978, BEB's unit costs rose by 28 per cent. During the first half of the period, when unit fuel costs were stable, they were more or less constant. However, since 1973 there has been a spectacular increase in coal and oil prices, and by 1978 expenditure per unit on these fuels was 53 per cent higher than it had been in 1968. Since they already accounted for 45 per cent of the industry's operating expenditure, this naturally had a considerable impact on its (total) unit costs. The rise in fuel costs has, to some extent, been offset by a reduction in staff costs per unit of output. Between 1968 and 1978, they fell by 9 per cent, due to the large rise in OMY and despite an increase of 24 per cent in the relative weekly earnings of the industry's manual workers. This rise has more than made good the (11 per cent) fall that occurred between 1963 and 1968 as a result of the Status Agreement and the cut-back in hours worked.

---

[7] United States Department of Energy (1978); Electricité de France (1978).
[8] United States Department of Energy (1978, 1979).

The rise in coal and oil prices would have had a smaller impact on BEB's costs if more electricity had been generated by nuclear power—as it would have been but for the serious delay in completing the advanced gas-cooled reactor (AGR) stations and the low level of output at the two that have so far been commissioned. Expenditure on fuel would also have been lower if the industry had imported more coal from abroad. During 1978-9, imports accounted for only 250,000 tonnes out of total coal deliveries of 78 million tonnes. A further 850,000 tonnes of foreign coal would have been taken under a contract for the supply of Australian coal but for a government scheme to promote the use of British coal. As a result, the CEGB, in the spring of 1979, was buying coal from the National Coal Board (NCB) at a price of around £22 per tonne—some of which was produced at a significant loss—and selling Australian coal, for which it was charged about £17 per tonne, to Electricité de France for around £13 per tonne. Although the CEGB was compensated for the loss that it incurred, it has been prevented from buying any more cheap coal from abroad. In the autumn of 1977, it could have signed up a long-term contract for 3 to 5 million tonnes of Australian coal which would have cost £2 to £3 per tonne less than British coal, and cheap supplies have also been available from Poland. Although imports might over the years have been built up to a substantial level, the direct saving would not have been enormous. However, the NCB would have been under pressure to control its costs and it would almost certainly have been forced to close down its high-cost pits.

Despite the steep rise in BEB's unit costs, its prices were no higher during 1978 than they had been in 1968. Moreover, during the intervening years they were lower, and in 1973 reached a point nearly 20 per cent below the 1968 level. Because of the divergence between costs and prices, there has been a substantial fall in the industry's gross profits. During 1968 they totalled £1,850 million and represented 40 per cent of revenue, whereas in 1978 they were £1,475 million, a margin of 25 per cent. However, during 1974 BEB's gross profit had been as low as £1,000 million, which constituted only 20 per cent of its revenue. This was because unit costs had shot up, due to the steep rise in the price of fuel, but there was no compensating increase in charges, due to the government's policy of price restraint. BEB's gross profits appear huge. However, when allowance is made for replacement cost depreciation, which amounted to £1,090 million, and for stock appreciation at £65 million, the industry only had a net profit of £320 million during 1978-9. In England and Wales the return on capital employed—net profit as a proportion of net assets at replacement cost—was a mere 1 per cent, and prices were 5 to 10 per cent below the industry's long-run marginal costs, as estimated with a real return of 5 per cent on investment.[9]

Although the sale of appliances and installation work account for only a small part of the industry's turnover, they are of considerable interest because they are the only area where BEB faces direct competition. If hire-purchase interest

[9] Price Commission (1979b, pp. 49, 56).

is disregarded, these activities sustained a net loss of over £8 million during 1978-9, after depreciation at historic cost and before allowing for stock appreciation. This represented a negative margin of $2\frac{1}{2}$ per cent on turnover. In England and Wales there was a small deficit on installation and servicing work, although major private contractors achieved a 4 per cent margin.[10] In their retailing activities, the Electricity Boards had a net margin of about $3\frac{1}{2}$ per cent, including hire-purchase charges but excluding supplementary depreciation. The corresponding figure for Currys, which is the largest electrical chain store in the private sector, was about $5\frac{1}{2}$ per cent.

4. The Coal Mines

Employment and Productivity

Between 1968 and 1973, employment at National Coal Board collieries fell from just over 400,000 to just under 310,000, and by 1978 the number had slipped to a little under 290,000; which meant there was an overall reduction of 28 per cent (Table 6.3). The drop in output was somewhat greater, and output per man-year fell by 11 per cent. However, there was a significant reduction in the number of shifts worked per week, and if allowance is made for this, productivity declined by 7 per cent. During the earlier part of the period there had been a small rise, and in 1970 output per equivalent worker was over 3 per cent higher than in 1968; but from 1971 to 1977 there was a steady decline, ignoring 1972 and 1974 which were affected by the national coal strikes. However, 1978 saw a slight recovery. The reduction in total factor productivity during the decade has been even larger than the fall in labour productivity. During 1978, output per unit of labour and capital appears to have been 13 per cent lower than in 1968.

Since 1968, some of the sources of productivity growth, which had made such a large contribution during the previous decade, have dried up. Between 1958 and 1968 the National Coal Board had the benefit of the large post-nationalization programme of modernization and reconstruction, but by 1968 expenditure on major projects had fallen to a very low ebb. For instance, during 1973-4 it amounted to only £14 million. Since then there has been a spectacular increase but relatively few schemes have been completed. And cutter-loaders and self-advancing pit-props, which had played such a large part in raising productivity, had already been introduced on a high proportion of faces by 1968. Nevertheless, the Board was anticipating that productivity would go on increasing at a rapid rate. In mid-1969 it forecast that output per man-shift (OMS) would increase from 42.5 hundredweight (cwt.) in 1968-9 to 75 cwt. in 1975-6, which is an increase of over 75 per cent.[11] (During 1975-6, it was

---

[10] Price Commission (1979b, pp. 33, 34).
[11] Select Committee on Nationalised Industries (1968-9, vol. 2, p. 479).

Table 6.3. National Coal Board collieries: output, productivity, and finances

	Output (1968 = 100)	Tonnage (m tonnes)	Employment (000)	Average number of shifts per week per wage-earner	Output per equivalent worker (1968 = 100)	Output per worker (1968 = 100)	Quantity of capital (1968 = 100)	Output per unit of labour and capital (1968 = 100)	Real unit operating costs[a,b] (1968-9 = 100)	Real unit staff costs[a,b,c] (1968-9 = 100)	Real staff costs per employee[a,b,c] (1968-9 = 100)	Relative weekly earnings of miners, October week (1968 = 100)	Real prices[a] (1968-9 = 100)	Revenue[a] (£m, 1978 prices)	Gross surplus[a,b] (£m, 1978 prices)	Gross surplus as % of revenue[a,b] (%)
1958	138.6	202.1	766.0	4.51	67.1	72.7	..	..	113.1	125.4	90.0	113.8	110.4	3,621	236	6.5
1963	126.1	191.2	588.7	4.23	84.2	86.0	..	..	104.9	112.0	96.4	106.6	109.1	3,201	394	12.3
1968	100.0	158.7	401.7	4.13	100.0	100.0	100.0	100.0	100.0	100.0	100.0	100.0	100.0	2,276	201	8.8
1969	91.5	145.7	360.4	4.09	102.9	102.0	98.2	100.8	100.3	98.8	99.7	96.4	98.2	2,057	140	6.8
1970	85.0	135.9	337.8	4.02	103.5	101.1	95.9	100.2	105.5	102.6	103.0	93.6	104.9	2,079	172	8.3
1971	83.9	135.7	332.8	4.12	101.5	101.2	94.0	98.8	131.3	121.5	97.2	98.2	112.1	1,742	−118	−6.8
1972[d]	66.9	108.4	317.9	3.47	97.8	84.5	92.0	91.5	121.5	120.3	118.8	100.8	112.3	2,063	28	1.4
1973	73.1	119.4	308.1	3.90	99.9	95.3	89.5	95.7	145.4	144.3	111.2	96.6	105.3	1,428	−369	−25.9
1974[d]	61.5	99.4	295.7	3.59	93.9	83.5	87.2	88.1	147.1	147.7	134.8	108.5	136.1	2,200	32	1.4
1975	70.8	116.8	299.6	3.99	97.7	94.9	86.4	94.1	164.1	165.1	146.7	117.0	158.6	2,426	137	5.7
1976	66.3	109.7	294.1	3.93	94.5	90.6	86.9	90.2	162.4	159.9	141.5	113.8	161.7	2,423	204	8.4
1977	64.3	106.5	292.1	3.93	92.2	88.4	88.7	87.4	169.1	165.2	143.4	112.9	165.7	2,420	167	6.9
1978	63.8	106.9	287.6	3.90	93.4	89.0	92.3	87.3	180.8	180.1	156.2	123.1	166.5	2,436	24	1.0

Note: The output index comprises 7 indicators (tonnes delivered to different types of consumer) combined with 1968 (and 1963) revenue weights. Allowance was made for the overall change in stocks. The per unit series were estimated using an index for the value of production at constant prices calculated with the wholesale price index for coal. The quantity of capital is the gross capital stock, as estimated by the CSO. Its figures contain some investment which has not yet fructified but must, in 1968, have contained capital at pits that had been closed prematurely. The gradual removal of these assets as their normal lives have expired should exert some downward bias on the estimates. A capital weight of 19.8% was used.

[a] Financial year (April–March) beginning during year shown, except for 1958 which is the calendar year.
[b] Pension fund deficiency payments and 'social costs' have been disregarded.
[c] Staff costs for all NCB employees.
[d] There were prolonged industrial disputes during the first quarters of 1972 and 1974.

scarcely higher than it had been in 1968-9, namely 44.8 cwt.) There was probably an element of wishful thinking, and even propaganda, in the NCB's estimate, and the Ministry was doubtful whether the rise would be as great. However, even it was expecting a substantial rise (55 per cent) and the Board did have solid reasons for believing that productivity would continue to grow. A number of possible sources of productivity growth had been identified, including retreat mining, new forms of mechanization, and improved versions and better use of existing equipment.[12]

Under the advancing system, which is normal in Britain, coal is extracted outwards in a series of strips parallel to the roadway along which the face was commenced. With the retreat method, tunnels are driven out and linked with a face. The coal is then extracted back towards the main roadway. With the advancing system progress is often interrupted because faults or other geological difficulties are encountered on 30 per cent of faces. Under retreat mining this does not happen because the panel of coal is proved before extraction commences. Moreover, it is unnecessary to maintain the tunnels in the region from which the coal has already been extracted—a task of some difficulty on an advancing face because of the need to hold up the sides on which the coal has been removed. On advancing faces it is usually necessary to cut holes, known as stables, at each end to accommodate the coal-cutter when it has finished a traverse, but under the retreating system the tunnel can be used instead. Consequently while an advancing face usually has a complement of eighteen to twenty men per shift, as few as eight men are required on a retreat face, though this gives a slightly exaggerated impression because it ignores preparatory work on the retreating face.[13] Moreover, in September 1968 the daily output per face was 35 per cent higher on retreat than on advancing faces. Some progress has been made with the introduction of retreat mining, but it has been less rapid than the Board had hoped. The proportion of output so obtained has increased from 4 per cent in September 1968 to 20 per cent in September 1978. However, by September 1978 only 14 per cent of all faces were using the retreat system, although geological conditions would permit 25 to 30 per cent of faces to be switched to retreat mining.[14]

This technique was expected to account for only part of the coming rise in productivity. Further mechanization was planned because, despite the use of cutter-loaders and self-advancing pit-props, mining had only been partly mechanized. A considerable amount of labour was absorbed in cutting stable holes, driving tunnels, and supporting their sides with stone packs. It was planned that the work would be mechanized by, for instance, using an extra power-loader to cut the opening. In 1968 the new methods had been applied to a very limited extent, but by March 1976 traditional stables had been eliminated

---

[12] Reid, Allen, and Harris (1973, pp. 35-6, 40-3).
[13] Harlow (1977, p. 211).
[14] Department of the Environment (1974, p. 12).

on over 70 per cent of all advancing faces, and by 1978 the ripping of side tunnels had been mechanized at one or more of the ends of a third of all faces.[15] Even more progress had been made with the introduction of machines to drive the tunnels for retreat faces and the roadways to new advancing faces. Despite this, the average daily advance has not increased and there has been a decline in the distance cut per man-shift devoted to this activity. The slow rate at which tunnels are driven explains, incidentally, why retreat mining has not been introduced more rapidly.

During the period 1968-78, a considerable amount of face equipment has, as planned, been replaced by superior machinery. Far more powerful cutter-loaders have been introduced and the new types of self-advancing pit-props can be shifted forward more rapidly. Conveyor belts which have a much greater capacity have been put in, and underground bunkers have been installed at pits where there is insufficient shaft capacity. The Board hoped that by such means it would be able both to increase cutting speeds and to reduce the amount of time that face equipment was out of action.

In spite of this, a sample inquiry showed that cutter-loaders ran for an average of only 1 hour and 46 minutes per shift during the early part of 1976. Even when allowance is made for the time (29 minutes) spent at the face ends, machines were only operated for 42 per cent of the time that was available (319 minutes) excluding travelling, refreshment, and preparation time.[16]

Why is machine running time still so short? One possible explanation is that underground travelling times have risen due to the growing distance between the face and the pit bottom. Between 1968 and 1974 there was a reduction of 18 minutes in the time that face-workers spent at their place of work, although there appears to have been little change since then. However, running time would not have been reduced by the same amount, partly because of the practice of cutting only two strips per shift and partly because machines only cut for one-third of the net time. Moreover, travelling time obviously cannot explain why this proportion is so low.

One reason is that some obvious and remediable causes of delay have been neglected. The face study in 1976 showed that 37 minutes were lost as a result of difficulties with the armoured flexible conveyor (AFC) onto which the cutter-loader deposits the coal and on which it is mounted. According to one of the Board's Area Directors, the AFC and related equipment 'represent only 6 per cent of the total face installation costs and statistics show that 20 per cent of lost production can be attributed to them. Improvement in design to reduce the delays due to AFC difficulties is long overdue'.[17] It is also noteworthy that 38 minutes were lost as a result of coal clearance delays, which seems a surprisingly high figure as the equipment is relatively simple. What has happened is that

---

[15] Harlow (1977, p. 210); Rawlinson (1976, p. 370).
[16] *Colliery Guardian*, p. 361.
[17] *Colliery Guardian*, p. 361.

attention and capital expenditure have until recently been concentrated on the face and that coal handling has to some extent been neglected.

However, it seems clear that there are other and more fundamental reasons why machine running time is so short and why so little coal is cut even when the cutter-loaders are working. In 1979 output per man-hour (OMH) was no higher than in 1976 when the Chairman of the Coal Board told the National Union of Mineworkers:

> we are not yet achieving the improvement in productivity which our massive investment programme would lead us to expect. Of course, a large part of the investment will take some time to materialise. But a good part is also spent on meeting current needs with the supply of adequate quantities of up-to-date machinery and equipment. Many pits have performed remarkably, but our overall performance is poor.[18]

There is little doubt that miners have lacked motivation, and the same may be true of managers. The Board has found that cutter-loaders are frequently stopped for no apparent reason, and managers report that often an excessively long period is taken to sort out the difficulties that do occur. Motivation would have been higher, and effort greater, but for the switch from piece-work to time rates, though this has at long last been reversed. The dropping of incentive payments was due to the National Power Loading Agreement of 1966, and took place as new faces were introduced.

At first the National Power Loading Agreement did not have any visible effect on face productivity, which continued to increase at a rapid rate. However, this may well have been because the widespread introduction of self-advancing pit-props was pushing productivity up. Searle Barnes (1969), who made some detailed investigations in Nottingham, found clear evidence that the Agreement was having an unfavourable impact, although it may have been more serious here than in other coalfields. When 17 day-wage faces were matched up with piece-work faces it was discovered that there was only one day-wage face which had a performance as good as that of its piece-work counterpart. On the remaining 16 day-wage faces, OMS was lower, often by a substantial amount, because their manning was greater and their output was almost always smaller. The Agreement appeared to have had an even more damaging effect on the productivity of those face-workers who were engaged on advancing the headings at either end of the face, and there were reports from several other coalfields that their performance was being seriously impaired.[19] Nevertheless the Board concluded a further agreement in 1971 by which those employed on other tunnelling were transferred from piece-work to day-work. This led to a reduction in their performance despite the widespread introduction of machinery.

By the end of 1971, the NCB had come to the conclusion that an incentive payment scheme was needed for face-workers. However, there was resolute

---

[18] National Coal Board Press Release, 7 July 1976, p. 7.
[19] Searle Barnes (1969, pp. 128–52, 166).

opposition within the NUM. Its leadership, which had become far more militant, was determined to secure a large wage increase on a national basis. The NUM's swing to the left was due partly to accumulated resentment at colliery closures and partly to the way in which miners had dropped down the earnings league. As a result of the National Power Loading Agreement, the position of face-workers, who play a key role in the union, had deteriorated particularly sharply.[20] The ending of piece-work also increased face-workers' sense of solidarity since they were now unable to increase their earnings through greater effort or their skill in negotiating improved rates for their teams. This was the background to the national coal strikes of 1972 and 1974, and helps explain why, in a national ballot in October 1974, the miners rejected a scheme for pit-based incentive payments. After this, wage restraint blocked any further attempt to introduce a proper incentive system, and it was not until early in 1978 that incentive payments were introduced through local negotiations, the miners having again rejected the proposal in a national ballot.

Between 1977 and 1978 there was only a small rise in productivity and the following year did not see any increase in the level of OMS. It would, however, be premature to conclude that incentive payments are ineffective or that productivity would have been no higher even if they had been introduced earlier. OMS appears to be rising now and there may have been special reasons why incentives did not lead to the expected growth in productivity. The industry has, due to the reduction in the age of retirement, been taking on a large number of new and inexperienced workers.

In any case there was probably another factor besides the absence of incentives that was and still is depressing efficiency, namely the cessation of colliery closures except in cases of exhaustion. Up to 1970 a large number of pits were shut each year because they made substantial and persistent losses. Moreover, major investment projects were only sanctioned at those pits that were listed as having good long-term prospects, and a pit with a poor financial performance was unlikely to be included. Hence managers and men had a powerful incentive for efficiency, as was shown time and again by the improvement which took place when pits were told that their future was in jeopardy.

The closure of pits on financial grounds more or less came to an end during 1970, when considerably more coal could have been sold if it had been available, and the Board also had production difficulties during 1972–3 and 1974–5 as an aftermath of the national strikes. What was more important was that the NUM, as a result of its leftward swing, began to oppose any closure that was not due to exhaustion, and it has over the years gained experience in mining engineering and become more skilful at arguing its case. The Board also became less willing to press closures because, as a result of the rise in energy prices, some unprofitable pits started to make more money. Moreover, it is difficult to

---

[20] Hughes and Moore (1972, p. 29).

combine an extremely optimistic attitude towards the industry's prospects with the closure of pits on the ground that they are grossly unprofitable.

Although, considered in isolation, the NCB's productivity performance looks very poor, it does not seem quite so bad when viewed in the light of what has been happening in other European countries. Poland and the USSR are the only major producers to have made large gains in productivity since 1968. Between 1968 and 1979 there appears to have been a rise of 73 per cent in OMS underground in Poland, though this figure is a little suspect. On the other hand, there was an increase of 28 per cent in France, of 21 per cent in Belgium, and of 15 per cent in Germany. The corresponding decrease in the UK was 7 per cent. Although the NCB was at the bottom end of the range, it should be borne in mind that Poland has been investing very heavily and that the continental countries had at the start of the period made less progress with the introduction of cutter-loaders and self-advancing pit-props. Moreover, in Germany and Belgium productivity has been stagnant or falling since the early 1970s. However, international comparisons also suggest that the Coal Board has considerable scope to improve its performance. During 1979 the output per man-hour for underground workers was 46 per cent greater in West Germany than in Britain, although our geological conditions are slightly more favourable.[21]

Costs, Prices, and Losses

Since 1968 there has been, as can be seen from Table 6.3, an enormous rise in the NCB's unit operating costs and their growth has been almost continuous. By 1978-9 collieries' unit costs were about 80 per cent greater than they had been during 1968-9. The increase has been so large partly because OMY has slumped and partly because of the substantial rise in the average weekly earnings of mineworkers relative to those of workers in manufacturing. Up to 1973, earnings in coal tended to rise less rapidly than in manufacturing, despite the national coal strike of 1972 and the large wage award that miners received as a result of the Wilberforce Inquiry. Early in 1974, after the second national coal strike, miners received another massive wage increase. This produced a more permanent rise in their earnings, and during 1978 these were boosted still further by incentive payments. Between October 1968 and October 1978, mineworkers' earnings rose by 23 per cent relative to manufacturing.

The massive rise in the Board's unit costs has led to a huge increase in its prices. By 1978-9 the price of coal was 66 per cent greater than in 1968-9, though most of this increase took place during 1974-5 and 1975-6. In 1968-9 collieries earned a gross surplus of £200 million, but because prices increased less rapidly than costs, there was a gross deficit of around £10 million in 1978-9, after allowing for £35 million of stock appreciation. The size of the industry's

[21] Eurostat (1980).

gross surplus depends on the way in which deficiency contributions to pension funds and what are termed 'social costs' are treated. In the NCB's accounts they are included in operating expenditure, although the Board receives a subsidy which covers part of the cost. However, they are best disregarded because they do not form part of the cost of producing coal during the year to which they appear to relate and would not be avoided if production were curtailed. Indeed, social costs are the expenditure which the Board incurs when it eliminates uneconomic capacity, such as the amounts paid to pension and superannuation funds to meet the extra cost of early retirement benefits awarded to redundant mineworkers. This is clearly a transfer payment rather than a resource cost. (It should be noted that the costs of closures and deficiency contributions have been disregarded not only here but wherever they are recorded as having occurred within the nationalized sector.)

On the other hand, it is necessary to allow for replacement cost depreciation, which has hitherto been ignored. The Board's estimates suggest that it amounts to about £185 million. If so, collieries had a net deficit of nearly £200 million during 1978-9. This was a particularly poor year but at no time since 1968-9 has the NCB managed both to cover its depreciation and to provide for stock appreciation; and in most years there have been large losses. It may be objected that, during the early years of the period at least, the Board had surplus capacity and misallocation would have resulted if depreciation had been met. If excess capacity exists, prices should be based on avoidable costs, which appears to mean that depreciation will not be recovered. The charge will in any case be excessive if depreciation is being provided in respect of surplus assets that will never be replaced. However, allowance has been made for this because capital has been written off and mining machinery usually has an opportunity cost. What is even more important is that coal is an increasing-cost industry, i.e. the greater the output the higher costs will tend to be as mines where the geological conditions are increasingly unfavourable have to be brought into production. It follows that even if the Board's high-cost pits only earn sufficient revenue to meet their avoidable costs, its low-cost mines will show a large gross surplus (or rent). This would more than cover its replacement cost depreciation.

But what happens is that the Board sustains a huge loss at its high-cost pits. Instead of closing them and basing its prices upon its marginal costs, the Board practises cross-subsidization and charges somewhat less than its average cost of production. The Board does not even have to recover the full cost of producing its deep-mined output because it is able to draw upon the handsome profit which it earns on its opencast production. This has always been relatively profitable but has, during the past few years, turned into a bonanza. Due to the rapid escalation in coal prices, there has been a large absolute increase in the revenue per tonne, which has not been matched by a corresponding rise in the absolute cost per tonne, primarily because this was considerably lower to begin with. During 1978-9 the NCB earned a net profit of about £95 million from its opencast output (before stock appreciation). On the other hand, a huge loss was incurred at those of the Board's pits that were unprofitable. I have roughly estimated from colliery output and employment data that the pits that made gross losses incurred an aggregate deficit of £180 million during 1978-9. In addition, their retention must have necessitated a substantial amount of investment, both directly in order that they might continue operations and indirectly because the NCB was put to the expense of purchasing equipment for its profitable pits which could instead have been transferred from loss-making collieries.

7 The Rationale for Marginal Cost Pricing*

MARTIN SLATER

1. Introduction

From 1945 to the present time, the principal UK energy industries have been in the public sector, and economic analysis of their activities has therefore proceeded within the context of the analysis of public sector enterprises in general. A central organizing principle in that analysis is that of 'marginal cost pricing', the idea that prices should in general be set equal to marginal costs rather than, as in the case of private sector firms, being set in order to maximize profits. In practice, marginal cost pricing has only been applied in varying degrees in different industries (most wholeheartedly perhaps in electricity, and more so in France than in the UK), but it is a reference point to which economists constantly refer in considering the desirability of various energy policies. This paper attempts to explain the reasons for, and trace some of the implications of, this deceptively simple principle.

Marginal cost pricing has a long history, dating back at least to Dupuit (1844), a French engineer in the public service concerned with the economics of road and bridge building. The idea was touched upon by Marshall (1890, Book V, Ch. 14), but resurfaces seriously in a famous article by Hotelling (1938), particularly in relation to railway freight rates. However, it was only after 1945, when widespread nationalization, particularly in Britain and France, significantly increased the area of applicability for principles of public sector pricing, that the modern interest in marginal cost pricing really took off. Significant early contributions were made in Lerner's *Economics of Control* (1944) and by a group of French engineers and economists working in Electricité de France (see Nelson (1964)). In the UK important early work was done by Turvey (e.g. 1968), again stimulated by the electricity industry (electricity is an industry which rather nicely conforms to the economist's stereotype, and so gains rather more than its fair share of attention). The USA, with no nationalized sector to provide a policy stimulus and where public utilities were subject to statutory regulation on an *average* rate of return basis, contributed relatively little to the debate at this stage. Modern textbooks which survey the current state of the art are Rees (1984c) and Brown and Sibley (1986).

The remainder of this paper is divided into three main sections: Section 2 covers the basic theoretical rationale; Section 3 considers various complications and qualifications which immediately arise in an imperfect world; and Section 4

---

* Martin Slater is a Fellow in Economics at St Edmund Hall, Oxford.

looks at the problems of ensuring adherence to such a rule in real-world institutional and political settings.

2. The Theory of Marginal Cost Pricing

2.1 General Principles

Economists see the problem of choosing the 'correct' prices for nationalized industries as simply a part of their wider concern for achieving an 'ideal' allocation of resources in the economy as a whole. By no means the smallest problem encountered here is to decide what 'ideal' means in a society composed of many individuals with conflicting interests. This question is widely debated, and there is no single comprehensive answer to it. However, for practical purposes there is a wide consensus in favour of the 'Pareto criterion', which states that one particular allocation of resources is to be judged less desirable than another if in the second, some people are better off, and nobody is worse off, than in the first. Any allocation that can be dominated in this way by another feasible allocation is obviously falling short of what could be achieved for the members of society. What we should look for at least is an allocation that *cannot* be dominated in this manner; such an allocation is said to be 'Pareto-efficient' (or a 'Pareto optimum'): from which it is impossible, in any way, to make one person better off without making another person worse off.

The Pareto criterion is only a partial answer to the question of optimal resource allocation, for it turns out that there will probably be many Pareto-efficient allocations for a given economy, varying particularly in the way their benefits are distributed. But economic efficiency (as opposed to the distributional, or equity, aspects) obviously has a strong claim to be at least a necessary part of a complete answer, and for this reason economists have considered its implications in some detail. Particularly, they have derived certain standard conditions about production and consumption that must be met if Pareto efficiency is to be achieved. One of these conditions is the famous one that all consumers' marginal rates of substitution (MRS—the rate at which consumers would be *willing* to give up commodity 1 for commodity 2) must be equal to the marginal rate of transformation (MRT—the rate at which producers *must* give up commodity 1 to produce more commodity 2). Further, it can be shown that this equality can be achieved in a decentralized market system via the price mechanism: utility-maximizing consumers subject to a budget constraint will act so as to bring their MRS between any two products equal to the ratio of their prices, while profit-maximizing firms under perfect competition will find that their profits are maximized when they set their MRT between any two products equal to their price ratio. Thus, so long as consumers and producers face the same set of prices, independent, self-interested actions will bring about the equality of MRS and MRT, and hence a Pareto-efficient allocation of resources.

$$\text{MRS}_{21} = \frac{p_1}{p_2} \qquad \text{(consumers' utility maximization)}$$

$$\text{MRT}_{21} = \frac{p_1}{p_2} \qquad \text{(producers' profit maximization)}$$

Therefore

$$\text{MRS}_{21} = \text{MRT}_{21} \qquad \text{(desired condition for Pareto efficiency)}.$$

Now this general principle can be applied to the particular case of a nationalized electricity industry (a monopoly) considering how to price electricity in a world in which the rest of industry is perfectly competitive. We assume that electricity is produced from coal, so that there is a MRT between coal and electricity given by purely technical considerations. Let us further assume that electricity is demanded for heating purposes, and it is also possible for consumers to use coal directly for the same purpose, so that consumers will have a MRS between coal and electricity, given by their tastes and other circumstances. To achieve a Pareto-efficient allocation of resources, we must arrange matters so that $\text{MRS}_{ce} = \text{MRT}_{ce}$.

Given the coal industry's price p_c, an electricity price of p_e will induce utility-maximizing consumers to set their MRS_{ce} equal to p_e/p_c, and hence arrive at a certain quantity of electricity demand. When the industry produces this quantity of electricity it will thereby determine its MRT_{ce}. If the resulting MRT_{ce} is not equal to MRS_{ce}, then we have not yet achieved Pareto efficiency, and clearly p_e is incorrect. Suppose that currently consumers' MRS_{ce}, at 2 units of coal per unit of electricity, is lower than the MRT_{ce}, at 4 units of coal per unit of electricity. The solution is to raise p_e, thereby reducing consumer demand for electricity; as consumers consume less electricity, their MRS_{ce} rises (assuming diminishing returns in consumption), and as the industry needs to produce less electricity its MRT_{ce} falls (assuming diminishing returns in production). Eventually we will come to a value of p_e which will equate MRS_{ce} and MRT_{ce} (at some value between 2 and 4, say 3) as required by the condition for Pareto efficiency. At this point we have

$$\text{MRS}_{ce} = \frac{p_e}{p_c} = \text{MRT}_{ce} = 3.$$

Looking at the second and third terms, we can see that the requirement is that

$$p_e = p_c \text{MRT}_{ce}.$$

MRT_{ce} is the number of units of coal that must disappear into the production process to produce one more unit of electricity. p_c is the price of those units, so the right-hand side as a whole is simply the marginal cost of electricity, and the equation simply states that the price of electricity must be set equal to its marginal cost.

Now suppose electricity can also be produced from oil. Then the same argument must imply that, for Pareto efficiency,

$$\text{MRS}_{oe} = \frac{p_e}{p_o} = \text{MRT}_{oe}$$

and hence

$$p_e = p_o \text{MRT}_{oe}.$$

Thus the marginal costs of all possible methods of producing electricity must be equalized at the price of electricity. If they are not—suppose $p_o.\text{MRT}_{oe} < p_e$ so that $p_o.\text{MRT}_{oe} < p_c.\text{MRT}_{ce}$—we should increase oil-fired production and decrease coal-fired until the two marginal costs are equal. It is clear that in so doing we will reduce the total costs of the electricity industry because we are replacing high marginal cost production by low marginal cost production. Thus a second implication of Pareto efficiency is that, where there are alternative methods of producing a given output, an enterprise must minimize the total costs of producing that output, calculated at prevailing input prices.

Thirdly, we can tackle the investment decision on identical lines. How much energy should be released for current consumption, and how much should be retained for investment (thereby enabling more energy to be produced next year)? If we extend our general equilibrium model in the Arrow–Debreu manner to include the time dimension (i.e. treat the same commodity in different years as *essentially different commodities with different prices*), we can find the optimal policy in exactly the same way as before. Considering 'energy in year 1' and 'energy in year 2', an intertemporal Pareto-efficient allocation will require that

$$\text{MRS}_{21} = \frac{p_1}{p_2} = \text{MRT}_{21}.$$

MRS_{21} is the consumers' *intertemporal* MRS of energy next year for energy now—how much more energy they would require next year to compensate them for the loss of one unit's consumption this year. Similarly MRT_{21} is the *intertemporal* MRT—how many extra units of energy we could technically produce next year for the cost of one extra unit of energy investment this year. Intertemporal Pareto efficiency requires the equality of these two quantities, and the correct selection of *intertemporal* prices, p_1 for energy this year and p_2 for energy next year, will bring this about.

However, consumers' and producers' decisions must be taken *now*, in year 1, so that p_2 represents not the actual price that will pertain in year 2, but the *present value* (in year 1) of that price. The second and third terms above therefore amount to

$$p_1 = \text{PV}(p_2)\,\text{MRT}_{21}$$

or

$$-p_1 + \text{PV}(p_2)\,\text{MRT}_{21} = 0.$$

This equation has the form of a present value investment calculation: p_1 is the cost of one unit of energy invested this year; MRT_{21} is the physical return in units of energy next year from that investment; and $\text{PV}(p_2)$ is the value of those units of next year's energy discounted to present value terms. Thus the equation says that the marginal unit of energy invested must just break even in present value terms. This reasoning can be extended to cover non-energy inputs and outputs as well. In general, therefore, any investment project that makes a surplus in present value terms should be carried out; any project that fails to make such a surplus should not.

If there is a clearly defined rate of interest r, then discounting to present value terms involves dividing by $(1 + r)$ for each year that any return extends into the future, and the above equation takes the familiar form of a discounted cash flow investment appraisal.

Thus by simple application of the necessary conditions for Pareto efficiency, we have derived three propositions:

(1) output prices should be set equal to marginal costs, calculated according to prevailing input prices;
(2) decisions between alternative methods of production should be made by minimizing costs, calculated according to prevailing input prices;
(3) investment decisions should be made by present value calculations, using market prices, optimally determined output prices, and the rate of interest.

Although we have demonstrated these propositions by rather stylized and simple examples, they hold good in much more general circumstances. Two points are worth noting about them. Firstly, although derived from a complex general equilibrium analysis involving the whole of the economy, they are very simple rules which may be put into operation by decision-makers at a fairly low level of decentralization, and they require only knowledge of market prices and the technical details of one's own industry.

Secondly, they amount to considerably more than just a pricing rule. In fact the approach is principally concerned with achieving ideal *quantities*, of inputs, outputs, and investment, and correct pricing is but one means to that end. In complicated real-world situations, as we shall see, what is required is not a slavish adherence to a marginal cost pricing slogan, but a feel for the underlying principles of efficient resource allocation.

2.2 An Alternative Partial Equilibrium Justification

For some applications, a simpler demonstration using a partial equilibrium diagram is useful. Figure 7.1 shows the demand curve DD for our industry's product. This ranks consumers in descending order according to their valuation of the product, i.e. how much money (which would otherwise be spent on other products) they are willing to give up to get hold of it. Set against this is the industry's marginal cost curve, which measures, at each level of output, the additional costs that the industry would incur by increasing output by one further unit. These costs represent extra inputs which have to be purchased, and which otherwise would have gone to the production of some other product. If the market system is working efficiently, competition for these inputs will ensure that their prices reflect in turn the prices that consumers would be willing to pay for the alternative products which the inputs might have helped to make. Thus the marginal cost curve ideally measures the 'opportunity cost', in terms of the value of other products forgone, of consuming more of our particular product.

A profit-maximizing firm would choose to sell 150 units at a price of £10 (for at this output its marginal revenue is equal to marginal cost). But the marginal unsatisfied customer would be willing to give up other products worth just under £10 in order to consume this product, whereas the marginal cost curve shows that effectively only £4 worth of other products would have to be

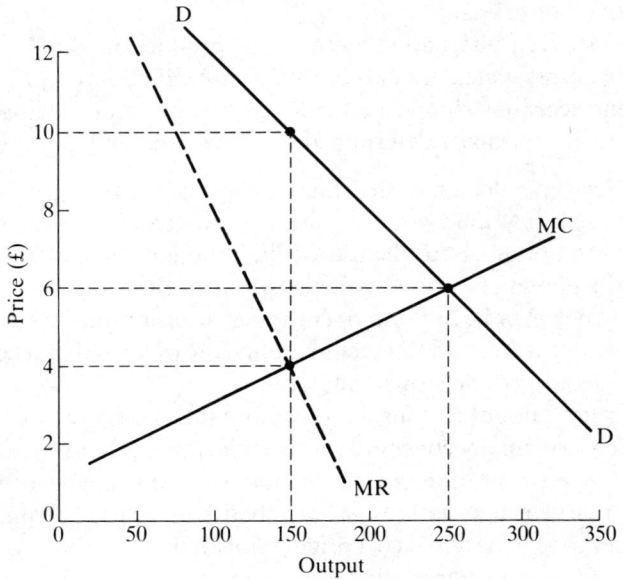

Figure 7.1 Partial equilibrium analysis

sacrificed in order to satisfy him. Therefore there would be a potential Pareto improvement in allowing this customer to be satisfied; hence the price is too high. This argument will hold as long as there are unsatisfied customers who value an extra unit of the product more highly than its marginal cost. Therefore the price should ideally be reduced to £6, allowing an output of 250 units. It should not be reduced below £6, because that would induce consumers to buy who value the product at less than the value of the products that ultimately have to be given up in order to produce it.

So again the desirability of a marginal cost price emerges, justified this time in terms of a *potential* Pareto improvement in a partial equilibrium analysis. An alternative way of expressing the same point is that a marginal cost price maximizes the sum of consumer surplus and producer surplus, the area between the demand curve and the marginal cost curve, which may be taken as a measure of the net benefit arising from this product. Figure 7.1 shows an increasing marginal cost curve; the analysis will hold also for a horizontal or even downward-sloping marginal cost curve, so long as the demand curve cuts the marginal cost curve from above.

Although this simpler justification is considerably easier to grasp intuitively than the earlier general equilibrium analysis, it does have the disadvantage that it keeps some of its most important assumptions hidden behind the scenes. In particular, there are many circumstances in which the financial marginal costs to the firm may *not* adequately reflect the true opportunity costs to society of further output of our product. Going back to the basic principles of the general equilibrium approach can help us to see more clearly what needs to be done when these assumptions are brought into question.

3. The Practice of Marginal Cost Pricing

The basic rules can thus be established without much difficulty. Most of the interesting problems, and therefore much of the literature, arise from the real-world complications and special cases that are met with when one attempts to implement the rules in practice. Some of the most important are considered briefly below (and arise in each of the industries discussed in this volume). The first three are problems associated with the concept of Pareto efficiency in general and are therefore criticisms of all price-based methods of resource allocation; the last four are more specific to public sector pricing.

3.1 Income Distribution

It is well known that for any given set of total resources and consumers' wants, there are in principle a very large number of Pareto-efficient allocations, each embodying a different distribution of welfare. Pareto efficiency therefore does not guarantee a desirable distribution of welfare, or even an acceptable one.

Thus marginal cost pricing in itself cannot guarantee a desirable distribution either.

In the field of energy pricing, a classic example of this is the case of low-income consumers. Suppose the marginal costs of electricity production are increased (by a rise in the international price of oil, say). Marginal cost pricing, looking at efficiency considerations alone, will require a corresponding rise in the price of electricity, to inform consumers of the greater costs the country now bears in providing electricity, and to encourage them to use less or to use alternative sources of energy. However, we are all familiar with press reports of old age pensioners dying from hypothermia because they are unable to meet their increased electricity bills. From these it would seem that the Pareto-efficient allocation has unacceptable distributional implications and that therefore the price must be maintained below marginal cost.

Although the first part of this conclusion may be correct, the second part does not necessarily follow. In fact this is obviously not a problem of electricity bills alone—the electricity bill could be met by failing to meet the food bill, in which case death would arise from starvation, apparently pointing the finger at food prices. Clearly the problem is not one of any particular price, but of generally insufficient income, and the best solution is to tackle the problem directly by increasing pensions rather than by distorting prices. This will be an economically more efficient way of meeting the problem, for electricity prices affect people other than pensioners, many pensioners do not heat by electricity, and pensioners may find that using the extra money to buy more insulation is a cheaper way of keeping warm than buying more electricity.

However, there is an important caveat to the above argument. Acting directly on the distribution of income is preferred because it does not introduce a distortion in the price for electricity. But the argument implicitly assumes that old age pensions can be increased without introducing a similar distortion elsewhere, i.e. by some form of lump-sum taxation on the more wealthy. In practice this is just not possible, and increased old age pensions must presumably involve increased income taxes or commodity taxes, thereby distorting the labour market or other product markets. Thus the optimal solution may involve a trade-off between distortions in different markets. The administrative costs of direct taxation and welfare benefits, and problems of tax avoidance and information gathering, may further increase the desirability of conducting at least a part of redistributional policy through distorting the prices of commodities. It might still be argued, though, that this might be better arranged by the industry charging strict marginal cost prices and the government levying varying indirect taxes or subsidies to meet the distributional objectives, than by confounding the two elements in a single distorted price. These and other arguments concerning distribution are considered in more detail by Dilnot and Helm (Chapter 3, this volume).

3.2 The 'Second-best' Problem

Another well-known difficulty with achieving Pareto efficiency is the 'second-best' problem. If *all* the marginal conditions are simultaneously met in *all* markets, we will have Pareto efficiency, but if for some reason they are not met in one particular market, it does not follow that maintaining them in all other markets will achieve the next best outcome.

An important instance arises in the energy sector because the National Coal Board (NCB) has never practised marginal cost pricing, but has tended to price coal on an average cost basis, effectively allowing low-cost pits to cross-subsidize high-cost ones (see Robinson (Chapter 17, this volume)). Furthermore, it is known that many pits produce their coal at costs considerably in excess of the price of imported coal. (Both these facts violate our second principle.) Therefore the NCB price does not reflect the true marginal opportunity cost to the economy of consuming more coal. Strict marginal cost prices for other forms of energy would thus leave coal relatively overpriced, giving too much of an incentive to consumers to substitute away from coal to these other forms; however, raising other energy prices comparably above marginal cost to reduce this incentive would only result in energy as a whole being overpriced relative to all other goods, producing a different set of distortionary effects. Furthermore, since coal is itself an input to the electricity industry, it would be wrong to base a marginal cost price of electricity (as in Section 2.1), or an investment appraisal of the relative merits of coal-fired and nuclear power-stations, on calculations that value coal simply at the NCB price (a potentially important consideration in the Sizewell inquiry (Department of Energy (1987)), for instance).

The ideal 'first-best' solution is always to remove the fundamental problem of the incorrect coal price, and this line is argued by Robinson and Marshall (Chapter 16, this volume). However, *if* nothing can be done about this, the 'second-best' solution is to adjust other energy prices in an optimal compromise between the various distortions noted in the previous paragraph. However, the information required for making this compromise is vast, particularly when, as is normally the case, not one but many industries may be pricing away from marginal cost. It would involve knowing the deviations from marginal cost for all industries, and all their own-price elasticities and cross-price elasticities of supply and demand. In practice we just do not have this information. Moreover, the need for such monstrous calculations defeats our very purpose, which is to provide a simple rule for conduct which can be easily applied at a relatively low level of decentralization within the economy with only a modest requirement for information.

This problem has led some economists to consider abandoning marginal cost pricing altogether. However, this is certainly an over-reaction. Firstly, although marginal cost pricing may not be the very best form of pricing, there is no reason to suppose that any other simple operational pricing system (e.g. average cost pricing) will do any better. Secondly, the quantitative significance of many of the distortions in other industries may be very small, with only a few very closely related industries being important. Thirdly, if the industries are indeed very close substitutes, as energy industries tend to be, marginal cost prices for gas, oil, and electricity will themselves put downward pressure on the coal price, bringing it closer to marginal cost itself. Thus the 'second-best' problem does not completely negate marginal cost pricing, but one must always be alive to the possibilities of other prices not truly reflecting opportunity costs, and be willing to make adjustments to the basic rules accordingly.

One *possible* approach is to accept that the high-cost pits survive only for social or political reasons, and that therefore the NCB price is not simply a payment for productive services but contains elements of disguised welfare payments. It would be preferable for these welfare payments to be made directly to the NCB by the government, thereby allowing the NCB to meet the import price and not rely on compulsion to retain its customers. As this does not happen, we must make the necessary adjustments and consider the import price as the true marginal cost.

This is only one *possible* approach, because the true answer would depend on an analysis of why this state of affairs had come about. The essence of the second-best approach is that there is some constraint that is preventing the attainment of the full optimum; the optimal reaction therefore depends on the nature of that constraint. Is it in this case a social one (the government wishes to protect miners), or a political one (the government does not wish to protect miners, but is forced to do so by union power or public opinion), or a balance of payments one (the government will not contemplate an increase in imports)? In all of those cases, is the government's concern only with the coal industry or with industry in general? Or is the distortion completely unplanned? In each of these circumstances, a different optimal policy may be appropriate. The papers by Robinson (Chapter 17) and Robinson and Marshall (Chapter 16) in this volume explore the coal problem further.

Taxation in its various forms produces the same complication. If prices include indirect taxes or import duties, these are not resource costs to the economy, and adjustments should be made accordingly. Important points to remember in the energy market are that electricity, gas, and coal are zero-rated under VAT, thereby already giving them a price advantage over non-energy products, while the oil market is subject to a complex array of taxes (see Devereux (Chapter 20, this volume)).

Another rather complex but topical second-best problem arises in connection with international trade. It is often argued that some other countries provide energy at subsidized prices to their manufacturing industries, giving them an unfair competitive advantage over British firms (for example, French electricity prices to industrial customers have been lower than equivalent UK prices, but

The Rationale for Marginal Cost Pricing

there is also argument over whether this is subsidy or the correct application of a more rigorous marginal cost policy in a heavily nuclear-orientated system). Should British energy prices be subsidized in response? In an otherwise ideal world, the answer would be no—if our trading partners wish to make us presents of underpriced imports, this is all to our advantage. However, this view presupposes perfect flexibility of our own market economy. If in practice the resources in our disadvantaged industries cannot be quickly redeployed in other directions, and if the running-down of these disadvantaged sectors is a process that cannot be reversed if the foreign subsidies are subsequently removed, some form of intervention may be justified. But the exact form of that intervention should be considered carefully, and it may be best achieved without affecting our own energy prices.

3.3 Externalities

As with all price-based mechanisms of resource allocation, marginal cost pricing will only achieve the desired result if all prices and costs represent true opportunity costs to the economy. But some costs and benefits do not fall directly on the firm that causes them. In such cases, prices give a misleading guide and need some correction.

An obvious example of such an 'externality' in the energy sector is pollution. Coal-fired power-stations produce air pollution and acid rain which affect people directly and damage crops and buildings. This is a cost to the economy (in so far as it occurs in the UK—exported pollution raises further issues) but it does not appear in the accounts of the electricity industry. Similarly, nuclear power-stations produce steady discharges of low-level radiation and have small but non-negligible risks of catastrophic discharges (see obviously the Sizewell Report (Department of Energy (1987))). Hence the true marginal cost of electricity may be higher than the electricity industry would calculate, resulting in underpricing of electricity. The above examples are of unfavourable externalities, but beneficial externalities are equally possible; in such cases there would be a tendency for prices to be set too high.

Externality arguments may be used to adjust both pricing and investment decisions. They may be seen as a particular type of 'second-best' problem where some important prices are wrong or do not even exist, e.g. the price of acid rain. One possible solution is therefore to attempt to supply those 'prices', either in the form of enforcing compensation for pollution damage, or in the form of a pollution tax. This brings the external costs into the firm's accounting system, so that calculations of marginal cost will now automatically include them. If these mechanisms are not available in practice, notional adjustments may be made in the form of 'shadow prices'—these are figures estimated to be the true social costs of inputs and benefits of outputs, and used in pricing and investment calculations even though the actual prices paid and received are different.

For example, in the 1960s 'shadow wages' of coal-miners were employed in calculations on the desirability of coal-fired versus nuclear power-stations. It was argued that although the actual wages paid to coal-miners were very high (as a result of strong union pressure and a uniform national wage system in which high-productivity pits cross-subsidized low-productivity pits), redundant coal-miners were unlikely to be re-employed for some considerable time, and therefore the opportunity cost to society of retaining them in the mines was very much lower (see Robinson (Chapter 17, this volume) for a modern-day appraisal of this type of argument). Hence the financial cost to the Central Electricity Generating Board (CEGB) of burning coal overstated the true cost to the economy as a whole. Investment calculations employing the resulting shadow price of coal recommended the construction of coal-fired power-stations, whereas on financial criteria alone, more nuclear stations would have been built.

A major difficulty with decisions taken on shadow price grounds (or less formal externality arguments) is that they will adversely affect the financial performance of the industry, unless the government compensates it. If financial performance is an important constraint on the industry, this will produce unpredictable harmful consequences on its other activities. An even greater danger is that vague, unquantified (and even quite precise but unverifiable) externality arguments can be employed to rationalize almost any decision. Both of these problems are strong arguments for keeping vague 'social' considerations outside the remit of nationalized industry managers. Government should attempt wherever possible to deal with externalities directly and to provide the right price incentives, leaving nationalized industries to concentrate on their own efficiency.

3.4 Short-run and Long-run Marginal Cost

One major difficulty in quantifying marginal cost lies in deciding the appropriate time-scale over which the additional output is to be provided. The extra costs of providing one more unit of output *today* may be greater or less than if we have some time to adjust to the new higher rate of output. This is particularly true where capital equipment is involved. To increase output on the current set of equipment may require overtime working, skimping of desirable maintenance, bringing back inefficient moth-balled plant, etc. at fairly high cost. However, if instead time is allowed to increase capital equipment to the appropriate level, these short-term expedients will not be needed, and the marginal cost (including the capital costs of the extra equipment) may be less.

The opposite may also occur, if starting from a situation of excess capacity. Here, in the short run it will cost little to supply extra demand: only materials and possibly some labour. However, if the high demand continues, eventually the existing machines will wear out and more of them will have to be replaced than would otherwise have been the case. Thus in this situation, the marginal cost of supplying additional output is lower in the short run than in the long run.

The Rationale for Marginal Cost Pricing

The difficulty for pricing, of course, is to decide which of these versions of marginal cost—short-run or long-run—should be used to determine the price. Both versions have their adherents.

The ideal policy would seem to be the following:

(1) charge a price in each charging period equal to the short-run marginal cost (SRMC), because SRMC reflects the true resource costs that have to be borne in this period;
(2) if SRMC > LRMC (long-run marginal cost), then steps must be taken to expand capacity; if SRMC < LRMC, capacity should be allowed to reduce;
(3) over time, as capacity approaches its optimal level, SRMC will also approach LRMC, so that price will also change, and eventually, when capacity has reached its optimum, price will be equal to both SRMC and LRMC simultaneously.

Figure 7.2 is a famous diagram illustrating this policy. The SRAC curves represent the average costs of three different sizes of plant that the industry might build, with their corresponding marginal costs SRMC. The envelope of all possible such SRAC curves gives the long-run average cost curve LRAC. LRAC must mathematically have a corresponding marginal curve, which is the long-run marginal cost curve LRMC. If the demand curve is DD and the existing plant is size 1, then the SRMC price is p_1 and the output q_1. But at this output SRMC > LRMC, so capacity must be expanded. As this new capacity comes on

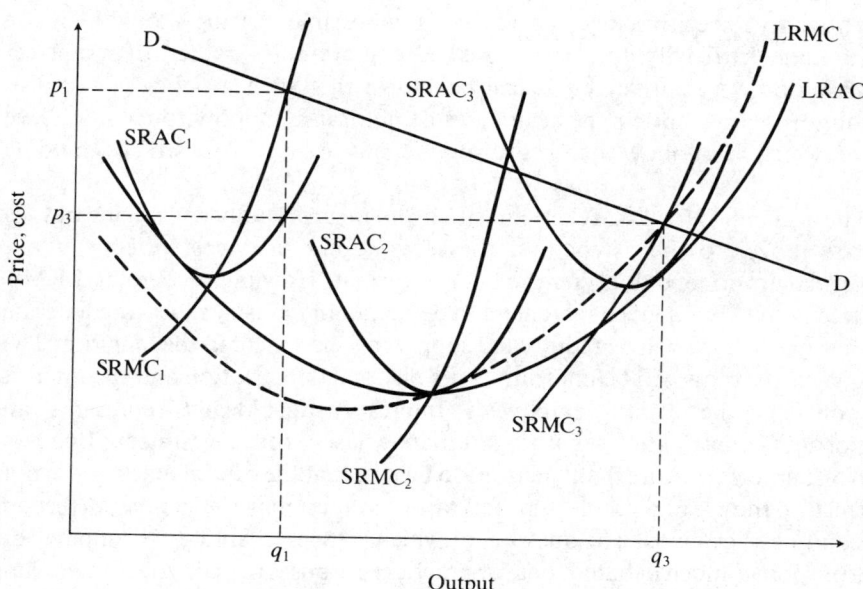

Figure 7.2 The ideal pricing policy

stream, optimal price will fall and output increase until p_3, q_3 is reached. As can be seen, p_3 is simultaneously equal to SRMC and LRMC.

This is therefore a SRMC pricing policy coupled to an optimal investment policy. However, in practice, when governments have espoused marginal cost pricing, they have tended to favour pricing wholly at LRMC. The reasons given for this preference centre around the greater stability of LRMC. In industries where capacity constraints are quite sharp, small fluctuations in demand might cause large swings in SRMC as the industry changes from being short of capacity to over-capacity. These fluctuations might give the wrong signals to customers for their own investment decisions; they might give rise to alternating large deficits and surpluses; and governments might in practice be more willing to make the price adjustments in one direction than in the other.

LRMC is the price that a perfectly competitive industry would charge in long-run equilibrium, and therefore this type of policy is in effect asking for the industry to be run as if it were such a perfectly competitive industry and not the monopoly it in fact is. Recently the new theory of 'contestable markets' (Baumol, Panzar, and Willig (1982)) has given an interesting twist to this analogy. 'Contestable markets' argues that even in the absence of large numbers of firms, perfect freedom of entry for potential competitors would enforce desirable pricing behaviour on apparently dominant firms. Of course, nationalized industries that have statutory monopolies or other non-statutory advantages do not exhibit perfect freedom of entry. Thus it has been argued that 'contestability' should be the real bench-mark, and that a useful policy might be to remove such barriers to entry wherever possible, thereby creating a direct incentive to correct pricing. Where that is impossible, pricing should be made to imitate artificially the conditions of contestability rather than perfect competition. In many cases, it can be shown that such a policy would also amount to LRMC pricing, but in cases of natural monopoly, the contestable price, which must allow the firm to survive, has to be LRAC, which would be higher.

The arguments for LRMC pricing are by no means wholly convincing but, as stated above, to the extent that they favour marginal cost pricing at all, governments do tend to prefer the LRMC variant. How in practice can LRMC be calculated? For a given increment in demand, an industry must calculate the extra capacity that will be required to supply this increment in the long run. The annual cost of this will be made up by its physical depreciation and the interest forgone on the capital employed. Interest again ideally represents an opportunity cost, that of the alternative uses for investment finance. Unfortunately there are many reasons to believe that the capital markets are not perfect, so the use of a simple 'market' rate of interest may well not be correct on 'second-best' grounds. The question of which is the relevant rate to employ for this purpose is much debated, but for practical purposes the UK government has decreed that it should currently be taken as 5 per cent in real terms (the required rate of return of the 1978 White Paper (HM Treasury (1978))). The two components of the capital cost must now be added to the marginal operating costs of the new equipment (SRMC) to form LRMC.

It is clear that there are significant problems of estimation and valuation here. These calculations are forward-looking, and concern the best technology currently available for building, not necessarily the existing technologies embodied in the current capital stock. The valuation of the investment itself may be problematic in times of rapid technical advance and/or inflation, particularly if there are no competitive markets in some inputs. Ideally we also require economically correct profiles of depreciation, not the simple rules of thumb that usually appear in accounts. Also in complex 'network' industries, new equipment may also have an effect in reducing the costs of supplying existing demand, and offsetting adjustment should be made for this.

The electricity industry will serve as a convenient illustration of some of these problems. Here there is a complicating peak-load problem (see Section 3.6), with electricity demand varying over the day and over the seasons. Nevertheless the industry is able to monitor the SRMC very closely half-hour by half-hour. The existing power-stations differ in their variable costs of operation; they may be of different ages, and use different fuels and technologies. They are ranked by these variable costs in a 'merit order', and are progressively switched in, starting from the lowest-cost one, as demand rises. At any time, therefore, the SRMC is the marginal operating cost (principally the fuel costs) of the highest-cost station then producing. However, the computation of the capital cost component of LRMC is by no means as simple.

In principle, new capital equipment will only be required if an increase in demand hits the peak period; increases in demand at other times can be accommodated by already existing unused plant (although this is further complicated in practice by the existence of several different kinds of peak). But any new capital equipment will shift the whole merit order upwards, replacing older less efficient plant in operation in all periods and resulting in savings of fuel costs. The CEGB deducts these fuel savings from the capital cost of the new plant to give its *net effective cost*. It is this which should be added to SRMC to give LRMC. The picture is further complicated by the fact that over the lifetime of the new plant, newer plant still will be built, pushing the new plant itself up the merit order and altering the subsequent fuel cost savings attributable to it. Computer simulation of the entire system for many years ahead is necessary to determine the precise value of the net effective cost (NEC). The estimation of NECs is therefore at the heart of the investment process in electricity, and the procedure is discussed in more detail by Jones (Chapter 12, this volume). Both the Monopolies and Mergers Commission report on the CEGB (1981) and the Sizewell Report (Department of Energy (1987)) include lengthy consideration of these calculations.

New plant can be of several different types, more or less suited to base-load and peak-load uses, and each will have a NEC. Only the type with the lowest NEC should be built, and it is this lowest NEC which determines the LRMC. As more of this plant is built, its NEC rises to meet that of the other types. In the industry's full long-run equilibrium, all NECs will be the same, indicating an ideal balance of the various types of plant. As the NECs rise towards the long-run equilibrium, the SRMC component will tend to fall as a consequence of less inappropriate plant being available to meet each particular time period's demand. If there are constant returns to scale in each plant, these two tendencies will cancel each other out.

However, in electricity it has been estimated that it might take more than thirty years for the industry to attain its optimal balanced capital structure, and the CEGB has argued that current prices should not be made to reflect such 'Utopian' considerations. The result was a compromise version of LRMC pricing which allowed deviations more in keeping with a SRMC approach in particular areas of over- and under-capacity. It has been argued elsewhere (Slater and Yarrow (1983)) that this resulted in relative overcharging of base loads. The real difficulty with this approach is that the choice of time horizon which defines these areas of deviation is essentially arbitrary, and there is now no clear link with the concepts of efficient resource allocation, which there would be if the rationale of SRMC pricing were completely accepted. This arbitrariness opens the door to manipulation to rationalize any particular pricing decision; the Monopolies and Mergers Commission (1981) criticized the CEGB for employing *different* time horizons in different parts of its pricing calculations.

Confusion over the precise meaning of LRMC has been responsible for much of the current political and managerial disillusion with public sector pricing. The economist's definition of the long run is quite clear: it is the length of time in which all relevant factors of production are variable. It is not some arbitrary length of time that allows some but not all investment decisions to be taken into account. LRMC is not simply SRMC with some vague addition for capital costs; it is not a price that will guarantee breaking even in the long run (as opposed to the vague idea that SRMC pricing will definitely not—Figure 7.2 shows that both these views are mistaken, because SRMC often exceeds LRMC); it is not an estimate of what the cost of energy will be in the year 2000. One can discern all these mistaken concepts behind public pronouncements and official documents. If the economist's long run proves to be too long in calendar time to tolerate, then this should merely reinforce the view that the correct policy is the SRMC one, and not some arbitrary redefinition of the long run.

But even here one might pause to consider. The thirty years quoted above for the electricity industry seems excessively long—surely it is not technologically impossible to adapt the capital structure much earlier. Furthermore one sees no sign of frantic current investment activity aimed at making inroads into this large capital imbalance. The CEGB's figures in the late 1970s and early 1980s showed great economic desirability of building more nuclear power-stations, but it was not in fact starting to build them, and its case for not including the cost advantages of nuclear power-stations in its prices was that in the foreseeable future (i.e. thirty years) nuclear power-stations would remain intra-marginal and therefore irrelevant for marginal cost calculations. Furthermore it was known that many other potential electricity investments had positive net present values but remained unimplemented. Thus the CEGB did not seem to be pursuing the optimal investment policy implied by the calculations. Why not? The answer presumably was again external constraints: political constraints on nuclear power-station building, financial constraints imposed by the government on available funds for investment. This leads us back once more to second-best problems; we need an explicit analysis of the form of these constraints and not an arbitrary fudge to get around them.

3.5 Financial Surpluses and Deficits

If prices are to be set at marginal cost, there is no necessary guarantee that financially an enterprise will just break even, in the sense of its total revenue covering its accounting costs. Breaking even is a matter of price being equal to average cost, and Figure 7.2 clearly shows that average and marginal costs often diverge. In the short run, excess capacity is likely to lead to deficits and shortage of capacity to financial surpluses. In the long run, the matter simply depends on the shape of the long-run average cost curve: if this is falling (increasing returns), LRMC will be below LRAC and hence there will be a deficit; if it is rising (decreasing returns), LRMC will be above LRAC and hence there will be a surplus. Only in the case of a horizontal LRAC (constant returns to scale) will an enterprise break even automatically. It is often said that nationalized industries, with their large economies of scale, typically fall into the increasing returns case and so have an in-built tendency to make deficits, but this is by no means necessarily the case; extractive industries like coal, gas, and oil, where expansion means that less and less favourable sources must be exploited, are more likely to produce surpluses with optimal policies.

It must be emphasized that such surpluses are not necessarily a sign of virtue nor such deficits a sign of vice—they are both the inevitable effects of correct marginal cost pricing in different technological circumstances. Ideally the Treasury should just absorb the surpluses and fund the deficits from general government revenue. However, this is not the way most governments see it, for one obvious reason: that it is impossible to distinguish deficits arrived at in this way from deficits simply due to internal inefficiencies. Understandably the Treasury is reluctant to give a blank cheque of this kind to industries, so usually some form of financial target is imposed.

With full information, a financial target could be set which led to no contradiction with optimal pricing, but more usually, with only partial information and governments keen to reduce in all possible ways their own overall deficits, meeting the target will involve some distortion. There are various ways to raise the money required while minimizing the losses from such distortions. Various forms of price discrimination may be practised which aim to take more money from intra-marginal consumers while continuing to confront the genuinely marginal consumers with prices close to marginal cost. The 'two-part tariff', where customers pay a fixed charge or a higher price on the first few units consumed and then a marginal cost price on the remainder, is a good example. A multi-product industry can adopt 'Ramsey pricing', whereby prices are raised most above marginal cost on products that are in inelastic demand. Such a policy can be shown to be best in efficiency terms, but the distributional consequences may be important, as the commodities worst hit are likely to be necessities. If possible, price distortions should only be made on final products; intermediate products should remain priced at marginal cost.

In such a system, where prices are driven by the financial target, targets may be seen to have three functions. Firstly, they can ensure that the opportunity cost of capital is covered; secondly, they can extract elements of rent from natural resources and intangible assets that an inadequate accounting system fails to flush out; and thirdly, they can act as a simple revenue-raising tax on the product of the industry. The government is currently proposing improved methods of accounting (HM Treasury (1986)) which could absorb the second motive into the first. It is also arguable that the third motive would be better served by levying VAT on the hitherto zero-rated products of the energy sector. This would be reclaimable by energy-using industries, unlike the present back-door taxation, with resultant improvement in economic efficiency. Distributional consequences, to the extent that they differ from those of the current system, may again be better handled through the income taxation and social security systems.

3.6 Peak-load Pricing

Some industries face demand that fluctuates in a regular periodic fashion. If, in addition, the product is impossible or expensive to store, this gives rise to the so-called 'peak-load problem'. If capacity is built to meet the peak level of demand, much of it will stand idle for most of the time. A better solution is to reduce price in the off-peak periods and to increase it at the peaks to encourage more efficient use of the capacity.

This is known as 'peak-load pricing', but it is obviously just a particular application of the general principles we have been following—the aim is simply to confront the consumer at all times with the true marginal cost of meeting his demand. As well as a pricing problem, there is also an investment problem: additions to total capacity have to be justified only in terms of their benefits to peak users, as off-peak users will make no use of the new capacity. This suggests the basic rule that the optimal pricing structure should involve charging off-peak users only the marginal operating costs, while peak users have to bear in addition the full capital costs.

In practice, peak-load problems can become quite complex. The fluctuation itself may be quite complicated: in electricity, for instance, there is a daily rhythm imposed on a weekly rhythm imposed on a seasonal rhythm. The timing and the height of peaks may be uncertain. There may be different types of capacity, better or worse suited to base-load or peak-period operation. There may be vintage effects, so that a power-station when new may be operated for most of the cycle, but as it ages and becomes relatively less efficient, it will be operated for progressively shorter periods, making calculation of its lifetime benefits difficult. But in principle all these considerations can and should be taken into account using the same general framework.

3.7 Exhaustible Resources

Another particular problem which occurs in the energy sector is the pricing of exhaustible resources, such as gas, oil, and to a lesser extent coal. What is the marginal cost of consuming another unit of gas this year? There are the marginal physical extraction and distribution costs, but these, at least in the short run, are likely to be small. But the really important cost is that the gas field will run out that much sooner. If at current rates of usage the gas field will be exhausted in 1995, the ultimate effect of taking one more unit now is that in 1995 one extra unit of gas consumption will have to be provided from somewhere else. It will have to come from a more expensive field, or be imported from a foreign producer, or be replaced by imported oil. In calculations it is customary to focus on one of the latter two options because the first has further knock-on implications for the other gas field. Thus the true opportunity cost of gas is the present value of the price of a unit of imported gas or oil in 1995. This causes considerable problems, firstly because it implies that the price of domestically produced gas should fluctuate in response to upheavals in the world oil market, which is unpalatable to many non-economists, and secondly because it requires a long-range forecast, and one should be quite rightly sceptical of current abilities in that direction. But there is no way that *any* sensible judgements can be made in this area without a view of the distant future—that is the very nature of the problem. This area is discussed in greater detail by Newbery (Chapter 2, this volume).

As usual, alongside the pricing problem is an investment problem, which here is the optimal rate of depletion of the resource. The rule here can be simply derived from our second principle: that all alternative ways of supplying the demand should have the same marginal cost. One obvious alternative way of supplying this demand would be to import gas *now* instead of waiting until exhaustion of the field; the cost of this would be simply this year's import price. If this is less than the present value of the import price in 1995 then this would represent a cheaper method of satisfying demand, and should be expanded until the two costs do become equal. This means leaving more gas in the ground at present and therefore extending the life of the gas field. The same argument can be applied to all intervening years, and so provides a complete time profile of extraction. This time profile will be found to maximize the net present value of the resource, in line with our third, investment, principle. However, it does again require forecasts of the import supply situation for the foreseeable future.

4. Policy

Despite the apparent logic of marginal cost pricing, there has been a considerable retreat from such principles in UK policy in the last decade. The 1967 White Paper on nationalized industries (HM Treasury (1967)) is generally accepted to be the high-water mark of 'economic' influence: it recommended the use of almost all the practices outlined above. But within a very few years it was equally generally accepted that the 1967 White Paper had failed. Inquiries were commissioned, and eventually a new White Paper was introduced in 1978 (HM Treasury (1978)), which laid heavy emphasis on financial performance and down-graded considerations of marginal cost to a very secondary importance. Although the White Paper itself argues that its financial controls are to be seen as consistent with optimal pricing and investment decisions, in practice this has simply not been the case: it is the initial setting by the government of financial targets and external financing limits that determines pricing and investment, not the other way round. We shall conclude by considering why marginal cost pricing has fallen from favour, and whether its eclipse is wholly justified.

Some would argue that the 1967 system did not fail, but that it was never really tried. From the start, political interference meant that pricing and investment decisions were rarely decided by the rules alone. Prices were frozen as part of anti-inflation policy, investment decisions were distorted for regional policy or as a result of other political pressures. Certainly the large nationalized industry deficits of the 1970s, which were the main cause of the government's dissatisfaction with the 1967 system, were paradoxically due much more to this factor than to any great adherence to marginal cost pricing. With the current government's declared less interventionist policies in general, one might expect that such political interference might have reduced, but this does not seem to be the case. The politician's interest in the affairs of nationalized industries is unfortunately much more fundamental.

The first weakness of the 1967 system was therefore that it did not meet the problem of 'government failure' (see Helm (1986) and Littlechild (1986)). It followed the economic consensus of the time in viewing government as an impartial arbiter, whereas in fact it is one of the important players in the game itself. The different levels of government have their own motivations, individual and party-political, bureaucratic, etc., and their own incentives and constraints, which typically have very little to do with economists' ideas of optimal resource allocation. In fact, inefficiencies in the allocation of resources are hardly very important politically by comparison with the odium of price increases, redundancies, deficits, surpluses, etc. Price (Chapter 14, this volume) gives some flavour of the fluctuating attitudes of governments to gas prices.

Can anything be done about this problem? One possibility often canvassed is to try to isolate the industries constitutionally from political influence, by setting up some holding company structure, with an autonomous supervisory nationalized industry commission. This sort of framework exists in Italy, and there is a British example in the BBC. But to some extent this negates one of the major political purposes of nationalization, which was indeed to give some political control over important industries, and it is perhaps naïve to think that such a separation would survive the pressures on it. One interpretation of privatization is that it is seen as necessary to remove the industries even further away if their autonomy is to be ensured.

However, political interference is by no means the only weakness. Even in the absence of such interference, it is clear that industries did not themselves fully implement the system. Not all industries even attempted to price at marginal cost, and only an insignificant proportion of investment was appraised in the recommended manner. There is an important difference between laying down some rules and ensuring that they are followed, and this was the second major weakness of the 1967 White Paper. It ignored what one might call the 'managerial' theory of the public sector firm. In the analysis of private sector firms, economists have for some time been alive to the possibility that the motives of managers may not necessarily be identical to those of their shareholders. Nationalized industry managers are presumably no different: they have no intrinsic interest in setting marginal cost prices, or in rooting out internal inefficiencies, and there is no very clear incentive system to steer them in this direction. It may be a rather complicated and tiresome business to calculate marginal costs and net present values, and if the result requires a major adjustment in price structure, thereby annoying those customers who lose without engendering comparable feelings of gratitude in those who gain, it may not seem worth while. As we have seen above, the Minister, who is the public sector counterpart of the shareholders, may not be too keen on such a result either, and so may well overrule it anyway.

The managers have a near-monopoly of the relevant information, so are quite well protected from external criticism. Bankruptcy is almost impossible; competition in the product market is excluded; the Minister has a nominal power to sack, but this is difficult to use in practice, and his motivations are complex (he is also often short-lived). The only real constraint is finance, and the 1978 White Paper relies strongly on this as a stick and a carrot. This is a move in the right direction, but the 1978 system is so crude, and it is conducted in practice with such disregard to considerations of microeconomic efficiency, that its results are still a long way short of desirable. Further refinement is necessary.

The final weakness of the 1967 system was that it did not take pains to ensure that the financial structures of nationalized industries were perfectly consistent with the optimal decision-taking framework. In its own terms, this was not unreasonable, for if the decisions were optimally taken anyway, the role of finance would be a purely bookkeeping one. But in fact, as the other mechanisms have collapsed, the financial constraints have come to the fore, and are now the principal influence on the decision-taking itself. Unless the form of these financial constraints is carefully calculated to induce decisions that improve resource allocation (and historically this has certainly not been the case), it is obvious that incorrect decisions will be forced.

Therefore what need rethinking are not the rules themselves, which as a logical exercise are obviously correct, but the incentive structures for implementing them. What kind of regulatory structures will best induce nationalized industry managers, or privatized industry managers, to take the decisions we want them to take? This is really what decentralized resource allocation should be all about, whether in the public or the private sector, and indeed was the way in which Adam Smith posed the question many years ago. In this light, marginal cost pricing should not be seen as simply a tool for nationalized industries alone, but as part of a more comprehensive apparatus for the consideration of resource allocation problems in general.

Part III

Electricity

8 Competition in Electricity Supply: Has the Energy Act Failed?*

ELIZABETH HAMMOND, DIETER HELM, AND
DAVID THOMPSON

1. Introduction

The 1983 Energy Act liberalized electricity supply in the UK. It aimed to establish a market in electricity by introducing competition from the private sector. As such it was an important step in Nigel Lawson's new market philosophy for energy (Lawson (Chapter 1, this volume)).

In practice, however, the legislation has failed to induce any significant new competition into the industry. Private generation of electricity has remained marginal, confined in the main to small by-product generation and to small-scale combined heat and power (CHP) schemes. There are no large-scale privately owned power-stations, and no one has started supplying electricity directly to final customers by renting use of the transmission system (as envisaged in the Act).

In this paper we investigate why entry has failed to take place. In particular we investigate two competing hypotheses. The first—'the efficiency hypothesis'—argues that little entry has occurred because the electricity industry is already highly efficient. Few opportunities for profitable market entry exist. The second—'the entry barriers hypothesis'—argues that the conditions under which entry can take place have not been 'fair' and, in consequence, entry has largely been prevented. Perhaps not surprisingly, we conclude that the truth lies somewhere between these two extremes.

Section 2 of the paper outlines in more detail the changes introduced in the Energy Act. Section 3 looks at potential entrants and the entry decision. Section 4 sets out the two hypotheses. We identify the price of electricity as a crucial decision variable, and Section 5 reports the findings of a detailed examination of

---

* This paper was first published in *Fiscal Studies*, February 1986.

At the time of writing this paper, Elizabeth Hammond was a Research Officer at the Institute for Fiscal Studies. Dieter Helm is a Fellow in Economics at Lady Margaret Hall, Oxford and a Research Associate at the Institute for Fiscal Studies. At the time of writing this paper, David Thompson was a Programme Director at the Institute for Fiscal Studies.

The research on which the paper is based was supported by the Economic and Social Research Council. The authors would like to thank John Kay, Michael Keen, John Maccadam, Catherine Price, and George Yarrow for helpful comments and suggestions on an earlier draft. They are also grateful for the helpful advice received from many individuals concerned with the electricity industry—from the electricity supply industry, from potential and actual private producers, and from the Department of Energy. The usual disclaimer applies.

the principles and practice of electricity pricing. In the final two sections we draw together our findings on the two competing hypotheses and draw out the implications for policy.

We do not provide a detailed account of the structure and characteristics of electricity supply (on which see for example Yarrow (1985) or Monopolies and Mergers Commission (1981)), but a brief outline at this point may help the reader. The public sector electricity supply industry (ESI) comprises the Central Electricity Generating Board (CEGB), the twelve Area Electricity Boards, and the Electricity Council. The CEGB generates power and supplies it through its transmission system to the Area Boards which take power from the transmission system at bulk supply points. Each Area Board supplies power to final customers through its local distribution network; it also collects revenue from customers and sells (and maintains) appliances. The CEGB charges the Area Boards for the power it supplies to them using a charging structure called the bulk supply tariff (BST).

The paper draws on the extensive published material on the electricity industry. This literature has focused primarily on cost and price structures, though there have been important recent papers on privatization and natural monopoly in the energy sector. Papers in the former category are referenced in the appropriate sections below. Amongst the latter, the reader is especially referred to Webb (1984), the section on the electricity industry in Vickers and Yarrow (1985), and Yarrow (1985). A parallel paper by Hammond, Helm, and Thompson (1985) on the privatization of British Gas is also heavily drawn upon.

2. The Energy Act

The 1983 Energy Act instituted two major changes in the statutes governing the industry. First, it abolished the monopoly dating back to the 1909 Electric Lighting Act, which had prohibited persons other than Electricity Boards from commencing to supply or distribute electricity, and to the 1911 Electricity (Supply) Act, which had restricted the establishment and extension of generating plants. Second, the Act required the incumbent Area Boards to publish tariffs ('private purchase tariffs') at which they would purchase electricity from private producers and, subject to technical feasibility, it required them to purchase power offered by private producers. The Act also required the ESI to allow private producers to make use of its transmission and distribution system in order to supply electricity directly to final consumers and to publish a schedule of charges for such 'use of the system'.

The Act was thus designed to be liberalizing, and it represented one of the last pieces of legislation that addressed the extent of competition rather than ownership in the government's privatization programme (see Kay and Thompson (1986) for a discussion). It focused directly on the entry decision, requiring an exposition of pricing principles as well as clarifying the right of private generators to sell power to the public electricity supply industry and to some customers directly.

However, it is worth noting that the relaxation of the monopoly related to the production of electricity as a 'main business'. Private generation was permitted before the Act, provided that this was for the company's own use or was a joint-production activity, and some private generators supplied power to the ESI. In practice, private generation both before and after the Act has been largely confined to CHP schemes, cogeneration, and small-scale generation based on waste products.

The evidence on private generation suggests that generation by the industrial sector has in fact been declining in recent years, although the volume of power supplied to the public sector does appear to have risen slightly since the legislation came into force (see Table 8.1).

Table 8.1. Electricity production by industrial and transport undertakings (gigawatt-hours)

	1980	1981	1982	1983	1984
Total production	18,476	17,926	17,266	16,960	16,479
Transfer to public supply	3,673	4,112	4,229	4,715	4,396

Source: Digest of UK Energy Statistics 1985.

Thus it can be seen that in practice very little has changed. What has altered, however, is the explicit exposition of prices and of the underlying principles, which has replaced the implicit arrangements made on an *ad hoc* basis between Boards and private producers prior to 1983. These principles affect the terms of entry by reducing the degree of uncertainty, but as we shall now show, these vary greatly with the type of potential entrant.

3. Potential Entrants and the Entry Decision

While the Energy Act is relatively simple, the entry decision itself is quite complex. In this section we identify the general form of this decision by considering it in the context of an investment framework. Next we show the potential impact of the incumbent on the components of that decision. Finally we indicate the problems that arise relating to the different types of entrant.

Investment Appraisal

There are effectively three different methods of entry which can be adopted:

(a) to supply to the ESI at the private purchase tariffs (PPTs) specified under the Energy Act provisions;

(b) to supply directly to final customers by making use of the ESI's transmission and distribution network;

(c) to produce power for one's own use in substitution for power supplied by the ESI.

What all entry decisions have in common, however, is the need to evaluate the future profitability of electricity production. A potential entrant using the traditional project appraisal techniques will attempt to estimate the future revenues from the project (the output multiplied by its expected price), subtract from this total the expected costs (largely fuel inputs and labour), discount these returns back to the present by some appropriate rate of return, and compare with the (discounted) capital costs.

Whichever of the three methods of entry is adopted, the expected future price of electricity (either the PPTs or the ESI's tariffs) is a crucial influence on the entry decision. This is indeed what the Energy Act concentrates upon, and we shall do likewise in this paper.

The Incumbent's Strategic Advantage

Each of the variables in the entrant's investment appraisal is uncertain, and the greater the degree of that uncertainty, the higher will the net present value of the project have to be for it to be undertaken (or the greater will the effective discount rate have to be). The incumbent can affect both the underlying variables and the degree of uncertainty. The incumbent can effectively be regarded as the ESI, although the Act does distinguish the roles of the CEGB and the Area Boards. The Energy Act must be interpreted and appraised according to the ways in which the ESI could in theory, and has in fact, acted to alter the expected costs and benefits of entry. Of course, the post-entry returns of the entrant will depend on how the incumbent reacts, and hence part of the entrant's estimates of future profits will be its expectations of these reactions.

Recent advances in the theory of oligopoly suggest a number of factors which may be very important in assessing future reactions by the incumbent (see Mayer (1985) and Vickers (1985) for a review). In particular these include the bankruptcy constraint, the existence of spare capacity, and the objective function of the incumbent. Let us briefly explore the implication of each of these factors.

First, the bankruptcy constraint. In the electricity industry, it is reasonable to assume this is asymmetrical. The ESI, being State-owned, does not face a binding bankruptcy constraint (although it is subject to public sector financial controls), while the entrant normally does. If the reaction of the incumbent to entry were retaliation, a price or non-price competitive war would be lost by the firm that first reached bankruptcy. Hence, in this case, there would be little incentive for effective entry, since that would result in financial disaster if retaliation took place. Second, spare capacity has been argued (Spence (1977)) to represent an important strategic threat to the potential entrant, especially where there are economies of scale. Its existence signals to rivals the potential for rapid output response to entry threats, as well as steering the existing incumbent towards a sales-maximizing objective in the short run. Dixit (1980) has shown that the incumbent's choice of capacity can be used to deter entry without, in this model, this resulting in idle capacity. A sales-maximizer is more likely than a profit-maximizer to respond to entry by retaliation, and hence the existence of the potential and actual incentive to defend market share reduces the entrant's expected profits. The third theoretical point reinforces the second. It concerns the new positive theory of the public enterprise. In a series of models, Rees (1984a, 1984b) has suggested that public enterprises are best considered as output-maximizers, with constraints concerning a minimum profit level or break-even requirement (in practice, a constraint established by the level of the external financing limit). This objective again signals to the potential entrant the possibility of price retaliation by the incumbent (subject to the tightness of the public sector financial controls).

When appraising the ESI, it is thus important to note that it does not face bankruptcy as potential entrants do (because the constraints imposed by public sector financial controls may be less binding); that it is characterized by considerable spare capacity; and that, subject to external financing limits, its statutory objectives are compatible with output-maximization.

What scope exists for the ESI to act to deter entry under the 1983 Energy Act? In part, this depends upon which of the three methods of entry identified above is adopted. Where the entrant chooses to supply directly to final customers, in competition with the ESI, the scope for deterrence is even greater than indicated by the above considerations. Additionally, the ESI may be able to influence (unfavourably) the levels of the charges made for 'use of the system' and thereby reduce the returns expected from entry. This additional problem clearly does not arise where the entrant is simply substituting own-generation for supply from the ESI. Where the entrant chooses to supply to the ESI at the PPTs, it might appear that the scope for an incumbent response to entry, at least in terms of the impact upon the expected prices faced by entrants, is significantly circumscribed. Effectively, the Energy Act could be regarded as a method of imposing a regulatory control which prevents the prices faced by entrants being reduced in response to entry. The effectiveness of this, however, depends critically on the specification of the tariffs that are paid to private generators and on the scope that exists for the level of these tariffs to be altered significantly in future years. This is an issue to which we turn in the next section, after first reviewing the type and extent of entry subsequent to the 1983 Act.

Methods of Production

We noted in the previous section that the level of private generation has been declining in recent years, although the 'export' of power to the ESI has been rising. No private producers have chosen to use the ESI's distribution system to supply directly to final customers. In practice, private producers use (or have considered) a spectrum of quite different methods of production with differing cost structures and incentives. These range over the undertaking of new power-station development, the purchase of an existing station from the CEGB, CHP projects, by-product generation in industry, and own-use generation. The principal differences are in the calculation of single versus joint costs, the time horizon of the project, economies of scale, and size of the capital finance required. Clearly the relevant factors influencing the decision to build the Severn Barrage are somewhat distinct from those of converting sewage gas into electricity. It is, we shall argue, no accident that entry has largely been confined to small projects of the latter type.

We investigated the extent of entry in two Areas (the North-East Area Board and the South-East Area Board). In the North-East, 9 proposals had arisen (by mid-1985) since the Energy Act came into force. Only 6 of these involved 'export' to the ESI. All 6 were put forward by other public sector bodies; 4 schemes involved utilizing gases from sewage plants and 2 were small hydroelectric schemes. The South-East showed a similar picture: a number of proposals have arisen for small-scale generation using gases from sewage farms or landfill sites.

A number of companies have considered the purchase of power-stations from the CEGB, but to date no transactions have been successfully completed. In each case, the power-station has been one which has already been scheduled for retirement and decommissioning by the CEGB. Typically maintenance work has been run down in anticipation of closure, and this means that substantial refurbishment would be required to continue operation. Furthermore, power-stations scheduled for closure might be expected to have comparatively high costs of operation. The scope for profitable operation by a new owner is therefore not substantial unless a very significant improvement in the methods of operation can be achieved.

Thus the major potential entrants following the 1983 Act have been firms considering the purchase of power-stations that the CEGB is currently closing down, CHP schemes, and by-product production. Entry in practice has been confined to small-scale schemes in the latter two categories.

4. The Two Competing Hypotheses

The fact that entry into the ESI has been minimal following the 1983 Act does not in itself establish that the ESI has acted to prevent entry, or that the terms of the Act do not properly inhibit the ESI from discouraging entry. For entry may well be prevented by the efficiency of the ESI: it may be the case that potential entrants cannot compete because of the superior performance of the incumbent.

There are thus two distinct hypotheses to explain the lack of entry—though of course the truth may lie somewhere between the two. Let us call these the 'efficiency hypothesis' and the 'entry barriers hypothesis'. Each of these will be examined in this section. Subsequently we shall look at the structure of the Act, and at the tariffs established under the Act's provisions, to see to what extent there is support for one or other of the two alternatives.

According to the efficiency hypothesis, the CEGB (in production and transmission) and the Area Boards (in distribution and revenue collection) are both productively and allocatively efficient. The potential entrants cannot match the ESI's performance, and therefore have not taken it on. Thus there is nothing wrong with the Energy Act. It has created a competitive threat, and will prevent any future monopoly abuse by the incumbent.

The alternative hypothesis is more critical of the Act, or at least its implementation. It claims that the entry conditions are not fairly set. The terms of competition leave the incumbent with advantages which result not from its inherent efficiency, but from its monopoly power—barriers to entry of one form or another prevent effective competition. The implications of this are that the Act needs substantial revision if effective competition is to take place.

Criteria for 'Fair' Entry Tariffs

The most important of the entry conditions is the price faced by entrants—in particular the tariffs (PPTs) at which the ESI offers to purchase electricity. What criteria can we use to specify a 'fair' level for these tariffs? To achieve economic efficiency the tariffs should be set in relation to the ESI's marginal costs. What this means in practice, however, depends critically upon the cost structure of electricity supply; we have considered three possible scenarios:

(a) constant returns to scale and optimal capital stock;
(b) non-optimal capital stock—spare capacity;
(c) economies of scale.

In the first scenario, short-run marginal costs (SRMC) are identical to long-run marginal costs (LRMC). Setting the PPTs at this level should, in the absence of non-price entry deterrence, result in entry taking place in cases where entrants can achieve higher productive efficiency than the incumbent ESI (and vice versa). The absence of entry in this scenario (and with tariffs appropriately specified) would indicate that the ESI is achieving at least the same level of efficiency as potential entrants and possibly better.

In the second scenario, SRMC is lower than LRMC. Setting the PPTs at SRMC would mean that an entrant who is potentially more efficient than the ESI may nevertheless be unable to earn an adequate return at the PPTs, and entry may be inhibited.

In the third scenario, setting the PPTs at LRMC may still preclude profitable entry on a small scale by a potentially more efficient private producer. The importance of sunk costs (and these are likely to be significant: consider for example the process of planning approval required for the Severn Barrage) will determine whether large-scale entry is likely to take place.

Entry Barriers Hypothesis

How might the incumbent distort the terms of entry? First, the ESI, being the principal (and, in the absence of use of the system, the sole) purchaser of privately generated electricity, might set the PPTs at an unattractive level. As we have seen, this turns out to be especially relevant where the existing spare capacity of the CEGB is taken into account. An output-maximizer with monopsony purchasing power would have little incentive to set attractive prices; indeed it might be argued that it would be economically inefficient to do so.

The second barrier to entry is that of uncertainty. If the ESI acts in such a way as to make the potential entrant less certain of the future price of his or her product, then the risk discount will be higher, and the marginal entrant will be deterred. Thus a crucial question in considering the 1983 Act is whether it has had an appreciable effect on the uncertainty (in particular that resulting from the discretionary power of the incumbent) surrounding the price that can be obtained in future years for privately generated electricity.

The classification of barriers to entry suggested by economic theory indicates that there may be a considerable number of other methods by which an incumbent might seek to reduce the incentives for entrants. In the case of electricity, the barriers relating to prices (and uncertainty about future prices) appear the most important, and in the sections that follow we shall examine in detail the determination of electricity prices.

Efficiency Hypothesis

But first we shall review the ESI's efficiency. This is clearly important when assessing whether the efficiency hypothesis can be supported. But because achieving improved performance is a major goal of liberalization policy, it is also important to assess how far, and in what particular respects, there is scope for improving the ESI's current performance.

We are concerned with both productive and allocative efficiency. Evidence on productive efficiency is hard to obtain, reflecting the difficulties of defining and measuring an appropriate indicator. The main sources of evidence are the investigations carried out by the Monopolies and Mergers Commission (MMC) into the CEGB and various Area Boards. On the CEGB, the MMC (1981) praised the 'technical efficiency and security of the Board's system'. However, the MMC concluded that there were serious weaknesses in the CEGB's evaluation of capital projects and that there has been over-investment in capacity. The MMC has generally reported less favourably on the productive efficiency of the Area Boards (see MMC (1983b, 1984, 1985)). The more significant criticisms, however, have generally related to activities that are affected by the Energy Act provisions only marginally (e.g. revenue collection) or not at all (e.g. appliance maintenance).

Allocative efficiency is, in principle, more readily assessed. Prices should be set systematically in relation to marginal costs (unless there are significant second-best effects). We are concerned both that the overall *level* of prices should reflect marginal costs and that the *structure* of prices should reflect the structure of costs—thus, differing costs of supply (whether differing geographically or by time period) should be reflected in price differentials.

We now analyse the evidence on the degree to which electricity tariffs do indeed reflect the principles of marginal costing.

5. The Price of Electricity

In this section we look at the principles determining the price of electricity. The price at which a potential entrant can sell electricity is the single most important entry condition. We are therefore principally concerned with the PPTs. But because the entrant can compete directly for final customers (by making use of the ESI's distribution system), we are concerned also with the ESI's tariffs and with the charges made to private generators for 'use of the system'. The central parameter in the determination of the price of electricity is the bulk supply tariff—the set of charges at which the CEGB supplies power to the twelve Area Electricity Boards for onward distribution to final customers. In the sections that follow we consider five issues.

(a) The first concerns the principles underlying the BST.
(b) The second concerns the *uncertainty* of the returns on investment in electricity generation which are earned over a number of years. Unless the entrant can make reasonable forecasts of future prices, a high (risk-related) return will be required to encourage entry.
(c) The third concerns the *level* of the PPTs. The Act requires that these are based on Area Boards' avoidable costs in purchasing electricity from private suppliers rather than from the CEGB. We consider the linkage between the PPTs and the BST.
(d) The fourth concerns the prices at which the ESI supplies electricity. As noted earlier, most private producers (both before and after the Act) generate power for their own use (in substitution for supplies from the ESI) as well as to supply to the ESI. To the extent that the prices charged by the ESI to such customers do not reflect LRMC, there will be a disincentive to generation for own-use.
(e) The fifth issue concerns the specification of the charges levied for making use of the ESI's transmission and distribution system.

The Bulk Supply Tariff and Marginal Costs

Following the 1947 Act and the White Papers on nationalized industries' pricing policy (HM Treasury (1967, 1978)), the bulk supply tariff is designed to reflect the long-run marginal cost of electricity. (See Electricity Council submission in Energy Committee (1984).)

The characteristics of the product affect the pricing structure. It is non-storable, and it varies in quality by the degree of continuity, by voltage, and by security of supply. The structure of the ESI's costs comprises the costs of producing electricity, the costs of transmission and distribution, and the costs of revenue collection and marketing. The CEGB's input costs are largely those of fuel, which in the last year before the miners' strike (1983–4) made up almost 60 per cent, with wages and salaries making up a further 10 per cent.

These various costs of the CEGB translate into the BST—the price at which the Area Boards buy electricity from the CEGB. The BST is itself made up of a number of charges, reflecting the characteristics of the demand and supply of electricity. Demand displays peaks and daily cycles, an annual profile by seasons, occasional rapid unanticipated variance requiring rapid response, and certain industrial demands which are both large and substantially predetermined. Because electricity cannot be stored, the tariffs are constructed to reflect the additional marginal costs imposed as demand shifts. On the supply side, as we have seen, fuel is the major cost. To the extent that there may be unanticipated movements in these prices, the pricing structure must also allow for adjustment.

The pricing system is designed to reflect costs at the margin of these fluctuations in demand and supply conditions. It therefore has a number of component charges, rather than one total value, which explain its principles and aim to indicate the real resource costs (and hence LRMC) to final customers.

The BST is in three parts:

(a) The system service charge: this is a fixed sum paid by each Area Board (see Table 8.2 for the amounts paid in 1985–6). The system service charge has two components. The first reflects the costs of the bulk supply points where each Area Board takes power from the CEGB transmission system; for all Area Boards these costs totalled £95 million in 1985–6. The second reflects the CEGB's unavoidable costs; these totalled £590 million for all Area Boards in 1985–6.

(b) Capacity charges: the BST contains a two-tier capacity charge (see Table 8.3). The peak capacity charge reflects costs of meeting the three highest peaks in demand during the winter period. These demands are, by definition, of fairly short duration and the peak capacity charge is based on the net cost of retaining old plant on the system for short periods of duty at the extreme peak. The basic capacity charge reflects the more persistent high winter demands and the more comprehensive measures which are necessary to meet the longer duration of capacity need; the basic capacity charge therefore includes the net cost of building new power-stations.

(c) Unit charges: these charges reflect the incremental costs of producing power (other than capacity costs). The structure of unit charges shows a pattern of extensive differentiation between different time periods (see Table 8.4). This reflects the different operating costs of plant used at different points in the demand cycle (see Hammond, Helm, and Thompson (1986) for a more comprehensive account). The unit charges are based upon a specified price for fuel, and the BST provides a multiplier for increases (decreases) in the unit rates to the extent that the actual price of fuel exceeds (falls below) the level specified in setting the unit rates.

Each of these charge components is derived from an interpretation of the long-run marginal costs of producing and supplying electricity. Yet the

Table 8.2. Bulk supply tariff: system service charge (£ million)

	1985-6
London	60.4
South-eastern	50.0
Southern	79.9
South-western	34.2
Eastern	82.0
East Midlands	62.6
Midlands	68.8
South Wales	33.2
Merseyside and North Wales	46.2
Yorkshire	71.7
North-eastern	41.7
North-western	61.4
Total	692

Source: *Bulk Supply Tariff 1985/6*, Central Electricity Generating Board.

Table 8.3. Bulk supply tariff: capacity charges (£ per kilowatt)

	1983-4	1984-5	1985-6
Peak	22	24	23
Basic	43	31	31

Note: The definition of the basic capacity charge was changed in 1985-6; the figures shown above are presented on the basis of the pre-1985-6 definition.
Source: *Bulk Supply Tariff 1983/4, 1984/5 and 1985/6*, Central Electricity Generating Board.

Table 8.4. Bulk supply tariff: unit rates, 1985–6 (pence per kilowatt-hour)

Times	Summer period (27 May to 29 September inclusive)	Periods other than summer (1 April to 26 May and 30 September to 31 March inclusive)
WEEKDAYS		
24.00–01.00	1.58	1.75
01:00–04.00	1.14	1.33
04.00–06.00	1.14	1.23
06.00–08.00	1.94	1.95
08.00–13.00	3.40	2.78
13.00–16.00	2.50	2.78
16.00–18.00	2.79	2.78
18.00–21.00	2.36	2.78
21.00–24.00	2.36	2.40
SATURDAYS, SUNDAYS, AND PUBLIC HOLIDAYS		
24.00–01.00	1.60	1.73
01.00–03.00	1.10	1.23
03.00–07.00	1.10	1.15
07.00–08.00	1.46	1.38
08.00–13.30	2.24	2.27
13.30–14.00	2.24	2.05
14.00–16.30	2.05	2.05
16.30–17.00	2.05	2.24
17.00–24.00	2.18	2.24

Note: In addition to the above rates, a peak surcharge rate of 2.38 p/kWh applies in the half-hour of highest system demand in the period 08.30–23.30 and in each immediately adjacent half-hour on each day except on weekdays in the summer period.
Source: *Bulk Supply Tariff 1985/6*, Central Electricity Generating Board.

apparent simplicity of the LRMC concept is deceptive. In particular, a number of well-documented problems arise in its application in practice. For example, how long is the long run and over what time horizon should the costs of new investment be considered? What are the cost implications of geographic location? What is the effect of government financing requirements? We next tackle each of these questions in turn.

How Long is the Long Run?

Economic theory differentiates between the short and long run according to whether capital is a fixed or variable factor. However, there is unfortunately no straightforward corollary in practice for this theoretical definition: some decisions are short-run operating decisions while others relate to long-run planning.

Competition in Electricity Supply: Has the Energy Act Failed? 169

The task in applying the concept to the production of electricity turns on the appropriate lifetime of a power-station, and on which sort of power-station (and hence technology) that calculation should be based. Should we take the costs of operating the most advanced nuclear station, the cost structure of French generation, that of the average current coal-fired station, of alternative energy sources, or what?

There have been a number of suggestions put forward in the literature. These vary from what is effectively an infinite time horizon examined by Slater and Yarrow (1983), to a short time horizon suggested by Rees (1983). In between there are effectively three others on offer. These are a thirty-year time span, the fifteen-year scale offered by Coopers and Lybrand Associates Ltd (1982), and the hybrid used in the BST of thirty years plus load management tariffs.

It cannot be decisively established which of these alternatives gives the most appropriate specification of marginal costs. In fact, the important point is precisely that the choice is open to debate; the principle of LRMC cannot, therefore, be interpreted to give an unambiguous specification of a level of costs.

Geographic Location

It will be noted that the BST charges at which each Area Board purchases electricity are identical. No allowance is made for the comparatively favourable (or unfavourable) location of particular Areas in relation to the location of the main sources of generation. This suggests first that there will be some allocative inefficiency in the structure of customer tariffs—differences in the costs of supplying electricity to different parts of the country will not be fully reflected in prices. Second, there are implications for the private purchase tariffs to which we return below.

External Financing Limits and Government Financing Requirements

If there exists ambiguity concerning the appropriate method of calculating LRMC, further unpredictability is introduced for the entrant by the relationship of government to the industry. Following experience of mounting deficits by public corporations in the early 1970s, and general macroeconomic requirements of funding public expenditure, governments have attempted to impose a system of cash limits on nationalized industries by setting them external financing limits (EFLs). These are effectively the amounts that each industry can borrow in the specified year. In the case of electricity and gas, they are negative; i.e. these industries are set targets for amounts that they must pay to the Treasury from profits. Now, if the revenue generated by charges based upon the LRMC of electricity is a given amount, say £x million, and rules are given for determining investment, yielding let us say £y million costs (using the required rate of return—RRR—principle after the 1978 White Paper HM Treasury (1978))), then the financial position of the nationalized industry is determined—at say £z million (where $z = x - y$). The EFL should thus, for consistency, be z.

But the government typically over-determines the system. It lays down LRMC for pricing, RRR for investment, and an EFL and financial target, say t. There is no guarantee that $t = z$, and in practice the almost inevitable conflict between the two is haggled out between the Treasury, the Department of Energy, and the ESI. An instructive insight into this process is provided by an inquiry conducted by the Select Committee on Energy in 1984 (an informative review of the Committee's findings is provided in *Public Money*, June 1984).

Forecasting the BST

We argued in the preceding section that the principles upon which the BST is based (relating charges to marginal costs) do not provide an unambiguous specification of the level and structure of the BST. In this respect, the potential entrant therefore faces uncertainty in forecasting the future level of the BST and electricity prices, uncertainty which is greater than that arising solely from the necessity of forecasting future input prices.

The extent of this possible uncertainty is clearly illustrated by the recent history of the BST. In 1984-5 the structure of the BST was significantly changed in response to the introduction of the Energy Act. The CEGB decided that some costs previously included in the capacity charge component of the BST were not properly regarded as avoidable. The system service charge component was introduced to recover these expenditures and a corresponding reduction was made in capacity charges (see Table 8.3). The significance of this change lies, as we explain in the next section, in the fact that capacity charges are a part of the basis of the private purchase tariffs offered to private generators, whilst the system service charges are not. In 1985-6 substantial changes were also made to the structure of unit rates (see Hammond, Helm, and Thompson (1986) for details).

It is interesting to note that these significant changes in the structure of the BST followed very shortly after a comprehensive review of the principles and implementation of the bulk supply tariff, which in conclusion 'reaffirmed its recommendation to the CEGB that the BST continue to be based on marginal costs. In considering the appropriate structure of such a BST it has found no major defect in the approach adopted for the present BST' (Electricity Council (1981)).

To add to the confusion for the potential entrant, the future relationship between the government and the electricity industry is uncertain. Privatization is openly discussed. The potential entrant must thus calculate the impact on the future price of electricity of the various options—complete privatization, regionalization, separation of Area Boards from the CEGB, and so on. It is even less clear than it was for gas in early 1985 what might be the form of regulation, the structure of competition, and the form of the BST (should it survive).

Competition in Electricity Supply: Has the Energy Act Failed? 171

It is thus clear that for the potential entrant, forecasting the BST is far from being a simple matter based on clear and unambiguous principles. Putting the concept of LRMC into operational form requires judgements about time horizons and investment profiles, which have in the past been matters of considerable dispute and which have led, as we have seen, to a series of both major and minor revisions. It is a reasonable bet that they will alter again, and hence the entrant faces uncertainty here.

But to compound these conceptual and operational problems, the impact of EFLs upon electricity charging structures provides a further source of uncertainty. Furthermore, with the prospect of privatization, none of these issues can be regarded with any certainty in the future.

In summary the potential entrant faces considerable uncertainty in forecasting the BST. In order to assess the prices he or she will face post-entry, an assessment is also required of the link between the BST and the private purchase tariffs, and between the BST and the tariffs charged for the supply of electricity by Area Boards. It is to these two issues that we now turn.

Linkage between the BST and the Private Purchase Tariffs

The principles upon which the private purchase tariffs are fixed are specified as being twofold in the legislation. First, they must not increase the prices payable by customers of the Board for electricity supplied to them by the Board; and second, the tariffs should reflect the costs that would have been incurred by the Board but for the purchase. This second principle in effect means that the private purchase tariffs reflect the avoidable costs of the Boards in purchasing electricity from private suppliers rather than from the CEGB.

The private purchase tariffs and the BST charges are set out in Tables 8.2 to 8.5. Comparison of the various tables shows that the structures of the two sets of tariffs are quite different. Our analysis shows, however, that in practice the two sets of tariffs are broadly equivalent, as we now explain.

The private generator is offered two alternative rate structures in the private purchase tariffs (see Table 8.5). Looking first at the '2-rate metering' structure, it can be shown that this is equivalent to a time-weighted average of the appropriate unit rates in the BST (see Hammond, Helm, and Thompson (1986) for details of the results reported in this section). The PPT is thus equivalent to the unit rate charges in the BST at a load factor of 100 per cent; no allowance is included for capacity charges. These are, however, included in the '3-rate metering' structure. Again it can be shown that the PPT rates relate to time-weighted averages of the relevant unit rates in the BST. In the case of the higher PPT rates offered in the November to February period, it can be shown that the amount by which these rates exceed the (weighted average of) BST unit rates is equivalent to the basic capacity charge (of £31 in 1985-6: see Table 8.3) averaged over the relevant hours. The PPT therefore offers to private generators operating at a 100 per cent load factor over the November to February period an

Table 8.5. Structure of private purchase tariffs (pence per kilowatt-hour)

	1985–6
3-RATE METERING	
Units supplied 00.30 to 07.30 hours	1.35
Units supplied 07.30 to 20.00 hours, Monday to Friday during months of:	
November and February	3.93
December and January	7.14
Units supplied at other times	2.47
2-RATE METERING	
Units supplied 00.30 to 07.30 hours	1.35
Units supplied at other times	2.52

Sources: *Purchase Tariffs*; various Area Electricity Boards.

amount equivalent to the BST basic capacity charge. This method of rolling up the capacity charge into a unit rate works to the disadvantage of potential entrants whose cost structure gives them a comparative advantage in supply during the periods of peak demand over which the basic capacity charge is measured. Equally, however, others may be advantaged (there is also a general provision in the Energy Act for private suppliers to negotiate individual terms). The PPT rates do not contain any allowance for the peak capacity charge in the BST, but there is provision for paying the peak surcharge on the unit rate. As noted earlier, no allowance is included for the system service charge on the grounds that the costs to which this charge relates are considered to be unavoidable. Subject to these omissions, it can be concluded that there is a clear relationship between the private purchase tariffs and the bulk supply tariff. However, the nature of the omissions (in particular the exclusion of the system service charge) indicates that the PPTs do not fully reflect LRMC.

One further point remains. We noted earlier that the BST rates charged to each Area Board are (apart from the system service charge) identical and no allowance is made for favourable (or unfavourable) geographic location. It will be clear from the above analysis that the same comment applies to the specification of the PPTs. There is no specific mechanism for encouraging the entry of private producers who are favourably located (and vice versa), although the provisions in the legislation to negotiate special terms could be adopted in principle. In theory the opportunity afforded to private generators to make use of the ESI's distribution system might provide a competitive cross-check, although our discussion below of the use-of-system charges indicates that this possibility will probably remain theoretical.

Linkage between the BST and Load Management Charges

The load management scheme provides a special scale of charges for certain groups of customers (in practice, medium to large industrial users). Under the scheme, customers who undertake to reduce their load on request from the ESI are exempted from some or all (depending on the precise nature of the load reduction offered) of the capacity charges and peak unit surcharge. In principle the pricing incentive enables the system's overall load factor to be improved and enables the CEGB to reduce its operating margin (the margin of capacity required to meet peaks in demand) and hence to reduce costs. The requirement to reduce load imposes corresponding costs on load management customers (either in terms of stand-by capacity or in terms of disruption to their production processes) and the decision about whether to go on load management will, at least in theory, reflect a balance of these two sets of costs. Specified in this way the load management scheme can be regarded as an efficient peak-pricing structure.

In practice, however, the number of load management calls (that is, the number of times consumers have been required to reduce their load) has been small, a reflection of the current level of spare generating capacity. In these circumstances the scheme becomes almost equivalent to offering to supply to load management customers at prices that reflect SRMC rather than LRMC. Indeed this was recognized as early as 1981 (Electricity Council (1981, p. 17)). The EFL requirement placed on the industry ensures that for ordinary tariff customers the tariffs charged remain close to LRMC-based prices, however. Interpreted in this light, load management can be regarded as a form of price discrimination. That is, whilst many load management customers would participate in the scheme if a larger number of load management calls were made, there are likely to be some who have only been attracted to the scheme by the low probability of a load call being made. Whether load management is an economically efficient form of price discrimination, or whether it shows undue preference, is an open question. (It might be argued, for example, that load management customers are those displaying the highest price elasticity in choice of fuel, but there appears to be no evidence on this, one way or the other.)

The importance of this effective price discrimination in the context of the 1983 Energy Act is that many load management customers are likely to be precisely those industrial users for whom own-generation, and the corresponding potential for export supply to the ESI, would be attractive. The availability of the load management tariffs therefore provides a disincentive to own-generation, as would the recently discussed proposals to provide even more favourable tariffs to large industrial customers (see Thomas (1985)).

Use-of-the-system Charges

The Energy Act has provisions enabling private generators to supply directly to final customers by leasing the use of the ESI's transmission network. We noted earlier that this might, in principle, provide a competitive cross-check on the absence in the private purchase tariffs of an explicit allowance for any locational advantages (or disadvantages) enjoyed by a private generator. And in principle, distortions in the structure of prices offered to final customers might be competed down to minimum levels by entry of this type.

The tariffs specified under the Energy Act for the use of the ESI's transmission and distribution system are based on a model that calculates the costs of a 'notional pathway' for transmission through the system (details of the tariffs are provided in Hammond, Helm, and Thompson (1986)). This highlights an immediate conceptual difficulty. When a private generator provides a power supply at a point on the network (say Wolverhampton) for sale at another point (say Bristol) then it is clearly inappropriate to think of power travelling between the two points in the way that, for example, a telecoms signal does. In practice the ESI's requirements to provide power at one point (the point of private supply) are reduced whilst at another point supply is increased (unless the private generator's customers are diverted from the ESI, in which event there is no change). The consequent change in the costs of providing and maintaining the distribution network (plus the net revenue loss to the ESI where demand is diverted) provides the appropriate conceptual measure of the incremental cost of system use that the private generator should in principle be charged.

We have not been able to determine whether the tariffs specified for use of the system correspond to this conceptual ideal (and we doubt whether this could be easily determined by anyone outside the ESI). A cross-check on the level of these charges might be provided by adopting a solution proposed in the parallel case of the gas industry (see Hammond, Helm, and Thompson (1985)). There it was suggested that if the industry were required to account for its own use of the system, as an internal profit centre, then the resulting level of profit (or loss) would give some indication of whether the use-of-system charges were being set at an appropriate level.

In any event, the option of using the ESI's system has not, to our knowledge, been adopted in a single case. Whilst suggestive, this does not necessarily indicate that the charges are set at too high a level. Economies of scale in revenue collection (see MMC (1985)) mean that private generators will find it difficult to compete with the ESI for smaller-scale customers whilst, as we noted in the previous section, many larger customers enjoy favourable tariffs under the load management scheme.

6. The Two Hypotheses Again

Why then has the Energy Act failed to induce entry into the ESI? To answer this question, we must return to our two hypotheses again—the 'efficiency hypothesis' and the 'entry barriers hypothesis'.

Considering first entry conditions, it is hard to avoid the conclusion that these are not appropriate, and for a number of reasons. The private purchase tariffs offered for the sale of electricity to Area Boards are required by the Act to be based on 'avoidable costs'. This leaves open the question of how these avoidable costs should be measured—should they be based on SRMC or LRMC, and how should these principles be implemented? In effect the apparently straightforward principles of the legislation have been left to the incumbent ESI to put into practice. We have seen that the PPTs have leant towards SRMC (because of the exclusion of the system service charge and the specification of the capacity charges in the bulk supply tariff). Two things follow from this. First, PPTs based on SRMC will give less of an incentive to entry than PPTs based on LRMC. Second, there is an inconsistency between the tariffs charged by the ESI (which are based on the full BST charges) and the prices offered to private generators.

Furthermore, the specified PPTs are offered only for a single year. Investments in electricity supply earn their returns over many years. The entrant faces uncertainty over his expected future revenue stream and this uncertainty is exaggerated by the scope that exists for the incumbent ESI to change the terms of the PPTs. The recent changes in the BST point to the realism of this uncertainty. In practice the LRMC concept that underpins the BST is capable of a number of differing interpretations—in particular with regard to time horizons and future investment profiles. And the interpretation adopted at any time is likely to be influenced (at the very least) by government financing requirements and EFLs. There is thus substantial uncertainty that is open to exploitation by the incumbent. And the increased risk of investment in an uncertain environment will inhibit entry.

In addition, the class of potential entrants who might substitute own-production for ESI purchases is affected by the load management scheme. This, we argued, is effectively a SRMC pricing arrangement. Large industrial consumers of electricity who are able to load-manage hence face lower electricity charges, reducing the incentives to produce their own electricity.

Thus for major potential entrants, the Energy Act has failed to set prices in a fashion that would induce market entry. The PPTs are not precisely related to LRMC, their future level is subject to significant uncertainty, and for the individual customer with scope for load management the purchase price is lower than for other classes of electricity consumers, hence discouraging own-production.

But because the entry conditions under the Act are far from ideal, it does not follow that they are collectively the cause of lack of entry. On the contrary, it is perfectly possible that economies of scale in production, as well as the ESI's productive efficiency, have made electricity supply unattractive to potential entrants. Indeed, as we have seen, there is considerable evidence to support the view that the ESI achieves high productive efficiency in generating power. It has also been suggested to us that economies of scale leave most potential private producers at a competitive disadvantage, although there are alternative views on this (see for example Yarrow (1985) and Energy Committee (1983)).

7. Conclusions

Thus we conclude that the Energy Act has failed to induce competition because both of the hypotheses are partially correct. The ESI's productive efficiency (and perhaps also economies of scale) do deter electricity production at the small-scale level. But entry barriers due to the incumbent's dominant position are hardly likely to encourage entry. There is scope for positive deterrence and this will discourage entry whether or not this scope is actually acted upon.

The Energy Act is flawed. In particular it leaves the incumbent ESI with effective control of price and entry conditions. In practice this has led, as we have seen, to the pricing conditions being set in a way which may discourage entry. It should be acknowledged that an argument could be made for saying that the current pricing conditions are economically efficient. By shading toward SRMC, the current pricing conditions inhibit entry whilst there remains excess capacity in the ESI.

But the type of judgement required here (effectively a balancing of the respective benefits of encouraging competition and of reducing excess capacity) is precisely that which should be taken by the industry's regulators. To leave this decision to the incumbent is simply to discourage clarity and provide the opportunity for entry deterrence. Furthermore it has led, as we have seen, to inconsistency between the PPTs and the ESI's tariffs.

There are a number of ways forward from the present position. The first of these is to tinker with the existing legislation. Amongst the most obvious changes that our analysis suggests are to make possible the offer of longer-term contracts and the clarification of the LRMC/SRMC criterion, especially with respect to load management customers.

The second option is to change the position of the incumbent with respect to private producers. Whilst the setting of PPTs and the terms of entry is left to the incumbent, the outcome will only be fair if either the ESI acts wholly altruistically or regulation by the Department of Energy is sufficient to deter anti-competitive practices. There is therefore a case for the regulation of the entry conditions to be carried out by a body external to the industry: either a newly created agency (see Yarrow (1985), Henney (1985), and Atkinson (1985) on proposals of this type) or, in the absence of a new agency, a body such as the MMC.

Both these options would enhance the effectiveness of the Energy Act. Nevertheless there must be considerable doubts as to whether this would result in effective competition in the electricity industry. The history of deregulation in sectors such as express coaching (see Davis (1984) and Jaffer and Thompson (1986)) and telecoms (see Vickers and Yarrow (1985)) points to the advantages held by a dominant incumbent when long-standing statutory protections are removed.

The creation of a market in electricity may therefore only be feasible if a more fundamental restructuring of the ESI is implemented. Certainly in the present situation, restructuring is required if competition is to be increased without adding further to excess capacity. It is clear, however, that a policy of restructuring has potential costs (for example, in changing the basis of co-ordinating the order in which different generating stations are utilized—the 'merit order') as well as possible benefits (see Yarrow (1985) and Webb (1984) for a discussion of possible options). But it is apparent that if the identified weaknesses in the ESI's performance are to be improved (in particular in relation to investment and over-capacity) then a reform that is more fundamental than the Energy Act is required. The closer examination of the costs and benefits of such reform is therefore of critical importance.

The Energy Act has brought real benefits both in liberalizing electricity supply and in codifying the conditions under which entry can take place. But, as in many other deregulated sectors, the advantages of incumbency (what might be called the incumbent's endowment) have acted to frustrate the objectives of liberalization. The simplicity of the Act has been misleading. It has effectively ceded regulation of the conditions of entry to the incumbent and, as in many other cases of self-regulation, this has not been conducive to the encouragement of effective competition.

9 The Potential of Incentive Regulation*

RICHARD SCHMALENSEE

1. Introduction

The idea for this paper began with the observation that incentive regulation is attracting a lot of attention these days from two very different groups of people—who are mostly unaware of each other's interest. Regulators see incentive regulation as a way of improving utility performance. Academic economists see incentive regulation as an example of an important class of long-term contracting problems—including defence procurement, Medicare reimbursement, coal supply, and managerial compensation. Regulators view incentive regulation as a possible modification of existing institutions; academics start with a clean blackboard and try to describe the best possible regulatory system.

I do not think a little communication between these groups can hurt much. I thus want to discuss what seem to me to be the most important lessons from academic work for the practical design of incentive regulation. To understand the arguments behind these lessons, it is useful to begin by comparing the regulatory status quo with two polar-case alternatives.

2. Frictions and Incentives

A few years ago, the success of deregulation in other industries led a number of commentators to advocate total or partial deregulation of the electric utility industry.[1] Most people would argue, however, that total deregulation of the industry would simply produce a set of unregulated monopolists, free from the threat of entry. And most would oppose such a change, on the grounds that the price of electricity would be set at outrageously high monopoly levels.

But, in the context of this paper, complete deregulation has one thing to recommend it. If an unregulated monopolist can reduce his costs by a dollar, his profits go up by a dollar, now and in the future. Complete deregulation would thus provide strong incentives to reduce costs. It would probably lead to higher

---

* Richard Schmalensee is Professor of Economics and Management at Massachusetts Institute of Technology and Special Consultant to National Economic Research Associates, Inc.
  This paper was first published privately for a NERA conference on electric utility, 12–15 February 1985, in *Surviving an Era of Changing Regulation*. A more detailed treatment of many of the issues discussed here and an analysis of recent US experience are contained in Joskow and Schmalensee (1986).

[1] For a discussion of these proposals, which concludes that their potential is very easy to exaggerate, see Joskow and Schmalensee (1983).

prices than the status quo, but costs would be likely to be lower. And there is no economic reason to suspect that service quality would be reduced.

Another alternative to the status quo is the perfect, frictionless regulation described in some introductory economics texts. In this regime, prices are set at each instant so that the regulated firm covers its actual costs and earns profits exactly equal to the opportunity cost of the capital it employs. Monopoly pricing is clearly not a problem here.

But incentives are a very serious problem indeed. Under perfect, frictionless regulation, the regulated firm has absolutely no economic incentive whatever to minimize its costs. Since managers must work hard and put unpleasant pressure on subordinates in order to hold costs down, it is easy to predict that costs would be higher than they are now, if regulation actually operated in this perfect, frictionless fashion. The service ethic would be the only force tending to keep the lights on and costs finite. Exactly this same problem arises when the Defense Department, or anyone else, uses cost-plus-fixed-fee procurement contracts.

In important respects, regulation as we know it is intermediate between these two alternative regimes. Commissions set prices, which remain fixed (except for some changes due to automatic adjustment clauses) until the next proceeding. The gap between prices and costs is much less than it would be under complete deregulation. If a utility can decrease its costs by a dollar, it can increase its profits by a dollar—but generally only until the next rate case. At that point, lower costs will be translated into lower prices. Incentives to reduce costs are thus less than under complete deregulation but certainly greater than under perfectly frictionless regulation. If prices were kept fixed for longer periods, incentives for cost reduction would be increased—but prices would be on average further out of line with costs.

These simple comparisons make several simple but fundamental points. First, incentives for efficiency are only present when prices do not track costs at each instant. Any incentive system must drive a wedge between costs and prices. In this sense, total deregulation is a great incentive system—though not nearly as good as effective competition.

Second, there is a trade-off between providing incentives for efficient production and ensuring prices that give incentives for efficient consumption.[2] The problem with unregulated monopoly from the point of view of society as a whole is that prices would be far above marginal costs. Buyers would consume an inefficiently small amount of power. If the intervals between price changes under conventional regulation were lengthened, it is likely that costs would be reduced. But input changes would drive costs further from prices on average. As in the case of monopoly, buyers would be led to inefficient consumption decisions—the difference is that prices might be too low and thus encourage too much use of power.

In the stable 1960s, William Baumol and others argued that regulatory lag

---

[2] This point is discussed at length in Schmalensee (1979, Chs. 7 and 8).

should be institutionalized in order to enhance incentives for efficiency.[3] A few years later, in the turbulent 1970s, these proposals sounded absurd, as rate proceedings became more frequent and commissions moved to automatic fuel adjustment clauses and other devices to keep prices in the same ballpark as utilities' rapidly changing costs. As this experience illustrates, the terms of the trade-off between incentives for cost reduction and incentives for efficient consumption depend on the environment in which the utility operates.

Third, as this experience also illustrates, all elements of any regulatory regime affect the incentives for efficiency that face regulated firms. Administrative practices affect regulatory lag; fuel adjustment clauses affect the incentives to economize on fuel costs; and policies toward plant abandonments affect incentives to make efficient additions to capacity when demand is uncertain and construction involves long lags. Every real regulatory system provides utilities with incentives of some sort, since every utility can take actions that affect its profits.

The title of this paper is thus somewhat incomplete. The real issue, both in theory and in practice, is the potential of *explicitly designed* incentive schemes, which can augment and modify the incentives provided by the aspects of regulation that have arisen by historical accident or administrative necessity.

3. Toward Optimal Regulation

I now want to describe the approach economists have taken to the analysis of optimal incentive regulation and efficient long-term contracting. Suppose that one could wipe the historical slate clean and design a regulatory system from scratch to optimize the performance of regulated utilities. What would such an optimal system look like?

To be a little more specific, suppose that a regulatory commission is interested in maximizing some measure of the performance of a regulated public utility, subject to the constraint that the firm receive enough profit to remain in business. If the commission knew enough, of course, it could simply dictate all the utility's decisions in order to maximize performance. But commissions in practice cannot manage the firms they regulate.

Commissions can generally observe a firm's actual costs, for instance, but they cannot directly observe the level of management effort or the quality of managers' decisions.[4] Actual costs will be determined both by managerial effort and decisions and by more or less random events beyond management's control. And commissions cannot disentangle with any precision the effects of these two

---

[3] See, for instance, Baumol (1967). For a general discussion, see Schmalensee (1979, pp. 119–21).

[4] Management audits do provide some information on how well-managed a firm is, but it is my impression that such information is far from perfect. In addition, it is unclear how one can translate an auditor's adjectives into the sort of quantitative decisions that are at the heart of incentive regulation in both theory and practice. For a discussion, see Schmalensee (1979, pp. 130–2).

influences on costs—or on any other performance measure. The heat rate or equivalent availability of any particular generating unit in any particular year, for instance, will depend in a complex way on a host of observable and unobservable factors, only some of which are subject to management control.[5] It is very difficult to imagine that a real commission will ever be able to say that exactly X per cent of the actual year-to-year change in any performance measure is explicitly due to management effort and decision-making.

Agency theory, or the *principal–agent* model, provides a general framework for dealing with problems of this sort. The basic problem considered in agency theory involves one party, the *principal*, who hires another party, the *agent*, to take actions on his behalf. The principal, here the commission, wants the agent, a regulated utility, to make some performance criterion, such as consumer welfare, as large as possible. The agent's actual performance depends on his effort and on random factors, and the principal can observe neither of these directly. The principal can make the agent's earnings depend on the observed level of performance, but he must provide earnings that are high enough on average to keep the agent viable—or, in this case, to keep the utility in business.

This general set-up has been applied to the problem of designing socially optimal regulatory regimes by several authors in the last few years.[6] The usual assumption is that the commission is interested in maximizing consumers' surplus or some other comprehensive performance measure that can be expressed in dollars. The commission can observe actual performance, which depends on three things it cannot generally observe directly: the effort devoted by the utility's management to reducing long-run costs; equipment failures and other random events; and the parameters of the utility's cost function.

The commission is assumed to make an incentive payment to the utility. The amount of this payment, which may be positive or negative, depends on observed performance according to an announced function. In the simplest case, the incentive payment is the firm's revenue, which depends on its output and the prices specified by the regulator.

The utility's profits are the sum of its incentive payments and its other net revenue. Other net revenue also depends on effort, random events, and its cost function. In the simplest case, other net revenue is equal to minus the firm's actual costs. The utility knows its own cost function. After it observes the random events, it selects the level of effort to maximize its profits.[7] The commission's optimization problem is to select a function relating incentive payments to observed performance in order to maximize the expected value of performance, net of incentive payments.

The commission makes this choice subject to two constraints. The first of these is the regulated firm's *self-interest*: the utility will act to maximize profits,

---

[5] See, for instance, Joskow and Schmalensee (1985).
[6] For a recent survey of this literature and good bibliography, see Sappington and Stiglitz (1985).
[7] Alternatively, management may be modelled as maximizing a utility function that is increasing in profit and decreasing in effort.

not net performance. The second is the *viability* constraint: the regulated firm's expected pay-off must be at least as high as it could earn elsewhere. That is, the utility must expect to cover the opportunity cost of its capital.

4. Limitations and Lessons

The principal–agent set-up mirrors some central features of real regulatory situations, but there are some obvious and important differences. First, it is not clear that regulators can generally either define or observe a single overall measure of performance. Regulatory commissions may not have well-defined objectives, and performance may be hard to measure along several dimensions. How, for instance, should the overall reliability of a large, interconnected electric power system be measured, and how should reduced reliability be traded off against lower rates? But the theory reminds us that without a well-defined objective, there can be no well-specified design problem. As I discuss further below and Landon (1985) stresses, this points to a set of serious practical problems.

Second, it is assumed in the principal–agent approach that the regulator can make positive or negative payments to the regulated firm, and those payments can depend in essentially any way on observed performance. But this substantially overstates the flexibility of real regulatory commissions. Public utility regulation involves setting prices and rates of return, not making or receiving cash payments. Formally, there are a set of important institutional constraints on the choice of incentive schemes that have been neglected thus far in the agency theory literature. It follows that any limits to incentive regulation that are identified by this analysis hold when these constraints are imposed.

Third, the theory assumes that managers' and shareholders' interests coincide, but this may not be completely true. Most incentive schemes in practice reward or penalize shareholders by operating on profits or allowed rates of return; management compensation is only indirectly affected. Even if regulators directly rewarded or punished managers, a firm's board of directors could undo this by adjusting managers' base salaries. If managers do not act to maximize the wealth of shareholders (who are their principals in another principal–agent problem), the effects of profit-based incentive schemes become quite complex.

Finally, regulation differs from most of the other contexts in which agency theory has been applied in one very important respect: there is no explicit contract between regulators and regulated firms. Thus commissions cannot generally bind themselves or their successors to follow any particular policies in the future. This difference has important implications for the design of socially desirable incentive regulation, as I shall discuss.

Despite these and other inconsistencies between theory and reality, the framework described above provides a sensible and intuitive way to structure one's thinking about the problem of designing incentive regulation. *Incentive regulation is used by regulatory agencies, acting as principals, to affect the performance of regulated firms, their agents.* Regulated firms know more than commissions, and commissions cannot directly make all managerial decisions.

5. Some Practical Principles

The principal-agent literature has not produced any formulas or guidelines comparable in simplicity and generality to those that have been found to guide utility pricing. There is nothing analogous to the principle that all prices should be based on marginal costs. And a good deal of the analysis is numbingly abstract. But it seems to me that the principal-agent literature and related work suggest five basic principles that should govern the design of real-world incentive systems:

(a) do not expect too much;
(b) permit high profits sometimes;
(c) use a comprehensive performance measure;
(d) weaken incentives when risk increases;
(e) put bounds on rewards and penalties.

In the remainder of this section, I discuss these principles and the reasoning that supports them.

Do Not Expect Too Much

There are two reasons for this. First, the regulatory status quo already provides incentives for efficiency. Explicit incentive schemes can only augment and modify those incentives; they cannot fundamentally alter the environment within which regulated firms make decisions. A management that performs poorly before explicit incentives are given may well do a poor job of reacting to those incentives. Moreover, regulated firms' managements are more secure against the threat of take-over than managements in unregulated sectors, and stockholders' ability to discipline inefficient management is correspondingly limited.

Second, agency theory shows that, even in principle, incentive regulation cannot generate optimal performance—it cannot match the incentives provided by competition as long as the regulator must keep the utility viable. On the one hand, incentives are necessary for efficiency, and the closer profits are held to the minimum necessary for viability under all conditions, the less scope the regulator has to provide incentives. If the firm is to be viable under all conditions even if the worst happens, the stick of bankruptcy cannot provide the same incentives it provides under competition. If follows that the carrot of high profits must be used; the utility must earn more than is necessary for viability under some circumstances. But, since excess profits come out of consumers' pockets, it is almost never optimal to give up so much control over profits as to give the regulated utility exactly the right incentives.

This lesson does not imply that incentive regulation is a bad idea. Well-designed incentive regulation can improve performance. And a well-designed incentive system can give a regulated firm much more reason to worry about costs than a regulatory regime that is in effect a cost-plus system.

Permit High Profits Sometimes

This lesson follows directly from the discussion above. Because the stick of large losses cannot be used to induce efficiency, the carrot of high profits must be employed. If incentives are to be powerful, it must be possible for a utility to earn very high profits if its performance has been exceptionally good. Moreover, the agency theory literature suggests that it may generally be necessary to pay for better performance by allowing returns in excess of the cost of capital on average, though the magnitude of this effect in practice is unclear.

This point should be reassuring in principle to utility managers and shareholders: well-designed incentive regulation should be on average more a bonus system than a penalty system, though it is designed to induce managers to work harder. But this point also raises potentially serious problems in practice, since commissions cannot sign explicit long-term contracts with utilities. Unless commissions can resist the temptation to lower rates when profits are high, no incentive system will work well. If utilities think they will be unable to avoid losses if they perform poorly (for whatever reason) but will not be allowed to earn 'excessive' profits if they perform well (for whatever reason), the incentive scheme lowers their allowed rate of return on average. One can expect utilities to react in an anti-Averch–Johnson fashion by under-investing, and the undesirable effects of this behaviour may well outweigh any desirable effects of better operating incentives.

Use a Comprehensive Measure of Performance

This is true whether incentive payments are explicit, as in the principal–agent analysis, or implicit, as in incentive schemes that tie the allowed rate of return to performance. The reason is simple: a regulated firm will act in its own self-interest and attempt to increase the performance measure on which it is graded. Thus, for instance, if the incentive scheme makes it profitable to increase generating unit reliability, reliability will be increased even if it means overspending on maintenance. As Landon (1985) discusses, rewarding low heat rates provides a number of perverse incentives. If distribution system outages are not punished, funds will be shifted away from distribution system maintenance. Only if incentives can be tied to a comprehensive performance measure can one be reasonably certain that the utility's performance will in fact be enhanced by incentive regulation.

As I noted above, regulatory commissions may find it difficult to formulate comprehensive performance measures; they will certainly find little useful guidance in the academic literature. Consider, for instance, economic cost per kilowatt hour (kWh), the measure that is implicitly rewarded by regulatory lag. Incentive schemes can easily be based on accounting cost per kWh, judged relative to historical standards or relative to the performance of other firms. Incentive regulation that sets a utility's rates on the basis of the costs of *other* utilities can make the much-discussed notion of 'yardstick competition' operational.[8] But accounting and economic costs may diverge for many reasons, including historical depreciation policies; and the problem of fairly comparing real, complex utilities with their own past performance or with the performance of other, similar utilities is very difficult indeed.[9] Moreover, unit cost is not fully comprehensive; outages are not considered, and unit costs can be reduced by actions that increase the probability of massive outages that rarely occur. Thus, accounting unit cost is not a fully satisfactory performance measure.

If no truly comprehensive performance measure is possible, taking into account such diverse aspects of performance as cost, reliability, technical change, and response to emergencies, performance measures should at least be chosen with an eye to minimizing utilities' incentives to make socially undesirable decisions that increase measured performance.

If utilities can make decisions that trade off generating unit reliability against heat rate, for instance, both should be considered in any performance measure. If construction costs are watched closely but fuel costs are passed through automatically, utilities have an incentive, in theory at least, to adopt technologies that are too fuel-intensive.[10] If this possibility is empirically important, it is correspondingly important to avoid treating these inputs asymmetrically in incentive regulation. The point is to choose performance measures that do not permit utilities to profit by socially undesirable actions with unmeasured adverse consequences. If this is not done, incentive regulation can worsen, not improve, utility performance.

Weaken Incentives when Risk Increases

The converse is that the more stable is the environment, the more scope there is to provide incentives that may drive prices away from costs. It may have made sense to lengthen regulatory lags in the stable 1960s; it surely made sense to reduce them in the early 1970s.

Agency theory makes clear that the exact features of any optimal incentive scheme depend critically on the commission's information and on the character

---

[8] For an interesting recent discussion of this possibility, see Shleifer (1985).
[9] See, for instance, Joskow and Schmalensee (1985) and Landon (1985).
[10] See, for instance, Joskow and MacAvoy (1975).

of the uncertainty it faces. In general, the regulator's choice of a function relating earnings to observed performance is complex and depends on the best and worst possible environments, and on the probabilities of all intermediate cases. No single incentive formula can be optimal—or perhaps even sensible—in all environments. The theory makes it clear that optimal incentive regulation in the 1960s differed from the optimum for the 1970s, which in turn differs from schemes that it would be optimal to adopt today.

To say more, we must consider the basic trade-off discussed above between the *sensitivity* of prices to changes in economic conditions and *rigidity* that provides incentives for efficiency. The more certain is the future, the lower the expected gap between prices and costs under any incentive system. It thus becomes less socially expensive to use rigidities and frictions to provide incentives for efficiency. In very uncertain times, on the other hand, a regulatory system with substantial rigidity will tend to produce prices that are undesirably far from costs. The reduction in incentives produced by automatic fuel-adjustment clauses may have been desirable in the 1970s, but not in the 1980s, and commissions seem to have moved away from such clauses recently.

All this means that no system of incentive regulation can be expected to be optimal for long periods of time. The process of incentive regulation must involve redesign, or at least adjustment of parameter values, when economic conditions change. And it is the future that matters, not the past, or even the present. If the past has been turbulent but the future is expected to be calm, strong incentives are optimal. On the other hand, if future demands and costs are very hard to predict with confidence, then it makes sense to weaken incentives in the interest of greater sensitivity, even if the recent past has not witnessed major shocks. There is no substitute for commission judgement here; incentive regulation does not lend itself well to detailed legislative treatment.

Put Bounds on Rewards and Penalties

In practice this is likely to mean placing upper and lower bounds on a utility's rate of return. There are two distinct arguments in favour of such bounds.

First, as I noted above, commissions cannot sign binding long-term contracts with utilities. Because of this, it is unreasonable to suppose that any commission would actually follow an incentive system that forced it to allow a firm under its supervision either to fail or to earn outrageous profits. There is thus no loss in excluding those possibilities in advance by bounding the actual rates of return.

Second, large, rapid changes in performance are more likely to be attributable to the environment than to utility management. Compare the effects of the oil shocks, for instance, with the effects of year-to-year changes in the quality of management decision-making. It is sensible to treat large, rapid changes in performance as mostly due to chance.

This can be accomplished by non-linear versions of the classic partial adjustment and sliding scale devices.[11] But it may be simpler just to make prices or rates of return completely insensitive to changes in the level of performance outside some band. That is, one might specify that the allowed rate of return will increase by 1 percentage point for every 4 per cent fall in costs for cost changes between -20 per cent and $+20$ per cent. If costs rise (fall) by more than 20 per cent, the allowed rate of return would be reduced (increased) by only 5 percentage points, no matter how large the cost change, on the theory that changes of this magnitude are most likely driven by the environment, not by management decisions. These numbers are only illustrative; the task of picking adjustment parameters and bounds is, in principle, quite complex and, again, reflects judgements about the importance of environmental uncertainty and management decisions.

6. Conclusions

Let me finally return to the question posed by my title: what is the potential of incentive regulation? In principle, incentive regulation has much to offer. Incentive schemes can move regulation from a regime based on historical accidents and administrative convenience to a system designed to optimize the performance of regulated utilities.

In practice, however, the potential contribution of incentive regulation seems relatively limited. Incentive schemes only add to the incentives for efficiency already provided by the regulatory status quo. Such schemes can make things worse if badly designed performance indices provide perverse incentives. And incentive regulation cannot reproduce competition; incentives mild enough to be credible are likely to have only mild effects on performance.

Still, incentive regulation is worth trying—carefully.

---

[11] For a discussion of these devices and their application in practice, see Schmalensee (1979, pp. 121–30).

10 Regulatory Issues in the Electricity Supply Industry*

GEORGE YARROW

1. Introduction

In the immediate future it is likely that two questions will dominate economic (and political) debates about the future development of the electricity supply industry (ESI): will privatization improve the industry's performance and should there be a substantial programme of investment in nuclear generating capacity? Underlying these questions, however, is a core set of microeconomic issues that has confronted policymakers for several decades. And since, in the process of change, it is easy to lose sight of them, it is upon some of these fundamental issues that the analysis of this paper will be focused.

The broad, microeconomic objectives of public policy towards the ESI can be broken down into three components; first, with a given capital stock and given demand conditions, that prices should be set to achieve an economically efficient use of available resources; second, that the flow of investment funds should also be such as to lead to an efficient allocation of capital between the ESI and other industries; third, that, whatever the final outputs of the ESI, they should be produced at the lowest feasible cost.

In practice, these three aims have frequently been overridden by wider macroeconomic policy goals. Thus, for example, at different times governments have intervened either to depress prices (as part of a general prices and incomes policy) or to raise prices (with the aim of reducing the public sector borrowing requirement). Nevertheless, in what follows the focus of attention will be upon systems of regulation designed to achieve microeconomic objectives and, to the extent that they are relevant, other policy considerations will be treated as a source of constraints on regulators.

2. Public Policy: The Recent History

2.1 The 1967 White Paper

In discussing recent trends in policy towards the ESI, the 1967 White Paper, *Nationalised Industries: A Review of Economic and Financial Objectives*

---

* This paper was presented at the conference 'Energy Policy' organized by the Institute for Fiscal Studies and the Economic and Social Research Council Industrial Economics Study Group in London on 1 May 1986.

George Yarrow is a Fellow in Economics at Hertford College, Oxford.

(HM Treasury (1967)), provides a useful starting-point. The main points of the White Paper are easily summarized. First, it was recommended that prices be set at long-run marginal costs, but that some adjustment towards short-run marginal costs should occur when it was clear that the existing level of capacity was suboptimal. Second, investment decisions should be based upon a test discount rate, calculated from the real returns available from low-risk private sector projects and therefore reflecting the opportunity cost of capital. Third, the pricing and investment guidelines were supplemented by a financial target for the industry which, in its original conception, is best viewed as an instrument intended to promote cost efficiency.

Thus, the three principal guidelines of the White Paper can be seen to follow fairly directly from the three microeconomic objectives. At this level of abstraction, perhaps the only doubt about the 1967 approach lies in the choice of long-run marginal cost pricing in preference to the short-run alternative since, particularly in industries where capital is very durable, and hence where excess or deficient capacity might persist for long periods of time, the latter would be the more natural initial bench-mark when seeking to achieve allocative efficiency. The point is, however, of relatively minor significance since (a) the problem is recognized in the White Paper and the role of short-run marginal costs is explicitly mentioned, and (b) there are information-signalling arguments *against* reflecting every movement of short-run costs in prices to customers.

The general problems arising from the 1967 approach to policy for the nationalized industries are well known and need not be discussed in any detail here (see Vickers and Yarrow (1985)). In effect the guidelines were an injunction to the industries to 'go away and sin no more'. Unfortunately, however, they were insufficiently precise to provide managers with an operational guide as to the distinctions between (from the policy perspective) desirable and undesirable conduct. That is, operational objectives were less than fully clear. More importantly, the White Paper did not establish effective systems for monitoring managerial behaviour, where by monitoring I mean the dual process of both measuring performance and rewarding or punishing managements in the light of their performance relative to the declared policy objectives. Thus, the 1967 White Paper was of only marginal relevance for some of the most central issues of regulatory policy.

Unlike most other nationalized industries, the ESI accepted the desirability of the marginal cost pricing principles enunciated in the White Paper and, ever since, has struggled with the problems of implementing a tariff structure based upon long-run marginal costs. As the subsequent experience has shown, estimates of such costs are, of necessity, highly subjective, in the sense that they depend upon a whole series of assumptions about demand and cost conditions in energy markets for several decades hence (see Jones (1985) and Slater and Yarrow (1985)). As a result, the ESI has effectively been in possession of significant discretion with respect to the choice of pricing policy, and it has been possible to select approaches that best meet objectives other than efficiency in resource allocation whilst still claiming that a marginal cost pricing approach was being pursued.

A similar point holds in respect of the investment guidelines. Given a particular test discount rate, it was all too easy to adjust demand and cost projections to justify particular aspects of an investment programme. Moreover, the long durations of projects imply that *ex post* evaluation of investment performance may yield little information that is relevant to current decisions, thus making it difficult to improve the incentives for efficiency in investment planning.

The absence of precise objectives, and of effective monitoring arrangements, did, in practice, open the door to the partial displacement of public policy objectives by managerial goals. Although it has proved difficult to construct convincing, general economic models of this process, some insight into the sorts of problems that may arise can be obtained by assuming that managers seek to maximize output, subject to whatever constraints are generated by the overall system of political control (see Rees (1984a)). In that case we can expect some bias towards over-investment: project appraisals are deliberately over-optimistic and the subsequent excess capacity can be used to justify lower prices, and hence higher sales volume and output, on the basis of short-run allocative efficiency considerations.

This type of argument can certainly be used in explaining the persistent excess capacity that has characterized the ESI over recent years, although other factors have obviously also been at work. The most notable of the latter have been the unanticipated declines in demand and increases in costs associated with the oil shocks of the 1970s. In the absence of sufficiently detailed *ex post* studies of the issue, the relative contribution of failures in regulatory policy to the excess capacity problem remains unclear.

Finally, before leaving the 1967 approach, it is worth drawing attention to one further area of weakness, associated with the financial target. This aspect of the guidelines was intended to promote cost efficiency, presumably on the assumption that a higher target would produce greater incentives for managers to reduce unit costs. However, quite apart from the problem of setting an appropriate level for the financial target—too low a value would put little pressure on managers to reduce costs and too high a value could, if it were infeasible, lead to demoralization—there is the difficulty that, in industries such as the ESI where price elasticities of demand are relatively low, a particular financial target might be met simply by raising prices. Again, given discretion over the choice of tariff structure, the response could be made to appear consistent with long-run marginal cost pricing by changing the assumptions upon which long-run marginal costs are estimated.

2.2 The 1978 White Paper

The 1978 White Paper, *The Nationalised Industries* (HM Treasury (1978)), went some way towards strengthening government control over the commercial policies of the industries concerned. Pride of place was accorded to the financial target, and marginal cost pricing was relegated to a more subordinate role. With respect to investment, the test discount rate for individual projects was replaced by a required rate of return criterion for investment programmes as a whole. In addition, each industry was required to produce a series of performance indicators that would be used to assess various dimensions of its internal efficiency. Finally, although not an innovation of the White Paper itself, the post-1978 framework of control has depended heavily upon annual external financing limits (cash limits).

Some of the developments brought about by the 1978 White Paper can unambiguously be classified as improvements on what went before. Thus, as noted earlier, financial targets are, on their own, very weak instruments for promoting cost efficiency. In recent years, however, the Central Electricity Generating Board (CEGB) has been set performance targets expressed in terms of a required rate of reduction in real controllable unit costs, and this procedure provides a much more direct method of attempting to influence the internal efficiency of the Board.

It can also be argued that the stronger governmental control that has been facilitated by the 1978 White Paper also serves to hinder the displacement of public policy objectives by self-chosen managerial goals. For example, the priority attached to external financing limits and financial targets can be viewed as a means of countering the over-investment and underpricing biases that are associated with output-maximizing managements. By the same token, however, the shift in emphasis has made it easier for governments themselves to use industries such as electricity supply to achieve short-term, non-microeconomic goals, particularly in the fiscal policy area. To the extent that these aims are frequently unstable, and are sometimes misguided, the longer-term damage to the industries concerned might well outweigh the benefits of tighter control.

More generally, the 1978 White Paper can be said to have failed to address the principal regulatory issues that have confronted successive governments. Thus, by focusing upon constraints on behaviour, industries such as the ESI have still been left in a position where their objectives are unclear. Indeed, in this respect, the 1978 White Paper could be regarded as a step backwards. Although lacking in operational guidance, the 1967 White Paper was at least clearly based upon the notion that the nationalized industries should be seeking to maximize economic efficiency; after 1978 it is no longer obvious that this is the case. Further, while the metering side of the monitoring task was improved by the introduction of a range of performance targets, there was no attempt to introduce more formal and more explicit systems of establishing incentives for better performance. That is, the rewards/punishment dimension of the monitoring function was given little priority.

2.3 The Energy Act

The most recent major piece of legislation of relevance to the regulation of the ESI is the Energy Act of 1983. Among other things, the Act sought to encourage new entry—or, more accurately, the threat of new entry—into the industry by removing certain statutory obstacles to the generation of electricity as a main business, and by laying down conditions under which the public corporations in the industry are obliged (i) to purchase electricity from private producers and (ii) to make their transmission facilities available to such producers.

This attempted opening-up of the ESI to competition is a potentially useful addition to regulatory policy instruments on a number of counts, of which two may be mentioned here. First, by imposing a market test on the industry, it can, in principle, provide information about the cost efficiency of the public corporations. If the internal efficiency of the public corporations is poor, we might expect to see this signalled by entry of new, more efficient producers. Such information could then be used by governments when setting future performance targets for real controllable costs. Second, unlike the system of control established by the 1967 and 1978 White Papers, the Energy Act implicitly contains a possible method of rewarding good, and punishing poor, performance. Assuming that managements prefer higher output and market share to lower output and market share, the mechanism through which this occurs is relatively direct. Both the output and market share of the CEGB will, provided the Act functions as intended, be dependent upon its cost efficiency relative to the cost efficiencies of other producers of electricity. Thus, for example, if the CEGB is inefficient, management would be punished by the losses in sales volume and in market share that would result from the entry of new firms.

3. Current Problems

3.1 Coal Pricing Policy

Throughout the period in which the ESI has attempted to implement marginal cost pricing policies, the rationale of the approach has been undermined by the nature of the industry's relationship with British Coal. The CEGB has been required by government to purchase the great bulk of its most important input from British Coal at prices that have been well in excess of the marginal opportunity costs of coal. In this way, the domestic coal industry has been protected from international competition.

As a result of coal pricing policy, the downstream messages to consumers about the opportunity costs of electricity have been, and still are, distorted. Arguments for marginal cost pricing in the ESI, based upon the desirable effects of the policy on allocative efficiency, are immediately invalidated if the industry's major cost component is itself distorted by suboptimal pricing in the upstream coal industry.

From a short-run perspective, the damages associated with the consequent allocative inefficiencies have been partially attenuated by the fact that, for the most part, the marginal plant in the various time-of-day and time-of-year (bulk) electricity tariff periods has been coal-fired. Thus, at least in those tariff components that are intended to reflect short-run costs (principally the energy charges), the intra-tariff deviations in relative prices from marginal opportunity cost levels have been more limited than would have been the case if, say, nuclear or oil-fired capacity had been operating at the margin. Nevertheless, electricity prices have been excessive relative to other products in the economy, gas being the most commonly cited bench-mark, with consequent losses in allocative efficiency.

A number of possible methods of correcting the underlying distortion have been suggested, including the following:

(a) allowing the CEGB to base its cost calculations on the marginal opportunity cost of coal, but to meet its existing financial obligations by raising more revenue from tariff components that are not linked to the volume of purchases, such as the system service charge;
(b) as in (a), but loosening the financial constraints on the industry to allow the average level of charges to fall;
(c) setting coal prices on a marginal cost basis and supporting the coal industry directly by increased subsidies from the public purse.

The first option is the least attractive because, whatever the structure of prices for electricity in bulk, at the retail level the Area Boards engage in a good deal of cost averaging when setting their own tariffs. Almost inevitably, the ESI's customers would continue to face tariff structures that biased their choices against electricity.

In respect of the other two options, there is little to choose between them on efficiency grounds: the difference is largely one of internal accounting, in the sense that it is a matter of how the red ink is distributed between the two industries. Given that it is the coal industry that is being protected, the most logical solution would be policy (c).

Despite their simplicity, it is unlikely that either of the two preferred options will immediately be implemented. Governments typically favour methods of industrial support and income redistribution that are less visible than direct financial subventions, as is witnessed by the prevalence of cross-subsidization in nationalized industries. Further, given the aims of reducing the capacity of British Coal and of improving its internal efficiency, there are political attractions in putting pressure on the coal industry indirectly, via an electricity industry that is attempting to reduce its input costs, rather than via reductions in direct subsidies. In this way, governments can distance themselves from some of the more visible, negative consequences of the adjustment. Hence, it is likely that correction of the distortion arising from protection of the domestic coal industry will proceed gradually, at a pace determined by progress in achieving the targeted structural changes in coal supply.

3.2 Financial Constraints

Since 1979 the financial constraints upon the ESI, particularly the external financing limit (EFL), have gradually tightened. In 1979-80, for example, the industry was allowed to increase its borrowings from central government to the tune of £232 million, whereas in 1986-7 it was required to *reduce* its net indebtedness to the government by £1,416 million. The only significant interruption to this trend occurred as a result of the 1984-5 coal-miners' strike, when the government chose not to raise electricity prices in line with the increased costs of generating electricity resulting from the temporary switch from coal to oil in input use.

That there has been a significant change in the EFL position since 1979 follows in part from the state of the industry's investment programme. Excess capacity, coupled with re-evaluation of nuclear options, has led to a period in which expenditures on the construction of new plant have been relatively low. However, the macroeconomic objective of reducing the public sector borrowing requirement has also had a significant impact on the levels of the financial constraints.

Indeed, the latter are such that prices to the ESI's customers appear to have been forced up beyond a point that can reasonably be defended on resource allocation grounds, even if coal is valued at the prices actually paid by the CEGB rather than at prices that more realistically measure its true opportunity costs. Nor does this conclusion depend upon an assumption that it is short-run marginal costs that should form the basis for calculating optimum prices (see Vickers and Yarrow (1985)). For example, the CEGB's own tariff structure, which is derived from a longer-term perspective, fails to generate enough revenue from its marginal cost based components to meet its financial obligations: the difference is made up by a system service charge which has recently accounted for a little less than 10 per cent of total revenue.

Since government finance has real resource costs—any feasible system of taxation will introduce some distortions into the market-place—it is, of course, possible to justify prices in excess of marginal costs on second-best grounds. That is, after making any appropriate allowance for income distribution effects, it can be argued that the most efficient level of electricity prices could well include some commodity tax element. Nevertheless, the structure of the implicit taxes on the ESI does suggest that economic efficiency could be improved without affecting the government's revenue flows. In particular, there is a strong case for avoiding taxation of intermediate goods, so as to avoid unnecessary distortions of input choices in the productive sector of the economy (see Diamond and Mirrlees (1971)).

With respect to electricity supply, the principle that there should be no taxation of intermediate goods is violated in two main ways. First, the supply price of coal is higher than its marginal opportunity cost. Second, the additional implicit tax component associated with stringent financial constraints on the ESI introduces a further distortion in the prices faced by electricity-using producers.

It is likely therefore that, with the same revenue yield to the exchequer, resource allocation could be improved by replacing the current tax elements in coal and electricity prices with value added tax at an appropriate rate. Again the obstacles in the way of this simple step are largely political. UK governments have been nervous about being seen to tax household energy usage, even though VAT on electricity is commonplace throughout western Europe: though less efficient, the less visible (to the public) system of taxation embodied in current arrangements has been preferred.

3.3 Potential Competition

Thus far, the response of private producers of electricity to the opportunities that the Energy Act was intended to open up has been disappointing: there has been little sign of significant new entry into the market. In itself, the absence of new entry does not necessarily imply that the Act has been ineffective, since it can be argued that the outcome only serves to demonstrate that the public corporations are relatively cost-efficient and that their performance cannot be bettered by the private sector (see Hammond, Helm, and Thompson (Chapter 8, this volume)). Changes in the structure of the CEGB's bulk supply tariff since 1983 do, however, give some grounds for concern that substantial entry barriers remain and that a more vigorous policy of liberalization will be required if effective competition is to be introduced into the industry.

In tariff years up to and including 1983–4 (the last year before the provisions of the Energy Act became relevant), the bulk supply tariff (BST) included a component called the service charge which was designed to recover the costs and expenses charged to revenue account in respect of the bulk supply points provided by the CEGB to meet the Area Boards' electricity requirements. The charge accounted for something of the order of 1 per cent of the CEGB's revenues from the Area Boards. The remaining parts of the BST were, in this period, claimed to correspond to the long-run marginal costs of producing electricity at various times of the day and year.

From 1984–5 onwards the service charge was replaced by a much larger system service charge, designed, among other things, to recover 'the unavoidable costs of the Board'. In effect, therefore, the CEGB revised downwards its estimates of marginal costs in 1984–5: costs that were previously classed as avoidable, at least in the longer term, were now designated unavoidable. The most interesting aspect of the tariff change, however, lay in the fact that the system service charge was distributed among Area Boards on the basis of the latters' electricity demands in 1982-3, and the individual Area Board charges were therefore made independent of their subsequent volumes of purchases. Thus, for a particular Area Board in later tariff years, the system service charge appeared as a fixed payment to the CEGB.

The relevance of this point for the question of new entry is that, under the terms of the Energy Act, Area Boards were required to offer to purchase electricity from private producers on terms that reflected the cost savings that would be possible if the Boards reduced their offtake from the CEGB. Since, given the new tariff structure introduced in 1984-5, Area Boards are not able to reduce their system service charges by switching purchases from the CEGB to private producers, the system service charge element in the BST is *excluded* from the calculations when the Area Boards fix the rates (set out in their private purchase tariffs (PPTs)) at which they offer to buy electricity from the private sector. The result is that, unless an Area Board completely disconnects itself from the CEGB, the average (per unit) revenue received by the CEGB significantly exceeds that received by a private producer for a load with similar characteristics.

Under the most favourable of the PPTs that are available, the disparity between the average prices paid to the CEGB and to a private producer can be estimated approximately from the proportion of the CEGB's revenues from Area Boards that is accounted for by the system service charge. These percentages have been as follows:

1983-4	1984-5	1985-6	1986-7
1.0%	7.8%	8.9%	9.8% (estimate)

Thus, as can be seen from the figures, there was a sharp upward movement in the relative importance of the service charge component of the BST in 1984-5, followed by further shifts in this direction in the last two tariff years. By 1986-7 the position reached was that private producers were being offered terms that were, at best, approximately 10 per cent less favourable than those extracted by the CEGB.

In defence of its pricing strategy, the CEGB can argue that, in contrast to its potential competitors, it is unfairly penalized by government policies with respect to coal pricing policy and to the industry's financial constraints, and that it is reasonable that rivals should face prices that reflect the Board's avoidable costs. Nevertheless, the ease with which the provisions of the Energy Act have been manipulated by tariff changes must provide grounds for concern. The fact is that the CEGB has been allowed to use discretionary powers to impede entry by reducing the average prices available to potential rivals. Moreover, irrespective of the actual tariffs set, the knowledge that the CEGB has such powers itself serves to deter entry by making entrants vulnerable to future tariff shifts and creating uncertainty about anticipated revenue flows.

Assuming that the aim is to promote fair competition in electricity generation, at some stage it will be necessary to tighten regulatory control over the private purchase tariffs (or their replacements) offered to new entrants. Ideally, this should be coupled with reforms that ensure that the existing public corporations (or their successors) are themselves not unduly penalized by other public policy constraints. The present arrangements do not constitute a satisfactory long-term option.

3.4 Investment

The final set of regulatory issues to be considered concerns the control of investment programmes in the ESI. At the moment, the provisions of the 1978 White Paper are nominally in force and the principal guideline to be followed is that new investment, taken as a whole, must yield a real rate of return of at least 5 per cent, which rate can be compared with the preceding level of 10 per cent (real) for the test discount rate of the 1967 White Paper. Like the test discount rate, this required rate of return (RRR) is designed to reflect the opportunity cost of capital and was set in the light of the expected returns on low-risk, private sector investment projects in the mid-1970s. In the event, it has been supplemented by more direct governmental control over investment programmes, principally via the mechanism of the ESI's annual external financing limit.

Since the mid-1970s, however, there has been a strong recovery in the profitability of the private sector, and the 5 per cent value now appears to be inappropriately low. Given the real returns on gilt-edged securities, for example, it is difficult to imagine any private sector company being satisfied with an *ex ante* return on projects of anything like this value, particularly when allowance is made for the systematic bias that tends to occur as a result of 'appraisal optimism'. Further, in the case of those public corporations where performance-related payment systems are not in place, governments may also want to take account of tendencies towards managerial output maximization by imposing more stringent criteria on investment appraisals.

If it is true that the implied cost of capital to the ESI is suboptimally low, one consequence is that, unless the position is corrected via EFLs, there will be a general tendency towards over-investment. The most important aspect of this tendency lies in the bias imparted to the choice of production techniques in electricity generation: more capital-intensive techniques will appear in a more favourable light than is warranted by underlying economic factors. In particular, the relative advantages of nuclear and coal generating technologies will be biased in favour of the former.

The choice of technique could also be affected by the suboptimalities in coal pricing policy discussed in Section 3.1. To the extent that it is the coal prices actually paid by the CEGB, rather than the marginal opportunity cost of coal, that are used in investment appraisals, a further bias towards nuclear technologies will also be imparted. This is just one example of a production distortion arising from the taxation of intermediate outputs.

The ultimate decisions about the type of generating technology to be adopted and the level of new capacity to be installed will, of course, also depend on a range of other factors, not least among which are the environmental side-effects and safety features of the alternatives. The point is not that coal-fired generating stations are necessarily superior to nuclear power-stations, but that the current, nominal framework of control for the nationalized industries provides a poor basis for even tackling the issue of the choice of technique.

4. Privatization

Discussion of regulatory issues in the ESI would be incomplete without some reference to the possible new approaches that might be adopted in the event that the industry is privatized. The available policy options depend partly upon the way in which the industry is transferred to the private sector and, in particular, upon whether or not restructuring will take place. At the time of writing (early 1987) everything remains to be settled but, for the purposes of discussion, it will be assumed that the existing structure of electricity distribution will be retained and that electricity generation will, at least initially, continue to be highly concentrated (i.e. there will be a 'dominant' generating company—the successor to the CEGB—perhaps facing one or more smaller 'competitors').

The likely features of the resulting, privately owned ESI can be summarized as follows. Shareholders rather than government would be responsible for the monitoring and control of managements and, as a result, it can be expected that managers will come to place greater weight on profit goals. The privately owned companies would have direct access to capital markets and the cost of capital would therefore reflect market rates. To limit exploitation of market power, retail prices, at least to domestic customers, would be regulated by a body similar in form to the Office of Telecommunications (OFTEL) and the Office of Gas Supply (OFGAS). Further, because of the difficulties in effecting a swift transition to competition in electricity generation, immediate and full deregulation of the bulk market would be unlikely to prove attractive. The Area Boards would therefore face price regulation in both output and input markets.

4.1 Regulation of Prices: General Problems

The most widely studied method of controlling the electricity prices of privately owned utilities is American cost-of-service regulation. This is based upon an implicit bargain between society and the firms concerned, in which the latter agree to supply electricity and the former agrees to allow prices that ensure that investors are, *ex post*, fairly remunerated for the capital they provide. Since prices are set to yield a fair return on capital, the method is frequently referred to as rate-of-return regulation.

The problems associated with rate-of-return regulation are well known (see Yarrow (1986)). For example, it provides only weak incentives for cost efficiency since there is a tendency for regulators to allow firms to recover their actual costs, whether or not the utility is actually producing efficiently. Moreover, by allowing utilities a designated rate of return on capital, it creates some incentives for over-investment: firms can increase profits by increasing their capital inputs (the rate base) beyond the point required by cost minimization. Finally, there is always the danger that society will break its part of the bargain: once utilities have invested in production capacity, they are unable to withdraw capital from the industry in the event that prices are set below levels that would yield a reasonable rate of return. There is therefore a temptation for regulators to choose lower prices either to please consumer interest groups or, in the event of excess capacity, to improve short-run allocative efficiency.

In the UK, regulatory policy has sought to bypass similar problems in the telecommunications industry by predetermining real prices for an initial period of five years. This has been done via a formula that constrains the annual increase in the price of a basket of services to be at least 3 per cent less than the increase in the retail price index. A similar approach has been adopted for gas, although, in this case, the formula also contains a term dependent upon the cost of supplies to British Gas and consumer prices can therefore be influenced by the company's purchasing decisions. Since it is also likely to be used if and when the ESI is privatized, it is clearly relevant to examine the strengths and weaknesses of this formula-based approach.

4.2 Indexation and Rate-of-return Regulation

At bottom, the UK system is simply a means of indexing allowable prices between relatively infrequent regulatory reviews. The advantage of the method is that, between reviews, the regulated, privately owned firm has incentives to improve its internal efficiency since, with predetermined prices, any reductions in unit costs will directly and immediately increase its profits. This suggests that, via the initial indexation formula, it might be optimal to predetermine real prices once and for all and to dispense with regulatory reviews entirely. Unfortunately, however, this would risk the consequence of a growing divergence between prices and costs, which in turn would eventually lead to supply failures (if prices were too low) or to allocative inefficiencies associated with monopolistic pricing (if prices were too high). Hence the need for periodic price reviews.

While indexation can be expected to have largely benign implications for operating efficiency, its impact on investment decisions is more problematic. Particularly on the generating side of the ESI, the long gestation lags of projects and the lengths of the economic lives of the assets together imply that the profitability of investment will be a function of the outcomes of a whole series of price reviews, stretching out for decades into the future. Investment decisions are therefore likely to be much more influenced by the anticipated outcomes of the reviews than by the initial indexation formula.

Because of this fact, it can be argued that the traditional weaknesses of rate-of-return regulation will inevitably reappear. Thus, suppose it is expected that regulatory reviews will bring prices back into line with actual costs to ensure that investors are fairly rewarded for the capital that they have supplied. There would be a tendency towards over-investment and, for some time before the date of the review, the incentives for internal efficiency would be diminished (since there would be an incentive to delay feasible cost reductions so as to obtain a more favourable decision on allowable prices).

4.3 Policy Credibility and Under-investment

Although the possibility of the above outcome cannot entirely be ruled out, there are strong grounds for believing that, in the case of the ESI, the more significant difficulties associated with the UK approach would derive from the policy credibility problem. UK public policy has been notably silent about the principles that will be used to calculate allowable prices once the initial indexation periods in the telecommunications and gas industries have come to an end. In itself, this silence creates uncertainty as to future outcomes and thereby tends to raise the cost of capital. More importantly, by failing to attempt to establish any sort of long-run bargain between society and private investors, it opens up some potentially damaging consequences associated with strategic behaviour.

To illustrate, suppose that, at the periodic reviews, regulators, when setting prices, seek to maximize allocative efficiency. In particular, suppose that prices are set so as to reflect short-run marginal costs. On the surface this might appear to be a highly desirable way in which to proceed. In practice, however, it would have severely damaging consequences for investment. Knowing that future prices would be influenced by short-run marginal costs, a profit-seeking management would have incentives to restrict the supply of capital since, by acting in this way, it would raise the marginal cost associated with any given output level and hence increase the regulated price of its output. In effect, investment would become a strategic instrument by means of which the regulated firm could restrict output, raise prices, and achieve monopoly profits. Moreover, it is easy to show that, in such circumstances, not only would output be suboptimally low but also the costs of producing the final output would be suboptimally high: cost inefficiency, as well as allocative inefficiency, would be a characteristic of the resulting equilibrium.

It was, of course, precisely this type of potential supply failure in the capital market that rate-of-return regulation was designed to avoid but, in so doing, it may have replaced one regulatory problem (under-investment) with another (over-investment). It should be noted, however, that, in practice, the biases in the latter direction may not be so great as simple economic models may suggest. In the 1970s there was some reluctance on the part of electric utility regulators in the US to allow cost increases, induced principally by higher inflation and the rise in input prices following the oil price shock of 1973, to be wholly passed through to customers. The possibility that society may not deliver on its side of the regulatory bargain is therefore a real one (even in an economy such as the US where there is a strong tradition of protecting private property rights) and the resulting (upward) effect on the cost of capital serves to offset the bias towards over-investment resulting from rate-of-return regulation.

There is therefore a danger that, by placing excessive stress on the known weaknesses of rate-of-return regulation, simplistic approaches to the regulation of privately owned electricity companies could lead to supply failures in the capital market. Put in its starkest terms, private investors may not be greatly attracted by the proposition that they should sink large sums of money into the building of facilities that will take many years to complete, will have a pay-back period measured in decades, and will have cash flows that are heavily dependent (a) upon the prices allowed by regulators who, beyond the initial indexation period, have an unclear brief and (b) upon the policy decisions of unknown future governments, including administrations likely not to be favourably predisposed towards private monopolies. It is quite possible, therefore, that private ownership could introduce a bias against capital-intensive projects in general, and against nuclear power generation in particular.

4.4 Government Revenues and the Coal Pricing Problem

In Section 3 it was argued that electricity prices have been distorted by suboptimal methods of raising revenue for the exchequer from the industry and of protecting the domestic coal industry. Privatization *per se* would do little to correct these deficiencies since they arise from aspects of regulatory policy, not from the structure of ownership of the industry.

For example, there would be a temptation for government to build relatively high electricity prices into the price control formula so as to be able to realize higher proceeds from the flotations of the public corporations. Given the preoccupation of fiscal policy with short-term financial flows, flotation would, if anything, intensify the pressures for higher prices, since the immediate effect of, say, a five year change in prices of 1 per cent on the public sector borrowing requirement would be much greater: government could immediately capitalize the incremental revenue flows by obtaining greater sales proceeds from the share issue.

Similarly, if the gas precedent was followed and the industry was allowed to pass increases in its coal costs through to final consumers, there would only be a limited incentive for a privately owned generating company to seek reductions in the price of its major input. On the other hand, if coal costs were excluded from the price control formula, any reductions in coal prices would not, at least until the first regulatory review, be reflected in final electricity prices and many of the distortions associated with an implicit tax on the intermediate product would remain.

In effect, establishment of a formula for regulating electricity prices in a privately owned industry requires the formulation, either explicitly or implicitly, of medium-term policies for taxation, coal pricing, and the electricity/gas price differential. Privatization does not remove the necessity for decisions on such matters, nor does it guarantee that existing deficiencies will be made good.

4.5 Ways Forward

The microeconomic public policy goals for a privatized ESI are, in principle, no different from those appropriate to the existing public corporations, namely short-run and long-run allocative efficiency coupled with production at least cost. It remains to consider methods by which, within the assumptions of this section (i.e. privatization of the public corporations in their current forms), regulatory effectiveness can be increased. Two policy instruments will be examined: liberalization and regulatory yardsticks (see Shleifer (1985)). The first of these is more relevant to the generating side of the industry and the second is more relevant to the services provided by the Area Boards.

Entry Again

The burden of the argument in Section 4.3 above was that, in the ESI, the implementation of a system of price regulation along the lines developed for the telecommunications and gas industries would introduce the possibility of significant supply failures with respect to the provision of capital to the industry. The threat of potential competition, however, goes some way towards mitigating this type of effect.

Suppose, for example, that a privately owned CEGB deliberately restricted investment so as to force up prices. Just as over-investment serves as a strategic barrier to entry (Dixit (1980)), under-investment would then open the door to new firms wishing to move into the industry. Provided there were no other major entry barriers, the existence of this threat, with its attendant implications for the prices, market share, and (ultimately) profit of the dominant producer, would provide some constraint on the pay-offs from restrictive policies (Hammond, Helm, and Thompson (Chapter 8, this volume)).

Thus, the promotion of conditions that ensure fair competition in electricity generation should be a central concern of regulatory policy for a privatized industry and priority should be given to strengthening the Energy Act. At this point it is worth stressing that increasing the scope for competitive forces and strengthening regulation of the industry are complementary policies. In the absence of regulation, the market power of a dominant generating company would enable it to engage in strategic behaviour to deter new entry and weaken the competitive threats. In the absence of potential competition, the CEGB would be more able to manipulate the regulatory regime to its advantage.

As the privatization of the British Gas Corporation shows, perhaps the major weakness of the current UK policy approach has been a failure to exploit fully the interactions between competition and regulation. The consequence of this failure is that the resulting policy regimes are likely to be less effective than they might otherwise have been in attaining the stated microeconomic objectives. It is therefore to be hoped that, should the ESI be privatized, there will be some attempt to improve upon what has gone before.

Yardstick Rate-of-return Regulation

In comparison with electricity generation, there is much less opportunity to enhance the role of product market competition in the distribution of electricity. The Area Boards *can* be regarded as regional natural monopolies and, given the levels of their sunk costs, there is little prospect that potential competition can provide significant incentives for improvements in allocative and internal efficiency. Hence, a much greater part of the policy burden must necessarily be borne by price regulation.

Fortunately, the structure of the distribution side of the ESI is conducive to the development of effective procedures for regulating prices. Recall that the main problems surrounding the regulation of a single monopolist are:

(a) the use, by the firm, of investment as a strategic instrument to influence the price decisions of regulators, leading either to over- or under-capitalization depending upon the criteria used to set allowable prices; and
(b) the difficulty of combining incentives for cost efficiency with the requirement that, to prevent monopolistic pricing or supply failures, prices do not move unduly out of line with costs.

Both of these problems can, however, be eased by making use of the opportunities opened up by the fact that there are twelve Area Boards, each pursuing economically similar activities.

Consider, for example, the implicit regulatory contract in which the totality of investors in privatized Area Boards is guaranteed a fair rate of return on capital, but in which the distribution of returns as between the groups of investors associated with particular Boards is made dependent upon the relative performance of the companies. Thus, if, at the time of a regulatory review, a given Board has achieved a higher rate of return on capital than the average for the twelve companies, it could be allowed to set prices in the subsequent period at levels that, in unchanged cost conditions, are expected to sustain above-average profitability. At the same time, the collection of allowable prices for the twelve companies could be such as to yield no more than a fair rate of return on the total capital employed in electricity distribution. Put simply, the *relative* prices of the distribution companies could be determined by the past relative performance of the companies, while the *average* level of retail electricity prices is determined by the fair-rate-of-return criterion.

It should be clear that, compared with single-firm rate-of-return regulation, this approach, known as yardstick regulation, has two potential advantages:

(a) the incentives for strategic over-investment are greatly diminished since each company is able to capture only a fraction of the profit gains that its behaviour would generate for the industry as a whole; and
(b) since better-than-average performance with respect to cost efficiency in one period would be rewarded with relatively favourable prices in the next, it is possible to introduce quite strong incentives for improvements in internal efficiency.

It should also be stressed that, unlike the regulatory systems for telecommunications and gas, the cost-efficiency incentives can be maintained even if the interval between pricing reviews is shortened. In effect, the regulatory system can re-create important aspects of the incentive structure of a hypothetical, workably competitive market in electricity distribution, combining pressures for cost reduction with prices that, over time, closely track movements in costs.

While yardstick regulation does not provide a simple solution to the policy credibility problem, it does partially alleviate some of the difficulties in this area. Thus, if credibility tends to lead to a bias towards under-investment, there will be some offsetting pressure in the opposite direction arising from the link between average prices and the industry-wide fair rate of return. It is also possible to lay down at the outset the details of the methods that will be used to fix allowable prices during the periodic regulatory reviews, thus reducing one type of uncertainty associated with the reviews. Finally, to the extent that it is expected that regulators will, in addition to rates of return, take account of short-run marginal costs when fixing allowable prices, the incentives for strategic under-investment will be reduced by the inability of individual companies to capture the full profit gains from their actions.

The most common objection to yardstick regulation is that the companies involved are never identical: each inevitably has idiosyncratic features that lead its own costs to be influenced by factors peculiar to itself. Thus, in comparing relative performance, it is impossible to distinguish between movements in the relevant indicator(s) (which, in the scheme discussed above, would be the rate of return on capital) that are caused by differences in efficiency and those that are caused by differences in extraneous, non-controllable factors.

The objection is, however, misguided. The information problem is common to virtually all incentive problems, and the fact that perfect measurement is impossible does not in any way imply that regulatory systems should not attempt to make the best possible use of the information that *is* available. Yardstick regulation, by utilizing the fact that the performance of UK electricity distribution companies *is* influenced by a number of common factors, would be a step in the right direction.

5. Conclusions

The traditional microeconomic objectives of promoting allocative and internal efficiency in the ESI remain valid, but the major policy failure has been an inability to establish regulatory structures that provide incentives for their attainment and that prevent them from being totally undermined by managerial goals and/or political expediency. Among the most pressing problems are: the allocative distortions caused by the combination of excessive coal prices and stringent financial constraints; ineffective monitoring (i.e. metering and rewarding) of managerial performance; an unduly low required rate of return; and the continued existence of avoidable barriers to entry into electricity generation.

While it is possible to deal separately with each of these problems, policy effectiveness will be improved if the interdependencies amongst them are recognized and if, therefore, a broad range of corrective measures are implemented. For example, simply raising the ESI's required rate of return may have little effect on investment decisions if methods of monitoring managerial performance are not altered, since it would continue to be fairly easy to revise demand and cost projections so as to obtain a more optimistic projected return. Similarly, promoting fair competition between the public and private sectors will be more difficult if both the cost of capital and the price of coal to the CEGB remain distorted.

Privatization is not a means of avoiding the difficult regulatory problems facing policymakers. Thus, although privatization might sharpen managerial incentives, there is a danger that one result could be more aggressive anti-competitive behaviour, including stronger actions to deter new entry. Nor does the fact that privately owned companies have direct access to the capital market imply that the resulting investment decisions will be efficient: as has been shown, there are grounds for believing that the investment level of a regulated private monopolist will be suboptimal. Hence, whatever the future structure of the ESI, regulatory policy will continue to be of central importance for the industry's economic performance.

11 The Role of Public Service Commissions in Facilitating the Development of Combined Heat and Power Generation in the US*

ALEX HENNEY AND DAVID THOMPSON

1. Introduction

Fundamental to the success of combined heat and power schemes is their ability to obtain fair terms of trade from the electricity supply industry (ESI) for the power generated. This article examines the contrasting outcomes of the UK Energy Act of 1983 and the US Public Utilities Regulatory Policies Act (PURPA) of 1978. In particular, it examines the role that leading state public service commissions, such as those in Texas, California, New York, and Massachusetts, have played in facilitating independent generation. This topic is but one aspect of competitive generation, which is currently debated in the US (e.g. see Joskow and Schmalensee (1983)).

2. The Impact of the Energy Act and of PURPA

The Energy Act 1983 was modelled upon PURPA which (*inter alia*) deregulated sales of power to utilities by independent generators who were either small power producers or who generated electricity as a joint product with heat (termed cogenerators). Both Acts require the public utilities to purchase power from independent generators at a utility's avoided costs, but the methods (and the spirit) with which they have been implemented have resulted in different outcomes.

In Britain there has been a negligible response to the Energy Act. The number of private generators has grown from 59 in 1983–4 to 78 in 1985–6. According to the Electricity Council, 'the increase in numbers has mainly been of smaller units of up to 5000kW'. This insignificant growth in output is regarded as 'an

---

* A version of this article was published as a memorandum of evidence in Energy Committee (1986c).

Alex Henney is a former chairman of the London Electricity Consumers Council and is an energy consultant. David Thompson was formerly a Programme Director at the Institute for Fiscal Studies.

The authors are grateful to Dieter Helm for helpful comments on an earlier draft of this article. Financial support for the research upon which this article is based came from the Economic and Social Research Council and from the Institute for Policy Research which supported a study tour in the US by Alex Henney.

Table 11.1. Private generation purchased by Boards (gigawatt-hours)

	1982	1984
By CEGB from Atomic Energy Authority	3,411	3,849
By CEGB from other producers (mainly National Coal Board)	294	160
By Area Boards in England and Wales	524	387
	4,229	4,396

Source: Tolley and Budden (1985).

Table 11.2. Sales by private generators to Area Boards

	Number	GWh purchased	Excluding largest 3 producers	
			GWh purchased	Average price (p/kWh)
1981–2	45	459	46.4	2.31
1982–3	45	864	55.5	2.18
1983–4	42	1,097	55.4	2.84
1984–5	52	343	94.4	3.46

encouraging sign' in which 'the Government will continue to take a close interest'.

As can be seen from Table 11.1, most privately generated power in fact comes from the Atomic Energy Authority. Whilst there has indeed been a growth in the power supplied by smaller private generators (see Table 11.2), this supply is still negligible when compared with public generation (amounting to less than 0.1 per cent of total supply in 1984–5).

In the US there has been a substantial growth in independent generation in response to PURPA. The rules that the Federal Energy Regulatory Commission (FERC) drew up to implement the legislation aimed to promote independent generation. A recent review for the US Department of Energy comments that the Act was 'viewed by Congress as a means of reducing the monopsony power of the utility and providing the independent generator with an assured market for its output on reasonable terms and conditions' (Pfeffer, Lindsay, and Associates (1986)). The Electric Power Research Institute estimates that between 1980, when the FERC promulgated its code governing independent generation, and mid-1985 applications for schemes totalling 16,000 MW (megawatts) had been filed with FERC (although not all would be installed). One estimate now puts the peak capacity of cogenerators in the US at 5 per cent of utilities' peak capacity.

By the end of 1985, 3,000 MW of firm capacity was on line in Texas and another 1,200 MW was contracted to be on line by the end of 1986 (Kepner,

King, and Edmunds (1985)). The President of the Houston Lighting Power Company has observed that the Company is the largest purchaser of cogenerated power in the nation today—15 per cent of its power is cogenerated. In California 607 projects were on line with a capacity of 3,034 MW as of 31 March 1986, and offers for another 652 projects with a capacity of 12,958 MW have been accepted (Public Utilities Commission of the State of California (1986)). While not all of the schemes will come on stream—the drop-out rate may be a half—the amount of independent generation is clearly substantial in a state with a maximum demand of 40,000 MW. The Californian Commission now refers to the 'cogeneration industry' (Public Utilities Commission of the State of California (1985)), and in the annual report and accounts of the Pacific Gas and Electric Company, one of the largest US utilities, the chairman reported that

more than 40 of our 100 largest customers now generate their own electricity on-site or plan to do so. These industrial customers use a process called cogeneration by which they generate electricity as part of another process. Some of these large users of energy supply all of their electricity needs and buy only back up power from PG & E.

Some of the development of cogenerated power has no doubt been assisted by tax incentives, and indeed a study for the US Department of Energy (Pfeffer, Lindsay, and Associates (1986)) cited a survey estimating that up to a quarter of schemes would not have been installed but for these incentives. Also purchase tariffs mandated by some states and commissions that are above avoided costs have promoted excessive levels of independent generation. For example, the Public Utilities Commission of California mandated rates for long-term contracts in September 1983, and suspended them in October 1984 because, with declining oil and gas prices, the basis of its long-term payments was too generous. Nonetheless a substantial amount of cogeneration has been genuinely economic, particularly in Texas.

3. Has the UK Energy Act Failed?

The different outcomes thus reflect in part the availability of tax incentives in the US, but a factor of central importance has been the difference in the ways the two Acts have been interpreted and implemented. There are two features of the UK Energy Act that are particularly important in explaining its failure to induce competition (see Hammond, Helm, and Thompson (Chapter 8, this volume)). First, the ESI—the incumbent monopoly—has been made responsible for the detailed determination of the terms of trade under the Act, including the computation of the avoided costs upon which the tariffs at which power is purchased from independent generators are based. Yet the calculation of avoided costs is a complex technical issue with scope for differences of opinion and judgement. To minimize the risk of bias, avoided costs should be determined in an independent manner to ensure that the issues are thoroughly scrutinized and determined equitably. The ESI, however, is not required to provide detailed public justification of its purchase tariffs. In practice, as the Energy Act was in the process of being introduced, the ESI restructured the bulk supply tariff (BST)—the internal transfer price at which power is supplied to Area Electricity Boards—in such a way as to reduce the elements in the BST that are designated as avoidable costs to the Area Board. This had the consequence of reducing the level of purchase tariffs offered to independent generators. However, the ESI was not required to provide evidence to justify this action. For this reason, it is not possible to judge whether the restructuring was efficient in economic terms. Nor is it possible, more generally, to determine whether the purchase tariffs are set at economically efficient levels. The purchase tariffs currently being offered raise a number of questions:

(a) There is no allowance for the higher avoided cost associated with supply at lower voltage levels, which avoid transmission losses, and hence should be reflected in a higher purchase tariff.
(b) There is no provision for geographically differentiated tariffs reflecting the different value of electricity at different points on the grid.
(c) The night rate purchase tariff, which is based on the BST night rate, is 10 to 20 per cent below the apparent marginal cost of energy. While this practice is justified by assertions about the economics of running part-loaded large machines at night, the reality may be that the BST has been lowered to improve the competitiveness of electric storage heating run off the Economy 7 tariff compared with gas.
(d) The industry has recently announced that it is introducing a reduced tariff for energy-intensive industrial users, some 170–200 customers accounting for one-third of industrial electricity usage, which will reduce their tariff by an average of 7 per cent, although some sites could see savings of up to 15 per cent. This tariff will reduce the incentive such users have to develop their own supplies and the competitiveness of any independent suppliers who might wish to sell to the energy-intensive users.

It is clearly fundamentally unsatisfactory for a private generator to have its prices determined by its main competitor. There is a provision for arbitration, but only on individual cases, not on the published purchase tariffs. It could be a lengthy and expensive procedure for the complainant. The ESI holds much of the relevant information and possesses significant financial strength; both factors are likely to deter potential complainants from pursuing a case through arbitration.

The second feature of the Act which militates against independent generators is that it makes no provision for long-term contracts at defined prices. Such contracts will be required by some, if not many, private generators to enable them to raise investment funding. In the opinion of the ESI, the terms of the Act preclude such contracts with Area Boards. It is suggested that if an Area Board signed such a contract, and if after a time the avoided cost elements in the BST were below the purchase price, then the Area Board would be acting *ultra vires*; while if the avoided cost elements were much higher than the purchase tariffs, the generator could claim that he was entitled to more. Prospective generators can, however, get a firm price contract from the Central Electricity Generating Board. However, such a contract is not subject to the provisions of Section 7 of the Energy Act, and the purchase prices that have been offered have generally been less than the private purchase tariffs offered by Area Boards under the Act. In contrast, the similar provision in PURPA prohibiting payment exceeding avoided cost has not proved to be an obstacle to long-term contracts in the US.

In summary, the Energy Act is flawed because it leaves the incumbent ESI with effective control of price and entry conditions. In practice this has led, as we have seen, to the entry conditions being set in a way which may discourage entry.

4. The Success of PURPA

In contrast, in the US, the roles played by some of the leading regulatory commissions (we cite New York, Massachusetts, Texas, and California) have lessons which might be adopted in the UK if there is to be a serious intent to develop independent generation. The commissions have adopted a positive approach to the development of independent generation. For example, in California the Public Utilities Commission has adopted a very supportive approach to independent generation which has included penalizing the Pacific Gas and Electric Company financially for failure to pursue cogeneration opportunities aggressively. The Commissions' identified objective is to act in the general public interest rather than to protect the interests of particular incumbent utilities. To facilitate economic independent generation, the commissions have:

(a) acted as independent arbitrators of the terms of trade;
(b) established the principle of non-discrimination between incumbent utilities and independent generators in determining the terms of trade;
(c) required standard long-term contracts;
(d) promoted competition in wholesale power generation.

We now consider each of these four factors in turn.

Independent Arbitration

All of the interested parties—utilities, cogenerators, utility customers—have a right to set out their views in open hearings subject to cross-examination, and all of the key data (including the utilities' system models) are on the table. The hearings by some commissions have proceeded in a series of steps, with scope for deliberation and further exploration of issues. For example, the New York Commission has undertaken several major hearings on the issues, commencing with the publication of a preliminary set of rules issued in January 1981. Meanwhile it started an investigation into the tariffs, charges, and rules for supply by and to private generators in Consolidated Edison's franchise territory, which it completed in May 1982. In its decision (State of New York Public Service Commission (1982)) it resolved a number of the issues, but left for further consideration the principle of calculating long-run avoided energy and capacity costs. This was undertaken in a hearing for all of the 7 major private utilities in the state. The hearing commenced on 28 November 1984, and adopted the following procedure:

(a) In January 1985 the Commission staff prepared a paper setting out their proposed methodology, and then spent 6 months in conjunction with the interested parties agreeing to the forecasting assumptions and developing a series of avoided cost estimates.
(b) The Commission staff published an analytical paper on 10 September 1985 incorporating their recommended avoided costs (State of New York Public Service Commission (1985)).
(c) The Commission invited all parties to comment on the paper in open hearing, and it produced its ruling on 27 March 1986. It then called for further work to develop standard forms of contracts.

Other commissions have followed similar procedures. The proceedings are based on facts rather than opinions. Furthermore, facts have to be quantified where possible, and nowhere was this more evident than in the analyses described above of avoided costs undertaken by New York Public Services Commission.

While different commissions have used different analytic approaches, and thus arrived at somewhat different answers, their procedures had the major benefit of allowing all to have their say and to see how decisions were arrived at.

Non-discrimination

The commissions have aimed to define the terms of trade in a non-discriminatory manner. For example, FERC has ruled that cogenerators should pay interconnection costs 'on a non-discriminatory basis with respect to other customers with similar load characteristics' and (contrary to the wishes of some utilities) 'back up charges shall not be based on the assumption (unless supported by factual data) that forced outages or other reductions in electric output by all qualifying facilities on an electric utility's system will occur simultaneously or during the system peak, or both . . . .'.

Both the New York and Massachusetts commissions have treated independent generators as symmetrically as possible to normal large customers, essentially regarding them as 'negative loads'. Thus if it is considered appropriate to charge customers for some aspect of service (for example, a capacity charge at certain times), then cogenerators should receive complementary credits.

Standard Long-term Contracts

All four commissions recognize the need for standard forms of contract and long-term contracts. Standard forms of contract are needed to reduce the time and legal costs of negotiating with utilities. This reduces the cost of entry and eliminates the opportunity for a utility to obstruct entry by requiring lengthy legal negotiations.

Long-term contracts are necessary for many independent generators to enable them to raise both short-term construction funding and long-term finance. The state of New York recognized this issue and provided in its Public Services Law that utilities must enter into long-term contracts to purchase and transmit power. A recent hearing in Massachusetts was initiated by the State Office of Energy Resources because the existing rules set by the Department of Public Utilities did not include specific provision for standard offer long-term contracts at fixed or semi-fixed rates.

The contracts are all authorized by the commissions. A feature of them is that they are diverse to suit the circumstances of different cogenerators and generation situations. Thus there are short-term contracts, which pay for energy but not for capacity, and long-term contracts with various payment terms. Some contracts provide for fixed prices over long periods, or for prices indexed to a fuel index or to a utility's actual cost, or for a minimum floor price with an adjustment dependent upon either a utility's short-run marginal cost or a fuel index. Some contracts provide for front-end-loaded payments. Further details are provided in Henney and Thompson (1986).

Competitive Bidding for Power Supply

Some commissions, including Texas and Maine, require cogenerators to bid competitively against incumbent utilities for contracts to supply power. In Texas, 'by mid-1984 it became apparent that the utility would be able to obtain real price competition among the cogenerators . . . the company formally started a negotiated bidding process for additional quantities of power . . .', and one contract was signed at 12 per cent below the utility's avoided capacity costs (Kepner, King, and Edmunds (1985)). Massachusetts is proposing to develop this approach. It intends to require the utilities to invite competitive bids for blocks of capacity, and to replace their capacity with cogenerated power if it is cheaper. In particular, if a utility wants to build a new power-station, it will have to estimate the cost at which it can supply additional power and, before receiving authorization to build, it will have to offer the capacity to competitive tender. This competitive tendering reduces the reliance that must be placed on forecasts of avoided costs by either a utility or a commission, and provides the obvious disciplines on potential developers against bidding either too high or too low. The procedure that the Department intends to follow is described in Henney and Thompson (1986).

5. Conclusion

The success of combined heat and power schemes depends crucially upon obtaining fair terms of trade from the ESI for power that is generated. The UK Energy Act has failed to ensure this. A recent report by the Select Committee on Energy (Energy Committee (1986c)) observed that 'the evidence appears to indicate that an Act intended to encourage the development of combined heat and power by the Electricity Supply Industry has actually had the reverse effect'.

We believe that important lessons can be learned from the successful implementation of the similar PURPA legislation in the US. Crucial to this success has been the part played by the state regulatory commissions both in acting as independent arbiters in the determination of the terms of trade and in ensuring that cogenerators are provided with long-term contracts for the supply of power.

How can this US success be imported into the UK? Eventually we would like to see the Electricity Council converted into an independent body, set up along the lines of a US-style commission, to regulate the industry and to deal with terms of trade for private generation. In the interim, we recommend that the Electricity Council sets up a panel comprising an independent chairman from outside the industry and two members of the Council or its staff, with the objectives of:

(a) conducting an open hearing (at which all relevant material would be available) on the appropriate terms of trade of private power generation. The hearing would lead to recommended terms of trade, including standard offer contracts;
(b) seeking a judicial review of whether Area Boards can or cannot sign long-term contracts. If they can, then the panel should devise suitable forms of long-term contract. If not, then it should invite Parliament to amend Section 7 of the Energy Act to explicitly require the Area Boards to enter into such contracts;
(c) adjudicating on complaints against Boards in setting terms of trade.

12 Risk Analysis and Optional Investment in the Electricity Supply Industry*

IAN JONES

1. Introduction

New electricity-generating plant may be ordered (a) to meet forecast growth in the demand for electricity; (b) at any given level of demand, to replace plant that is no longer physically operable for engineering and/or safety reasons; or (c) to reduce the costs of meeting a given level of demand.

Given the statutory obligation of many electricity supply undertakings to provide security of supply to a certain level, investment for reasons (a) and (b) is essential, whilst potentially cost-reducing investment is optional.

For an organization such as the electricity supply industry seeking to minimize the discounted costs of meeting expected future demand for electricity, a necessary condition for undertaking optional investment is that the present value of the operating or variable costs of those existing plants that would be displaced by the new plant exceeds the present value of the total costs, including capital charges, of new plant. An equivalent formulation of this necessary condition is that the present value of the operating cost savings yielded by the new plant exceeds its capital cost, or that the net present cost of the investment project is negative. Under certain circumstances, for example, if the expected capital costs of new plant and the operating savings it is expected to yield do not vary with the passage of time, then existing capacity should be replaced as soon as the net present cost of doing so is negative. However, the expected net present cost of electricity-generating projects depends importantly upon the expected operating efficiency of the plant, and hence the savings of primary fuel burnt at less thermally efficient capacity which would be displaced by the new plant and which is likely to be 'calendar time dependent' (Marglin (1963)). A negative net present cost is then no longer a sufficient condition for replacement, although it remains a necessary condition. The optimal point of time at which existing capacity should be replaced is when the estimated net present cost of doing so is minimized. The investment appraisal problem may therefore be characterized as one of investigating the choice between a series of mutually exclusive projects; that is to say, the replacement of a given unit or units of existing capacity at a number of possible alternative dates in the future.

* This article was first published as 'The application of risk analysis to the appraisal of optional investment in the electricity supply industry' in *Applied Economics*, 1986, by Chapman and Hall. Ian Jones is a Senior Consultant at National Economic Research Associates (NERA).

Operationally, an investigation of this problem in the context of the electricity supply industry involves two steps. First, an assessment must be made of whether the expected net present cost of new capacity for commissioning at the earliest possible date in the future will be negative. If so, the second step involves an examination of whether the net present cost will become more or less negative if the commissioning of the capacity is deferred.

Using evidence in recent reports by the Monopolies and Mergers Commission (MMC) (1981) and the House of Commons Select Committee on Energy (SCE) (Energy Committee (1981)) and in material submitted to the Sizewell B Public Inquiry by the Central Electricity Generating Board (CEGB) and other parties, this paper illustrates the use of risk analysis in assessing how the net present cost of the Sizewell B project might vary if its commissioning date was deferred. The paper is therefore complementary to earlier work by Evans (1984), who has applied probabilistic decision analysis to estimating the benefits of the Sizewell B project if constructed at a specified date in the future.[1] At the same time it examines an issue, that of appraising optional investment, neglected in the existing literature on the economic appraisal of electricity-generating projects which has tended to concentrate on the choice of what was characterized earlier as 'essential' investment (for example, Bates and Fraser (1974)).

The remainder of the paper is organized as follows. Section 2 reviews economic appraisal procedures in the electricity supply industry with particular reference to current CEGB practice. It discusses the Board's appraisal of optional investment projects and the material submitted on the issue of deferment to the Sizewell B Inquiry. Section 3 presents a risk analysis of the deferment decision.

2. Economic Appraisal Procedures in the Electricity Supply Industry

As with many major electricity supply organizations, the CEGB's plant-ordering decisions are informed by the application of investment planning models of the kind described, for example, by Turvey and Anderson (1977). A two-stage planning process is involved. In the first stage, a 'global' programming model is used to develop a background plan representing in very broad terms an optimum pattern of system development for perhaps forty to fifty years ahead, given the expected capital costs, operating performance, and economic lifetime of alternative types of generating plant, and the forecast

---

[1] Evans specified a set of triangular subjective probability distributions for a number of key determining variables. A Latin hypercube sampling procedure was used to derive randomized values of the parameters, which were then applied in a series of runs of a system planning model. The analysis also treated the Sizewell decision as an enabling decision which would (or would not) permit the construction of a series of nuclear generating units. Benefits were therefore estimated not simply in respect of the project in its own right but of a programme of nuclear construction.

prices of fuel inputs.[2] The background plan is often sensitive to assumptions regarding the maximum rate at which certain types of plant may be commissioned, and planning studies may specify several different background plans corresponding to differently specified planning constraints.

What are sometimes referred to as marginal or incremental studies (see Turvey and Anderson (1977)) are then used to 'fine-tune' the background plan and to prepare detailed appraisals of individual investment projects. An important function of marginal studies is to represent the characteristics of specific choice situations in greater detail than is necessary (or indeed feasible) in global studies. For example, in the present context, an important issue is whether and to what extent technical progress is expected to affect plant performance; in global studies, the technical characteristics of alternative types of plant are usually assumed to be invariant with respect to the plant commissioning date (Evans (1981)).

Net Effective Costs and Net Avoidable Costs

The nature of the calculations undertaken in marginal studies is illustrated in Figure 12.1. Figure 12.1a shows forecast load duration curves in some future year, i, D^0 and D^1. If Q^0_{mi} is the system maximum (instantaneous) demand (in units of, say, gigawatts (GW)), then total system annual energy demand in units of GW hours is equal to

$$\int_0^{Q^0_{mi}} D(Q_i) dQ_i. \tag{1}$$

A load increment is represented as

$$\int_0^{Q^1_{mi}} D^1(Q_i) dQ_i - \int_0^{Q^0_{mi}} D^0(Q_i) dQ_i \tag{2}$$

where D^1 is the load duration curve with and D^0 the load duration curve without the increment of demand.

It is assumed for simplicity in what follows that there are three types of plant available to meet demand: nuclear (n), coal-fired thermal (t), and gas turbine (g), with running costs per hour in year i of R_{ni}, R_{ti}, and R_{gi} respectively. As shown in Figure 12.1c, $R_{ni} < R_{ti} < R_{gi}$. We also assume that the planning background plant mix (Q^0_{ni}, Q^0_{ti}, and Q^0_{gi}) is sufficient to meet the load increment at the system maximum demand ($Q^1_{mi} - Q^0_{mi}$).[3]

Figure 12.1b shows the integrated load curves corresponding to the demand

---

[2] For an account of such a model, see Evans (1981).
[3] The CEGB's own incremental studies do not, in fact, make this assumption. See Bates and Fraser (1974).

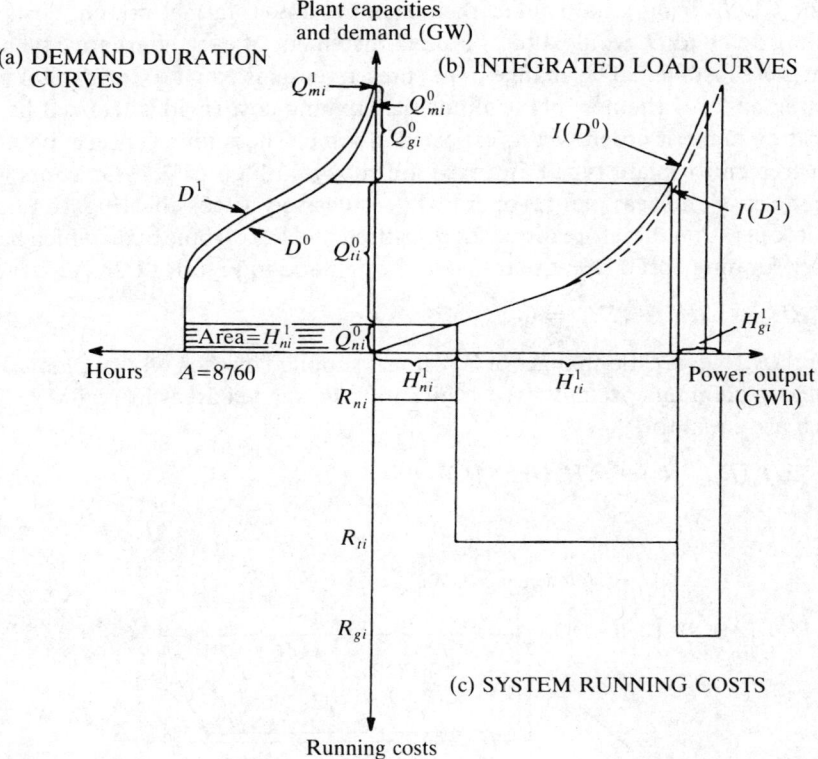

Figure 12.1 The relationship between system demand duration curves, plant outputs, and system running costs

duration curves D^1 and D^0 (Turvey and Anderson (1977)). H^1_{ni}, H^1_{ti}, and H^1_{gi} are in units of GW hours and are equal, respectively, to

$$H^1_{ni} = \int_0^{Q^0_{ni}} D^1(Q_i)dQ_i \tag{3a}$$

$$H^1_{ti} = \int_{Q^0_{ni}}^{(Q^0_{ni}+Q^0_{ti})} D^1(Q_i)dQ_i \tag{3b}$$

$$H^1_{gi} = \int_{(Q^0_{ni}+Q^0_{ti})}^{Q^1_{mi}} D^1(Q_i)dQ_i. \tag{3c}$$

Finally, Figure 12.1c shows the total running costs incurred in meeting the demand D^1_i from the set of plants in the background plan (Q^0_{ni}, Q^0_{ti}, and Q^0_{gi}).

The CEGB's marginal studies then examine how total net present costs of meeting demand D_i^1 would alter if a small increment of each alternative type of plant were installed. The change in net present costs is equal to the capital and running costs of the new plant minus the running cost savings realized by its operation in merit order over its expected lifetime. The running cost savings for an increment of plant type n in year i are shown in Figure 12.2. Given its low running costs, nuclear plant is operated as intensively as possible (Figure 12.2a) and its operation therefore displaces coal-fired and gas turbine plant which have higher running costs. Because total energy demand in Figure 12.2b is given,

$$\Delta H_{ni} + \Delta H_{ti} + \Delta H_{gi} = 0 \tag{4}$$

where ΔH_{ni} etc. are the changes in GW hours supplied by each type of plant. The change in running costs involved is given by the shaded areas in Figure 12.2c, which are equal to

$$[\Delta H_{ni}(R_{ni} - R_{ti}) + \Delta H_{gi}(R_{gi} - R_{ti})]. \tag{5}$$

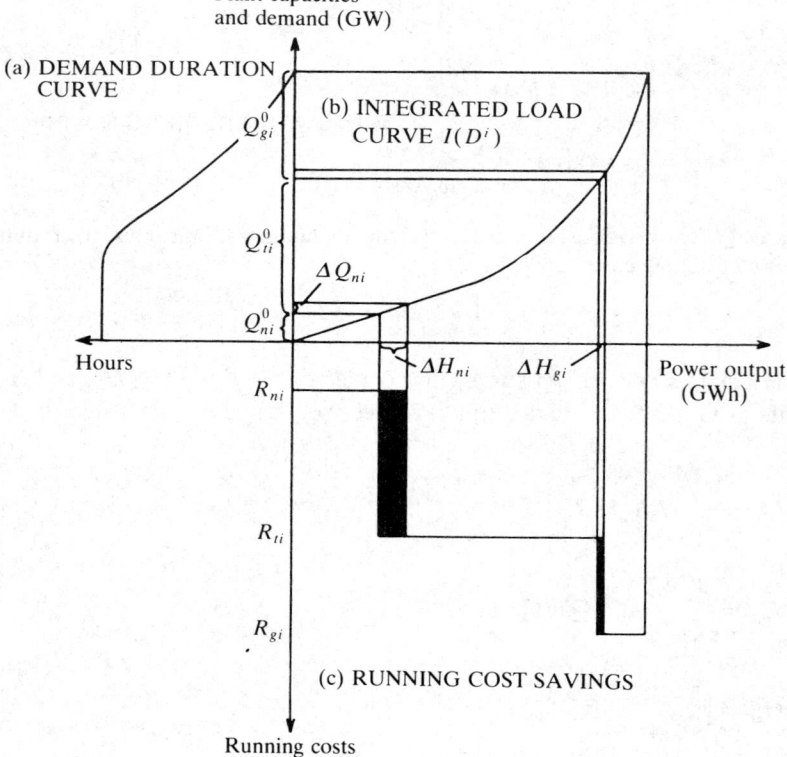

Figure 12.2 Running cost savings for an increment (ΔQ_{ni}) of nuclear plant on base load

Running cost savings are calculated by reference to the forecast demand duration curves and the background plant mix over the lifetime of the project.

Thus net present costs for plant type n are equal to

$$C_n + \sum_{i=0}^{i=k} [\Delta H_{ni}(R_{ni} - R_{ti}) + \Delta H_{gi}(R_{gi} - R_{ti})]\delta_i \tag{6}$$

where C_n = project capital costs;
δ_i = a discount factor;
k = expected operating lifetime.

Figure 12.2 can also be used to show how variation in the background plant mix affects the running cost savings and hence the net present cost of the project. Suppose at one extreme that the quantity of nuclear plant in the background plant mix was permanently constrained to the level given by Q_{ni}^o in Figure 12.1a. Then any further nuclear capacity would be operated as intensively as possible throughout its life and would realize correspondingly high running cost savings

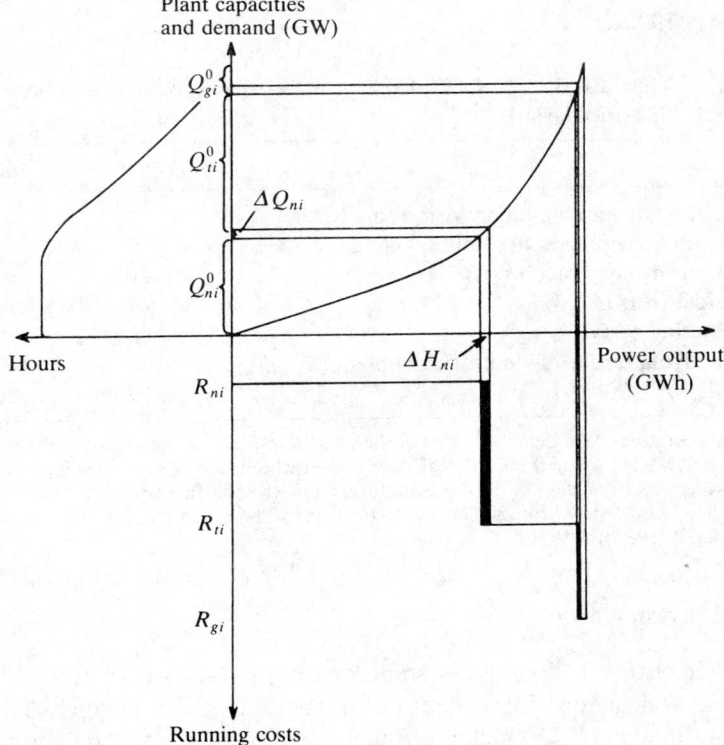

Figure 12.3 Running cost savings for an increment (ΔQ_{ni}) of nuclear plant not on base load

(Figure 12.2). If, on the other hand, a relatively high level of nuclear investment were assumed in specifying the background plant mix, then an increment of nuclear plant introduced early in the planning period would eventually be operated less intensively and the running cost savings would be correspondingly smaller (Figure 12.3).

The CEGB's economic appraisal documents express the resulting net present costs of investment projects in units of pounds per kilowatt (£/kW) per annum. These are known as the net effective costs (NECs). This is done in order to compare between projects with different output capacities and different expected economic lives. The use of equivalent annuity values strictly requires that each alternative investment opportunity consists of a sequence of identical projects evaluated to a common terminal date.[4] In practice, the potential degree of error introduced into the evaluation process by such an assumption is small given the durability of typical generating projects and the current rate (5 per cent real terms) of discount applied in CEGB appraisal.

Table 12.1 shows the main components of the NECs of pressurized water reactor (PWR) and coal-fired plants as currently estimated by the CEGB (1982b, Table 14 (revised)). The figures refer to a planned commissioning date in the early 1990s.

Table 12.1. The CEGB's estimates of the net effective cost of future stations (£/kW p.a., March 1982 price levels)

	PWR	Coal-fired
Capital charges at station and provision for transmission equipment, decommissioning plus interest during construction (C)	91	49
Inclusive fuel costs (F)	28	112
Other operating costs (O)[a]	12	8
Fuel saving from displacing less efficient plant (S)	160	143
NEC $= (C+F+O-S)$	-29	$+26$

[a]The 'other operating costs' item is the sum of the direct non-fuel operating costs of the new plants and what are referred to by the CEGB as marginal overhead costs (including, for example, an allocation of research and development expenditure) minus the non-fuel operating cost savings elsewhere on the system arising from the operation of the new plant in merit order.
Source: CEGB (1982b, Add. 6).

Optional Investment

Optional investment also allows some existing plant to be withdrawn from operation. The savings from plant decommissioning are not included in the CEGB's estimated NEC which is estimated in respect of an increment of load;

---

[4] The degree of approximation involved is discussed in Rees (1974).

instead, a separate estimate is made of the costs that would be avoided through retirement of marginal plant at about the date the new plant would be commissioned. As calculated by the CEGB, these net avoidable costs (NACs) consist mainly of the avoidable costs of operating and maintaining the plant, including repair and maintenance costs, the fuel cost savings realized by its continued operation (arising from the displacement of plant lower in the merit order), and an allocation of marginal system overhead costs. Like the NEC, the NAC is expressed in units of £/kW per annum. The CEGB currently expects the NAC of plants that would be candidates for decommissioning in the early to mid-1990s (the earliest possible commissioning date for Sizewell B) to be about £14/kW per annum (CEGB (1982b)).

Denoting $P(\text{NEC})$ as the present value of the costs and benefits covered by the NEC calculations, and $P(\text{NAC}_i)$ as the present value of the costs saved by plant decommissioning in year i (consequent on the optional investment), the total net present cost of an optional investment project for commissioning in year j, NPC_j, may be written as

$$\text{NPC}_j = P_j(\text{NEC}) - \sum_{i=j}^{i=j+n} P(\text{NAC}_i) \qquad (7)$$

where $(j + n)$ is the first year in which essential investment is required to be commissioned.

In the 1960s and 1970s CEGB investment appraisal procedures were primarily concerned with the choice of generating plant to meet requirements stemming from the growth of demand of electricity. In the circumstances, the preferred investment alternative would be shown by the type of plant with the lowest NEC. The activities of the Monopolies and Mergers Commission and the Select Committee on Energy, and now the Sizewell Inquiry, have required the CEGB to consider how it should evaluate the case for optional investment.

At the time of the MMC and SCE reports, the CEGB appeared to believe that the replacement of elderly plant would be economically justified as long as the NEC of the preferred investment alternative (shown by the NEC of PWR plant in Table 12.1) was less than the NAC associated with the maintenance of existing marginal capacity (MMC (1981, paras. 5.15 and 5.123); Energy Committee (1981, para. 64)). We may infer that a comparison of this kind, supplemented by the results of sensitivity tests on estimated NECs of investment projects, formed the basis for the CEGB's view reported in the MMC and SCE reports that a 1,500 MW (or more) per annum programme of investment in nuclear plant would be justified on purely economic grounds from the late 1980s onwards. Neither the MMC nor the SCE commented on the possible invalidity of a simple comparison of NECs and NACs as the basis for decisions on capacity replacement, although both bodies, the MMC in particular, made serious criticisms of the CEGB's investment appraisal procedures and concluded that in the light of their findings, the Board had failed to demonstrate a robust case for its preferred programme.[5]

The MMC Critique

In particular, the MMC criticized the CEGB's practice, in both internal documents and material supplied to external agencies such as the Department of Energy, of focusing on so-called basic estimates of NECs, which were not derived from central or most likely estimates of all the determining variables. In preparing these basic estimates, the Board distinguished between background variables, such as the prices of primary fuel inputs, and technical or plant-related variables, such as construction time and cost. The basis for this distinction, so the Board argued, was that whereas the future values of background variables could be treated as strictly exogenous with respect to any actions that it might take, the values of plant-related variables could be affected by the actions of the Board or its suppliers.

The basic NEC estimates incorporated forecast values of background variables which were intended to represent central estimates. Basic estimates of plant-related variables, on the other hand, were in 'the nature of targets which should be feasible to achieve under ideal circumstances—that is to say, if the design of plant is settled, and if there is committed management and a compliant labour force' (MMC (1981, para. 5.94)).

The CEGB sought to justify the use of basic NECs incorporating target values on two grounds. First, it suggested that there was no rational basis for the formation of a central or most likely estimate for the variables in question. The Board also suggested that if it were seen not to be 'planning for success' (i.e. not using target values in its appraisals) then there would be a danger of self-fulfilling prophecies. In respect of construction times, for example, the Board suggested that 'if you start putting longer [than target] periods in, longer periods will occur' (MMC (1981, para. 5.94)).

In the CEGB's appraisal documents, the basic estimates of NECs were supplemented by a range of sensitivity test data. These explored the effects of variations around the forecast mean values of the background variables, and of failure to achieve the target values of the plant-related variables.

The Commission made several detailed technical criticisms of the CEGB's analysis of background variables (notably primary fuel prices) and of its sensitivity test material. More fundamentally, the Commission also argued for a complete reorientation of the Board's approach to investment appraisal so that

---

[5] The programme was referred to in a statement delivered to the House of Commons by the Rt. Hon. David Howell, MP, then Secretary of State for Energy, on 18 December 1979. As so often with British government statements of this kind, bets were hedged. Whilst describing the 1,500 MW programme as a 'reasonable prospect against which the nuclear and power plant industries can plan', the Secretary of State added the qualification that 'the precise level of future ordering would depend on the development of electricity demand and the performance of the industry'. The text of the statement appears in Annexe A in Energy Committee (1981).

it focused upon outcomes derived from central estimates of all the relevant determining variables. Several points were made in support of this contention. First, the Commission argued that central estimates could be made in respect of plant-related variables, if the CEGB were prepared to devote the necessary time and resources to the task. In this context the Commission drew attention to techniques which would allow a more systematic examination of project risks within the basic structure of critical path analysis used by the CEGB in project planning. The Commission also noted that whilst the analysis of the Board's previous experience would play an important part in the formation of expectations about the most likely outcome, account should properly be taken of how management action might realistically lead to an improvement over past performance. However, central case estimates incorporating judgements of this kind should be clearly distinguished from *targets* 'which would represent an improvement on the most likely level of performance, in the context of motivating line managers and providing incentives to contractors and their workforces' (MMC (1981, para. 5.162)).

Second, the Commission noted that in seeking to justify a programme of nuclear investment on cost-saving grounds, the Board was obliged to compare the NEC of such plant either with the NAC of maintaining fossil-fired plant, or, in certain circumstances, with the cost of life extension (also of fossil-fired plant). Because the CEGB estimates of NAC and of life extension were based on central estimates of the relevant variables, it was important to develop equivalent central NEC estimates (MMC (1981, para. 5.164)).

Third, it was argued by the Commission that unconscious appraisal optimism might be introduced into investment appraisal if the starting-point for assessment of risk and uncertainty represented a highly favourable and unlikely combination of circumstances. This risk would be reduced if the starting-point for the presentation of investment appraisal results were the central estimate of project NEC: 'The tendency, even in the 1979-80 Development Review, to describe the basic NECs as "central estimates" is indicative of a risk which would be avoided if the estimate in question really were "central" in the normally understood sense' (MMC (1981, para. 5.165)).

In the light of its critique of the Board's methodology, the MMC asked the CEGB to re-estimate the NECs of nuclear and coal-fired plant for a commissioning date in the early 1990s using an alternative set of values for the plant-related variables from those used by the CEGB itself. These values fell some way short of the target levels used by the Board in computing basic NECs. Although the MMC did not make the claim itself, the reader could perhaps be forgiven for thinking that the values specified corresponded to what in the MMC's view were central or most likely outcomes (MMC (1981), Table 5.13). Given these revised assumptions, the NECs increased from $-£18/\text{kW}$ per annum to $+£18/\text{kW}$ per annum for advanced gas-cooled reactor (AGR) plant and from $+£22/\text{kW}$ to $+£25/\text{kW}$ per annum for coal-fired plant (MMC (1981), Table 5.14). The revised estimate of the nuclear NEC therefore exceeded the CEGB's then central estimate of NAC (of £10/kW per annum) by £8/kW per annum, and so not even the necessary condition for investment to replace existing capacity was satisfied.

The Sizewell Inquiry Material

The documents produced by the CEGB for the Sizewell Inquiry (CEGB (1982a, 1982b)) indicate that it has responded to some of the MMC's criticisms of its investment appraisal methodology. In particular, the CEGB states that its appraisals are based on parameter values for plant-related variables which are central estimates and not targets (CEGB (1982b)). Of more immediate interest, the Sizewell material indicates that the Board now recognizes that an estimated NEC less than the NAC is not a sufficient condition for investment in advance of when capacity is needed to meet demand (CEGB (1982a, para. 9.14); CEGB (1982b, paras. 20–3)).

The CEGB has examined the economics of the Sizewell project against a wide range of outcomes regarding the demand for electricity, the level of primary fuel prices, and the development of generating plant mix. It estimates that the NEC of the project for commissioning at the earliest possible date (in the early 1990s) will be negative even under the least favourable circumstances of low demand growth and fossil-fuel prices. Given future levels of electricity demand and fuel prices which fall towards the middle of the range of possible cases considered in the CEGB appraisal documents, the estimated NEC ranges from −£29/kW p.a., assuming a relatively rapid growth in the quantity of nuclear plant in the background plant mix, to −£69/kW p.a., if it is assumed that there was no further investment in nuclear plant after the Sizewell project had been completed (as shown in Figure 12.3).

On the CEGB's estimates, the necessary condition for commissioning new plant at the earliest possible date is therefore clearly satisfied. The supplementary condition, that NPC_j (as defined in equation 7 above) is less than NPC_{j+1}, is examined by the CEGB in three stages.

First, a comparison is made between P_j (NEC) and P_{j+1} (NEC) (as defined above). Because $P(NAC_j)$ is positive,

$$P_j(\text{NEC}) < P_{j+1}(\text{NEC}) \tag{8}$$

would be a sufficient economic justification for proceeding with the project. However, even if

$$P_j(\text{NEC}) > P_{j+1}(\text{NEC}), \tag{9}$$

immediate construction would be justified if

$$[P_j(\text{NEC}) - P_{j+1}(\text{NEC})] < P(\text{NAC}_j). \tag{10}$$

Estimation of the costs covered in the NAC calculations therefore forms the second step in the CEGB's marginal analysis of deferment; and the final step is to compare $P(NAC_j)$ and $[P_j(NEC) - P_{j+1}(NEC)]$ (if the latter is positive).

Figure 12.4 shows the profile of expected costs and benefits for the Sizewell project over its lifetime. If the project capital costs (including interest during construction) are expressed as an annuity from the commissioning date (year j), then the effect of deferment from year j to year $(j+n)$ on project net present costs may be expressed as

$$\Delta NPC_{j+n} = P(A) - P(B) - P(X) + P(Y) \tag{11}$$

in terms of the areas shown in Figure 12.5. Thus deferment reduces the present value of capital and operating costs (by an amount equal to $P(Y) - P(X)$), but also leads to a net loss of fuel-saving benefits (equal to $P(A) - P(B)$).

Because of the settling-in process expected for most types of plant, which means that it does not immediately attain its settled-down output rate, the loss from deferment of fuel-saving benefits in the early years of operation extends over a longer period than the term of deferment itself (as shown in Figure 12.5).

The CEGB's calculations of the effect of a one-year deferment on those elements of Sizewell's net present cost covered by the NEC calculations are shown in Table 12.2. In addition to the costs and benefits of deferment on the Sizewell project itself (shown in Table 12.2), the CEGB's deferment calculations also estimate the effects of deferment on a notional successor station. These follow-on effects are estimated to yield a saving from the deferment of Sizewell

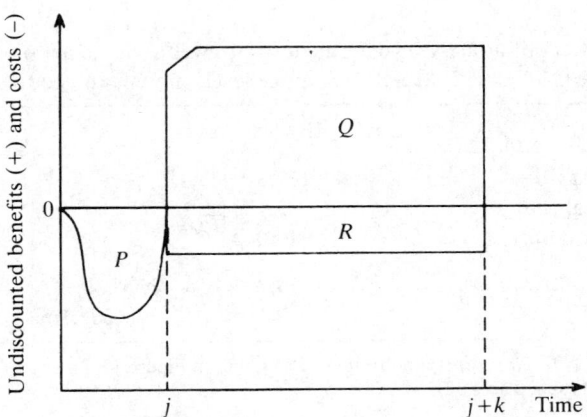

Notes: j=commissioning year; $j+k$=retirement year; P=construction costs; Q=reduction in fuel costs at other plants; R=own fuel and operating costs minus reduction in non-fuel operating costs at other plants.

Figure 12.4 Cost–benefit profile of a generating plant investment

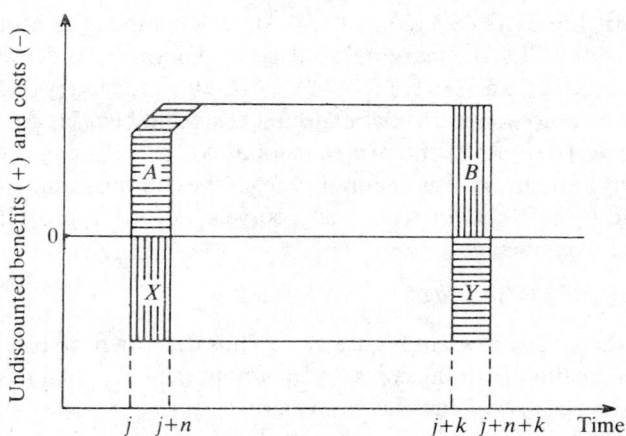

Figure 12.5 The effect of deferment

of about £5 million (present valued to 1991). Subtracting this figure from the £67.1 million shown in Table 12.2 yields an estimate of the increase in net present costs resulting from deferment of about £62 million. To these costs must be added the cost of deferring the decommissioning of elderly plant, which the CEGB estimates at about £13 million. These calculations therefore suggest that a one year's deferment would increase the net present costs of the Sizewell project by about £75 million, and support the CEGB's case for proceeding with construction as soon as possible.

Table 12.2. Effect of a one-year deferment of Sizewell B on its net present costs as estimated by the CEGB (£m, March 1982 price levels present valued to end-1991)

	Early years	Final year
Change in fuel savings	162.3 ($P(A)$)	9.9 ($P(B)$)
Change in capital charges	93.1 } ($P(X)$)	16.5 } ($P(Y)$)
Change in other costs	10.0	1.3
Subtotals	+59.2	+7.9
Total		+£67.1m

Note: The items A, B, etc. correspond to items A, B, etc. in Figure 12.5.
Source: CEGB (1982b, Table 33).

Although the estimated NEC of the project varies quite widely as a function of the rate of nuclear plant commissioning after Sizewell (for reasons explained earlier), the estimated change in net present cost resulting from deferment is almost invariant to the background plan. This is because the calculations are dominated by the early year fuel savings and capital cost items, neither of which are affected by the subsequent rate of growth of nuclear capacity.

3. A Risk Analysis of the Deferment Decision

The CEGB's evidence to the Sizewell Inquiry includes much sensitivity test material showing the effects on the NEC of the project for commissioning at the earliest possible date of variations in a wide range of determining parameters. The economics of project deferment are subject to the same uncertainties, but the CEGB's evidence contains only one sensitivity test (see Table 12.3) and that shows a combination of effects which together would reduce the estimated cost of deferment to zero. In itself this is highly unsatisfactory; no explanation is given of the probability status of the outcomes illustrated and it is possible that given the hypothesized combination of events, the NEC for commissioning at the earliest possible date would exceed the NAC of retaining existing plant in operation, so that not even the necessary condition for optional investment would be satisfied.

Table 12.3. Conditions for zero cost of deferring Sizewell B by one year (scenario 'C')[a]

Change in basic assumption	Change in cost of deferment (present valued to end-1991, £m)
Fossil fuel prices assumed to remain at current levels up to and including 1995–6	−60
A reduction of 7% points in the average annual availability of Sizewell B up to and including 1995–6	−10
A 6% increase in capital cost of station	−5
	−75

[a]The term 'scenario C conditions' refers to values of major exogenous variables (UK economic growth, world market prices of fossil fuels, UK real exchange rate) lying towards the middle of the range of outcomes considered in the Board's scenario studies.
Source: CEGB (1982b, Table 36).

In order to throw more light on the nature of the uncertainty surrounding the Sizewell deferment decision, we first of all describe the results of a wider range of sensitivity tests (such as the CEGB has provided in respect of the estimated NEC of the project). These illustrate how variations in the values of certain key determining variables affect the costs and benefits of deferment.

The Key Variables

The CEGB's deferment calculations presented in Table 12.3 were in respect of a commissioning date in the early 1990s. However, the time-scale for the Sizewell project has slipped considerably since the CEGB's main economic evidence was published in November 1982. The government's decision to approve the project was not announced until March 1987. Even if there is a prompt start on site, it is unlikely that commissioning will occur until late 1995, given the CEGB's views on the likely construction time. We therefore begin by re-estimating the cost of a one-year deferment from a commissioning date in the mid-1990s, given values for the determining variables similar to or consistent with those assumed in the CEGB's deferment calculations (see note to Table 12.4).

The result is shown in Table 12.4. The only difference from the CEGB's deferment calculations reported in Table 12.2 is that the estimated value of the early year fuel savings item is slightly higher so that the overall cost of deferment is also marginally increased.

The estimated cost (or benefit) of deferring the Sizewell project is simply the difference between the potential fuel cost savings that are forgone and the capital and other operating costs that are avoided. It follows that the loss from deferment would be lower if the costs of fossil-fuels were less, or if smaller quantities of fossil-fuel were displaced, and also if the capital and other operating costs were higher.

The cost of deferment is also highly sensitive to the rate at which plant performance might be expected to improve with the passage of time, a topic which has received relatively little attention at the Sizewell Inquiry.

A key, though unstated, assumption in the CEGB's deferment calculations, shown in Table 12.2, is that the plant performance does not alter as a result of deferment, implying that the undiscounted costs and benefits of the project remain virtually unchanged if it is deferred.

Elsewhere in the CEGB's evidence to the Inquiry, however, we learn that in fact the Board predicts a continuing improvement in the performance of the PWR plant that it wishes to order for construction at Sizewell, so that units commissioned four to five years after Sizewell are expected to achieve a settled-down annual average availability of 70 per cent, compared with the figure of 64 per cent assumed for Sizewell (CEGB (1982e, para. 177)). This would represent an annual rate of improvement of nearly 2 per cent.

The term 'availability' describes the relationship between the *potential* power output of the plant and the power output it would have produced if operated continuously at design rating. The economic significance of improved availability is therefore that the plant's power output, and hence the fuel-saving benefits it would realize (through the displacement of plant with higher operating costs), would be increased.

The improvement in availability projected by the CEGB appears, moreover, to be the outcome of 'generic' design improvements realized by the plant

Table 12.4. Effect of a one-year deferment of Sizewell B on its net present costs (£m, March 1982 price levels present valued to end-1991) for commissioning in 1995

Change in fuel savings[a]	165.3	⎫
Change in capital charges	93.1	⎬ early years
Change in other costs	10.0	⎭
	+62.2	
Other costs/benefits	+15.9	
	+78.1	

[a]The change in fuel savings was calculated as follows. CEGB (1982d, Table 8) gives world market prices for steam coal and crude oil (both c.i.f. NW Europe) in 1995 as $85/tonne and $45/bbl respectively. CEGB (1982c, Figure 2) gives the CEGB's view on the path of the real exchange rate under scenario 'C' conditions, implying a rate of $1.50:£ in 1995. This was used to express world market prices in terms of UK currency. Delivered prices were then obtained by adding £2/tonne to the world market price of coal to represent the capital charges of constructing modern high-capacity import facilities which are not currently available in the UK and by using the same assumption as made by the CEGB on the relationship between crude and H.F.O. prices. CEGB (1982b, Table 30) gives the quantities of coal and H.F.O. displaced by Sizewell in the early years of settled-down operation as 1.8 and 0.9 mtce, respectively. Sizewell would therefore save some £212m of fossil-fuel inputs, at a cost of £30-40m in nuclear fuel cycle costs (CEGB (1982b, Table 14)), yielding a net saving of approximately £177m. However, because the plant is not assumed to reach its settled-down performance immediately, the loss of fuel-saving benefits through deferment is slightly less than this.*

*Suppose that the plant was expected to displace aV worth of fossil-fuel in its first year of operation (with $a < 1.0$) and V worth thereafter (at settled-down output), then the present value of fuel savings lost through deferment would be:

$$\frac{aV}{(1+r)} + \frac{V(1-a)}{(1+r)^2}$$
$$= \frac{V}{(1+r)}\left[\frac{(1+ar)}{(1+r)}\right] \quad \text{(A)}$$

With $a < 1.0$, the term in square brackets in expression A is less than unity, and expression A is less than the value of fuel savings lost through deferment if settled-down performance was achieved immediately (equal to $V/(1+r)$).

manufacturers or by the transfer of additional operating 'know-how' from other utilities using similar plant, factors which are strictly exogenous to any decision concerning the Sizewell project itself.

Although the CEGB's case does not offer any empirical evidence to back up its engineering judgements on the likely rate of technical progress, some support can be found in the existing literature on 'learning by doing'. In particular, Joskow and Rozanski (1979) have examined the effects of learning by doing on the performance of nuclear plants, using a measure of plant performance (the power or load factor) that is closely related to the concept of plant availability.[6] These authors sought to explain the variation in plant performance in a sample of nuclear plants in the USA and elsewhere in 1975-6 as a function of plant vintage, operator experience, and plant size. With respect to the vintage effect, they concluded that there is 'evidence of an industry learning curve, with technological improvements increasing the ultimate capacity factors of new plants at the rate of 5% per year'.[7] At the mean capacity factor for their sample (about 60 per cent), this rate of improvement represented a 3 percentage point annual gain in availability. However, an inspection of the authors' regression results shows that the size of the 'technical progress' coefficient is highly variable with respect to the specification of the estimating equation and also that even in the authors' preferred equation, the standard error on the coefficient remains relatively high (the 95 per cent confidence interval is 5 per cent ±4 per cent p.a.). Moreover, Joskow and Rozanski's sample includes plants of several design vintages and supplied by more than one manufacturer. For reasons discussed by Klein (1977), we would expect the rate of technical progress estimated in a sample of this kind to be higher than the expected rate of improvement within any single design vintage, such as the 4-loop Westinghouse reactors with design ratings in excess of 1,000 MW which the CEGB wishes to construct at Sizewell.

The economic implication of the CEGB's expressed technological expectations is that if the project were deferred, its performance would improve and its undiscounted benefits would be higher, as shown in Figure 12.6. The change in those elements of net present cost covered in the NEC calculations resulting from deferment would be given by

$$\Delta \text{NPC} = P(A') - P(B) - P(C) - P(X) + P(Y) \qquad (12)$$

where the areas B, X, and Y correspond to those identified in Figure 12.6. To this would be added the other items (chiefly the cost of maintaining existing capacity) identified earlier. It may be assumed, as a first approximation, that these other items are invariant with respect to the rate of technical progress in plant performance. The difference between equations 11 and 12 is equal to $P(C) + P(A) - P(A')$ which represents the increase in the present value of fuel savings realized as a result of improved plant performance gained through deferment.

Sensitivity Test Results

Table 12.5 shows the sensitivity of the cost of deferment to variations in certain

---

[6] The term 'load factor' refers to the relationship (usually expressed as a percentage) between the output *actually produced* by a plant and the output it would have produced if it had been operated continuously at its full design rating over some specified period of time. The load factor is equal to or less than the availability, falling below it as the plant is operated on part-load.

[7] The term 'industry learning curve' refers to the rate of improvement in the plant designs supplied from manufacturers as distinct from the learning curve reflecting increased experience on the part of plant operators.

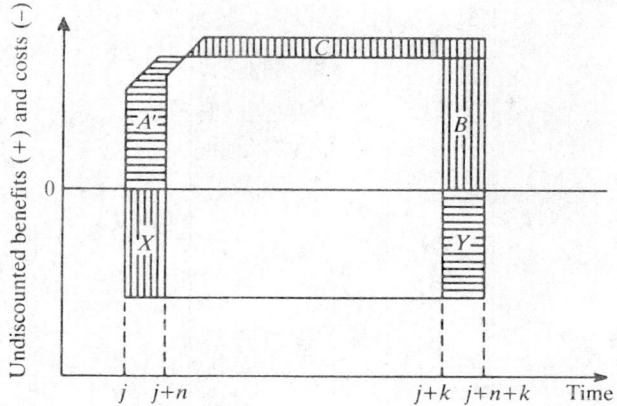

Figure 12.6 Deferment with improved plant performance

key parameters. In each case, the alternative parameter values chosen for the sensitivity tests represent the author's subjective estimate of the lower (i.e. least favourable to the project) bound of a 95 per cent confidence interval for the variable in question. This table highlights the sensitivity of the deferment calculations to variations in the world market prices of fossil-fuels, the real effective exchange rate, and the rate of improvement in availability.

Sensitivity analysis of this kind is a useful device for identifying the main sources of risk. However, if estimates are available of the probability distributions of key determining variables, then formal risk analysis (Evans (1984) and Hull (1980)) may be applied to yield an expected value for deferment and also the probability that deferment would produce a reduction (or an increase) in the net present cost of the project.

Drawing on the author's evidence to the Sizewell Inquiry (Jones (1983)),[8] subjective probability distributions are specified for the values of the following variables in 1995:

(a) the world market price of steam coal (WMPC);
(b) the world market price of crude oil (WMPO);
(c) the sterling real effective exchange rate (R).[9]

---

[8] For a review of subsequent evidence to the Inquiry, see Sizewell B Public Inquiry Transcript, Day 200.
[9] The real effective exchange rate for sterling (RERUK) may be defined in the following way:

$$I(RERUK)_t = I(MERUK)_t \times \frac{I(PUK)_t}{I(PW)_t}$$

where $I(RERUK)_t$ = an index of the real effective exchange rate in period t;
 $I(MERUK)_t$ = an index of the market effective (currency basket) exchange rate in period t;
 $I(PUK)_t$ = an index of the UK price level (GDP deflator) in period t;
 $I(PW)_t$ = an index of the world price level (in principle incorporating the same weights as used in estimating MERUK).

Table 12.5. Sensitivity test results on deferment (£m, present valued to end-1991)

Parameters	CEGB assumption	Alternative assumption	Effect on deferment	Revised effect of deferment on project NPC
World market coal price	$85/tonne	$55/tonne	−34	44.1
World market crude oil price	$45/bbl	£25/bbl	−39	39.1
Sterling real effective exchange rate	$1.50 = £1	$2.00 = £1	−50	28.1
Capital charges (p.a.)	£93.1m	£116m (+25%)	−23	55.1
Availability level	64%	58%	−10	68.1
Rate of increase in availability	0% p.a.	2% p.a.	−32	46.1

Notes: CEGB estimate of cost of deferment (from Table 12.4): 78.1.
Expressed in terms of March 1982 $US (sterling real effective exchange rate, 1980, $2.21 = £1).

A distribution is also specified for the rate of increase in availability (GA) based on the earlier discussion. For all these variables the specified probability distributions are substantially different from those underlying the CEGB's own assessments (reported in Table 12.4). The CEGB expected world market prices of both steam coal and crude oil to increase in real terms by approximately 2 per cent from 1980 onwards, and the real effective exchange rate of sterling to decline by 2.6 per cent between 1980 and the turn of the century. By contrast, the modal outcomes for WMPC and WMPO in the present analysis assume annual rates of increase of approximately 1 per cent and 0.5 per cent respectively; for R the modal outcome assumes a decline of approximately 1.5 per cent p.a. The subjective probability distribution specified for the increase in plant availability yields an expected value for this parameter of just under 1 per cent p.a., a considerably lower figure than that assumed by the CEGB.[10]

Unlike the risk analysis reported in Evans (1984), it is not assumed that the determining variables are all independent of each other. In particular, it is assumed that the sterling real exchange rate (R) is itself jointly determined by the world market price of crude oil (WMPO) and a set of exogenous factors (X), including the future level of UK coal and natural gas output and the rate of growth of productivity in the traded goods sector of the UK economy relative to that achieved in the traded goods sector of the UK's trading partners. Although the nature and extent of the oil price effect on the exchange rate are not well understood, experience to date suggests that the sign of $\partial R/\partial \text{WMPO}$ is positive (see Jones (1983)). Thus the expected value of the change in project net present costs arising from deferment (EVD) may be written as

$$\text{EVD} = [\overset{+}{\text{WMPC}}, \overset{+}{\text{WMPO}}, \overset{-}{\text{GA}}, \overset{-}{R}(\overset{+}{\text{WMPO}}, \overset{+}{X})] \tag{13}$$

where WMPC = world market price of coal;
    GA   = rate of increase in plant availability.

The signs of the derivatives are as indicated. Whilst the direct effect of an increase in WMPO is to increase the loss from deferment, its indirect effect through the exchange rate works in the opposite direction.

Table 12.6 shows the values of R associated with specified values of WMPO and X. Table 12.7 shows the subjective probability distributions for the WMPC and GA variables.

---

[10] The justification for this is that whilst there is reason to expect some continuing improvement in plant performance as a result of both learning by doing and continuing research and development effort, the 2 per cent rate projected by the CEGB would appear to imply a level of lifetime availability well in excess of the performance even of existing oil- and coal-fired plant of mature design. Several authors have drawn attention to the grossly over-optimistic projections of nuclear costs and levels of performance that have characterized the industry in the past. See, for example, MMC (1981) and Komanoff (1981).

Table 12.6. Subjective probability distributions for WMPO and X and associated values of R (probabilities in parentheses, R values expressed in terms of $ per £)

		Performance of UK economy (X)		
		Low (0.25)	Medium (0.5)	High
WMPO ($/bbl)	33 (0.3)	1.5 (0.075)	1.6 (0.15)	1.7 (0.075)
	36 (0.6)	1.6 (0.15)	1.7 (0.3)	1.8 (0.15)
	45 (0.1)	1.8 (0.025)	1.9 (0.05)	2.0 (0.025)

Table 12.7. Subjective probability distributions for WMPC and GA (probabilities in parentheses)

WMPC ($/tonne, c.i.f. ARA March 1982 prices)	$60 (0.3)	$70 (0.6)	$85 (0.1)
GA (% per annum)	0 (0.4)	1 (0.4)	2 (0.2)

Notes: C.i.f. UK prices are derived by adding £2/tonne (see Table 12.4).
Combining the subjective probability outcomes in Tables 12.6 and 12.7 yields a set of $9 \times 3 \times 3 = 81$ possible outcomes.

Results

The first issue for consideration is whether, given a continuation beyond the mid-1990s of rates of increase in the international prices of steam coal and crude oil of about 1 per cent p.a. and a decline of the real exchange rate of about 1 per cent p.a., the expected NEC of the project for commissioning in the mid-1990s would be less than the NAC of retaining existing plant in operation.

We have examined this issue by estimating the present value of the fuel cost savings realized by the plant over its lifetime (using analytical methods) and comparing these with the costs of constructing and operating the plant shown in CEGB (1982b). It is estimated that if the plant remained on base-load operation throughout its expected 35-year life, its NEC would be significantly negative (approximately −£23.4/kW p.a.). The calculations suggest, however, that further optional investment in nuclear plant after Sizewell would be justified only to the extent that such plant could be operated on base load (see Figure 12.3) unless construction costs were reduced or plant performance improved.

The cumulative probability distribution for the effect of a one-year deferment (from 1995 to 1996) on the expected net present cost of the Sizewell project is shown in Figure 12.7. The results may be summarized as follows:

(a) There is a 56 per cent probability that a one-year deferment would lead either to no change or to a reduction in the expected net present cost of the project. If there were a benefit to deferment (i.e. the net present cost is reduced) then

the maximum reduction in net present cost would be approximately £40 million.
(b) Conversely, there is a 44 per cent probability that a one-year deferment would increase the net present cost of the project. The maximum increase in net present cost would be approximately £45 million.
(c) The expected value of a one-year deferment would be to reduce the net present cost of the project by approximately £2½ million.

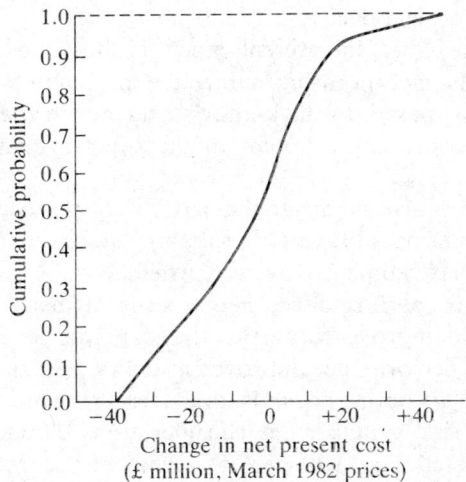

Figure 12.7 Cumulative probability distribution for the effect of a one-year deferment on the expected net present cost of the Sizewell B project

The expected cost of deferment is some £80.5 million less than the figure of £78.1 million in Table 12.4. The difference between the two estimates can be disaggregated into two elements:

(a) the effect of lower fossil-fuel prices and a higher real exchange rate, given an assumption of unchanged plant performance, which reduces the cost of deferment by some £63.1 million;
(b) the effect of allowing for the possibility of improved plant performance at the given (lower) fossil-fuel prices and higher real exchange rate, which further reduces the cost of deferment by some £17.4 million.

We also note that if plant availability were to increase at 2 per cent p.a., as the CEGB expects, then a one-year deferment would allow a reduction of approximately £30 million in the net present costs of the project, at the given fossil-fuel prices and sterling real exchange rate.

4. Conclusions

In contrast to its earlier practice, the formal decision algorithm used by the CEGB to assess the case for deferring the Sizewell project is sound. However, its application in practice is flawed both because the results reported in the Board's economic documents do not take account of the possibility of technical progress in plant design anticipated elsewhere in the Board's evidence to the Sizewell Inquiry, and because the CEGB does not adequately explore the uncertainty surrounding the deferment decision.

In examining these issues, the present paper has illustrated the application of formal risk analysis incorporating subjective probability distributions for certain key variables, based on the author's evidence to the Sizewell Inquiry. The possibility of some improvement in plant performance has also been assumed.

The results of this analysis indicate that a decision to proceed with Sizewell's construction as soon as possible would probably reduce the CEGB's generating costs as the CEGB has claimed. However, when account is taken of potential improvements in plant performance, there is a slightly better than even chance that deferment would improve still further the economic benefits expected from the project. If plant performance improved at the rate predicted by the CEGB, then there would be very strong grounds for deferring the project unless demand need or 'strategic' risk management considerations dictated an early start. Finally, we have noted that a neglect of technical progress in plant design characterizes many of the planning models used to inform decision-making in the electricity supply industry. Our results suggest that estimates of the costs and benefits of deferment based on such an assumption may be seriously misleading.

13 Electricity Supply in Europe: Lessons for the UK*

DIETER HELM AND FRANCIS McGOWAN

1. Introduction

Privatization of the UK electricity supply industry (ESI) is firmly on the political agenda. As with British Gas, the problems of competition and regulation are being faced after the decision to proceed with the sale. But unlike with British Gas, the government has proposed a degree of restructuring prior to privatization in its recent White Paper, *Privatising Electricity* (Department of Energy (1988)).

The proposed change in ownership and structure follows on from the liberalization of the first Conservative government. The 1982 Lawson *market doctrine* (see Chapter 1, this volume), which sought to create a market in energy and use competition as a method of regulation, was effected through the 1983 Energy Act. This Act set the terms of electricity sales and use of the system by the private sector, and was designed to encourage new private sector entrants into the industry. Though these measures have fallen short of expectations (see Hammond, Helm, and Thompson (Chapter 8, this volume)), they created a climate for further radical change.

Such developments prompt a number of critical questions on ownership, organization, regulation, and competition. What difference does ownership make to performance? What are the relative merits and shortcomings of vertical integration? How far should distribution of electricity and other energy sources be horizontally combined? How far should the sector be regulated, and should the form of regulation concentrate on cost or price factors? Is there scope for competition and entry and, if so, how should it be managed?

In seeking to answer such questions, international comparisons are typically made with US experience: the US ESI has been widely researched, providing a

---

*Dieter Helm is a Fellow in Economics at Lady Margaret Hall, Oxford and a Research Associate at the Institute for Fiscal Studies. At the time of writing this paper, Francis McGowan was a Research Officer at the Institute for Fiscal Studies; he is now a Research Fellow in the Energy Group at the Science Policy Research Unit, University of Sussex.

This article has arisen out of research into the structure of energy provision in a number of European countries, which included interviews with officials from government and industry in France, the Federal Republic of Germany, the Netherlands, and Sweden, as well as with the officials at the International Energy Agency, the European Commission, and in the UK. These primary sources are supplemented where possible by the secondary literature available. The paper was first published as Institute for Fiscal Studies Working Paper 87/10 (1987). The authors are grateful for comments on an earlier draft from David Thompson and Bill Robinson. The usual disclaimers apply.

considerable basis for comparative work. Yet a better basis might be the European ESIs: in terms of institutional development, scale, and technology they are closer to the UK than the US is. They also illustrate a diversity of ownership and regulatory structures, providing evidence of how different systems perform. The aim of this paper is to see what insights European ESIs might provide to the questions that reorganization poses for the UK. The analysis focuses on four countries—France, Germany, the Netherlands, and Sweden.[1]

Section 2 of the paper examines the patterns of ownership in European ESIs, identifying three stylized models to which our case studies conform. In Section 3, we focus on organizational structure, particularly the costs and benefits of vertical and/or horizontal integration. Section 4 deals with regulation, in particular the informational problems that the regulator faces, and the potential for competition. In both Sections 3 and 4, we illustrate how these issues arise in our case studies. Finally, the paper draws out the implications of the European evidence for the future development of the UK ESI.

2. The Importance of Ownership

The diverse patterns of ownership in the European ESIs provide a basis for examining some of the consequences of alternative structures for the UK. It would be expected that, if ownership matters, the differing structures would result in differing performances.

The development of European ESIs derives in part from industry-specific characteristics such as the prevailing technical conditions, fuel sources, the geography of distribution and production, and the natural monopoly element. The historically given political attitude towards State participation in the economy has also influenced the pattern of ownership and organization. The relative merits of different types of ownership cannot be assessed without reference to these factors. We therefore first consider the historical origins of the current diversity in European ownership structures.

The Second World War represented a turning-point in both the underlying characteristics of the industry and the political development of the State. Accordingly, while rationalization has been a constant factor in the industry from its inception, the war and subsequent reconstruction saw the process accelerate as national grids were established (as distinct from the pre-war period when many countries were still extending supply and local networks were being created).

---

[1] On the US ESI, see Joskow and Schmalensee (1983). Most British literature has focused on the national experience with little reference to other countries, much of it prompted by the current government's liberalization and privatization policies; see Hammond, Helm, and Thompson (1985 and Chapter 8, this volume), Vickers and Yarrow (1985), and Yarrow (1985). Aside from one or two works, notably Corti (1976) and Lucas (1985), most comparative work is conducted by international agencies, such as the EEC and the International Energy Agency (IEA).

Electricity Supply in Europe: Lessons for the UK

For expositional purposes, we can identify three broad stylized categories into which most can be fitted. These are the *unitary integrated State-owned system*; the *mixed dominant State incumbent system*; and the *decentralized mixed ownership system*. These three stylized models we have identified were in large part political responses to rationalization: replacing the market by nationalization at one extreme; maintaining the *status quo ante* of decentralization largely intact at the other; and the growth of a State incumbent's role in an otherwise mixed and regionalized system.

Notwithstanding their differences in approach, in enacting rationalization all three systems involved considerable concentration and an increased participation by public authorities (at least until the 1980s when policies of privatization and increasing competition may have halted, if not reversed, that trend). We first discuss the development and present structures of our case studies, before evaluating their relative performance.

Unitary Centralized Systems: Nationalization

By far the most dramatic change occurred in the UK and France where the pre-war decentralized systems were centralized into unitary ones, using ownership as the vehicle for that change through the nationalization of the sector. Prior to this, both countries' ESIs had consisted of a myriad of small producers and local distributors (see Hannah (1979) and Hicks (1938, pp. 108-9) on the UK, Virole (1986) on France). Nationalization provided an opportunity to take advantage of new developments (particularly growth in demand, technical advances in transmission, and hence greater economies of scale) and to extend the network to more remote areas.

The nationalization process in both the UK and France did, however, derive from more than simply rationalization considerations, and these other objectives have had a marked influence on the subsequent development of the UK and French ESIs. Nationalization also sought to address other market failures, most notably monopoly and the social costs of the relative decline of the coal industry. The latter issue has been a dominant feature of UK electricity policy, ultimately imposing on the CEGB a long-term requirement to purchase coal at above marginal cost.

Nationalization also formed part of a wider energy policy in guaranteeing and co-ordinating supply. Risks were pooled, demand was forecast centrally, and strategic considerations were met at the government level. It constituted an important prerequisite for more general central investment and infrastructural planning. Reconstruction was widely believed to require a continued active role for the State, as it had done during the war (Helm (1986) and Cairncross (1986, pp. 465-9)). The supply of power to industry and to the domestic market was an important part in that process in both the UK and France.[2]

---

[2] For a useful encapsulation of this, see the Economic Survey of 1947.

From a system of decentralized utilities, the 1947 Electricity Act reorganized the UK ESI into an integrated industry. It built on the partial rationalization that had taken place during the inter-war period: the 1926 Electricity (Supply) Act created a national grid owned by the Central Electricity Board which also controlled (but did not own) generating capacity. The near collapse of the system after the war exposed market failures which enhanced the Labour government's plans for nationalization of the sector as part of its broader economic plans. The 1947 Act created a 'British Electricity Authority' for generation and transmission, and 14 Area Boards for distribution, a structure which remains broadly intact (Hannah (1979, p. 349)).

The French pre-war ESI had been even more decentralized than the UK's. Although the government in the late 1930s had made tentative moves towards tariff harmonization and rationalization of the industry, it remained localized and in private hands. In 1946 there were over 150 production companies, more than 1,000 distributors, and 80 companies operating transmission lines. Nationalization brought almost all production and distribution, as well as the transmission system, under Electricité de France (EdF) as part of a broader policy of public ownership, including the coal and railway industries (these retained their own electricity capacity but sold any surplus to EdF). The pre-war distribution structure was retained in so far as the joint administration for gas and electricity was continued (Lucas (1985, p. 10)).

Decentralized Systems: Retaining the Pre-war Structure

Most other European countries chose to rationalize within existing ownership structures, largely because of the relative strength of the incumbents *vis-à-vis* central government. Thus the Dutch system has been, and continues to be, decentralized largely as a result of the relative weakness of the central state *vis-à-vis* local tiers of government, while in Germany, the federal character of government also permitted the retention of localized utility provision. This section outlines the development of these countries' industries.

The Dutch Case

The Dutch ESI, along with most of the country's public utilities, has been traditionally the concern of the local authorities (de Ru (1985, pp. 322–3)). Historically, they have incrementally developed their role in the ESI, as private companies were regulated out of the sector and the municipalization of services increased around the turn of the century. Concentration has occurred but it has been within the pre-existing framework, with communes banding together to form utility companies.

The division of ownership largely follows the level of operation. The bottom tier of local government, the communal authorities, is mainly responsible for distributing electrical power to domestic consumers. There are currently around 80 electricity distribution utilities in the country, made up of one or more communes (compared with 250 in the 1940s and 500 in the 1920s). The production of electrical power is largely in the hands of companies set up by the provinces and the larger municipalities. There are currently 13 of these (compared with 50 in the 1940s), accounting for just under 90 per cent of the country's electrical needs.[3]

Dutch provincial producers generally supply power to consumers in their area either directly or through the communal distributors. They co-ordinate their activities across the country within the SEP (Samenwerkende Elektriciteits-Produktiebedrijven or the Electricity Producers' Co-operative) which is in charge of day-to-day transmission of electricity between regions and between the Netherlands and other countries. The SEP owns the 380kV lines and the national control centre; all other lines and switching equipment belong to the major producers. The SEP was formed in 1948 to ensure a steady supply of power throughout the country.

There has been a steady shift towards public ownership as the proportion of private production (some of which is sold to the network) has declined. This tendency has reinforced the position of the local authority utilities and has not involved the central government in an ownership role. However, the State is playing an increasingly active role in the reorganization of the sector.[4]

The German Case

The German system is similarly decentralized, again largely as a result of the federal character of government and local autonomy. As in the Netherlands, there is little direct role for the federal government and no direct State ownership of production, transmission or distribution facilities. Federal participation has been limited to holdings of the assets of partially publicly owned industrial companies such as VEBA. The federal government has, however, reduced its shares in these companies as part of its privatization programme.[5]

Nonetheless, public participation in the sector has developed through the growth of mixed companies at the Laender (regional), municipal, and communal levels. While there is also considerable private investment in many of

---

[3] *Elektriciteit in Nederland 1985*, an annual review of developments in the Dutch ESI, illustrates the decline in utility numbers.
[4] *Financial Times* Netherlands supplement (21 October 1986) and International Energy Agency (1986).
[5] The principal holding companies owned by the federal government have been VEBA and VIAG, the former concentrating on energy provision and the latter on industry. Both have substantial shares in some of the largest utilities in the country. On the history of these and other government holdings, see Turot (1970, pp. 16–17). As with the bulk of State holdings, they are constituted and left to operate as private companies, with government influence primarily exercised through representatives on the supervisory board. For an interesting discussion of the role of government in State-owned enterprises, see National Economic Development Office (1977) and the chapter on Germany in Centre Européen d'Entreprise Publique (1984). On the government's plans for privatization, see *Bulletin: Gesamtkonzept für die Privatisierungs- und Beteiligungspolitik des Bundes*, 26 March 1985. On the sale of VEBA, see Himmelmann (1985) and 'Veba sell off to raise DM3b.', *Financial Times*, 26 January 1987.

the larger companies, wholly private companies accounted for only 3 per cent of sales in 1985. By contrast, wholly publicly owned companies were responsible for 25 per cent of sales, and mixed companies (defined as less than 95 per cent publicly and less than 75 per cent privately owned) accounted for another 71 per cent.[6]

In the post-war period, the trend has been towards larger mixed ownership companies, mainly at the expense of the smaller private producers and some public distributors. There has also been substantial rationalization in the sector: the number of producers and distributors combined has fallen from over 3,000 in the inter-war period to under 1,000 now. Of these, 600 are large enough to merit inclusion in official statistics.

Although there are more than 300 production companies, the bulk (around 60 per cent) of West Germany's energy needs are met by 8 of them (Bongaerts and Kraemer (1986)). Most of these have emerged out of the private producing companies, such as the coal and steel cartels of the late nineteenth and early twentieth century. As they developed their production and distribution capacities, they came into competition and conflict not only with each other, but also with municipal councils which produced or distributed power for domestic consumption (Lucas (1985, pp. 178 ff.)). The restructuring of the large private companies with municipal capital effectively defused much of this rivalry between public and private companies, while the 1935 Energy Act confirmed the principle of monopoly supply backed up by subsequent regulations on concession areas, placing an obligation on utilities to provide power to all consumers within their areas. This law has remained in force ever since.

While production is focused in 8 large companies mostly owned and operated at the Laender or inter-Laender level, distribution is mainly carried out by municipal or council governments (and is more likely to be wholly publicly owned). As in the Netherlands, these companies are often engaged in the provision of other energy services with certain larger towns developing combined heat and power (CHP) programmes.

Given the scale of operations of the major producers in Germany, there is greater self-sufficiency within regions than in the Netherlands. The inter-Laender transmission network is therefore looser than in other countries; the Deutsche Verbundgesellschaft (literally, the German Interconnection Association) is owned by the largest producers and provides transmission between their concession areas and with other countries.

Overall there is a trend towards greater concentration of the German system, with the larger utilities tending to merge with smaller regional production and distribution companies in their areas. The integration of these small companies has been encouraged by regulatory authorities, to reduce the perceived technical and allocative inefficiencies of small-scale operation.

---

[6] The main industry association, Vereinigung Deutscher Elektrizitaetswerke, provides detailed data on the industry. See its 1985 report *Die Oeffentliche Elektrizitaetsversorgung*, p. 43. For details of who owns the utilities, see Commerzbank (1985).

Dominant State Incumbents within a Mixed System

The third pattern lies in between these two extremes: here, rationalization has been obtained within existing structures but certain key functions have been taken up by the State. In such a system, ownership becomes an indirect method of planning and regulation. The Swedish ESI exemplifies this.

As in other countries, pre-war electricity was produced either as a by-product of industry or through a local utility owned either privately or by the local government. However, due to the quickly recognized potential of hydro-power in providing Swedish needs, and the State's ownership of land, the State played an important role at a much earlier stage in development. Its principal vehicle was, and remains, Statens Vattenfallsverk (literally the State Waterfalls Authority, or Vattenfall).[7]

Prior to the 1940s, Vattenfall's activities were mainly confined to producing power (it has always been the largest single producer in the country), but as networks grew from the local to the national level (because most of the power was in the north and most of the population was in the south of the country), the State came to play a greater role. This was confirmed in the 1940s when the Swedish government made Vattenfall responsible for the transmission grid, and has grown since as Vattenfall has been the main investor in new capacity and taken on expired rural distribution concessions.[8] By contrast, private investment has been regarded as low and deficient in recent years relative to the public sector. The reasons for this investment failure are complex, and we return to them below.

Aside from the growth of public participation, the structure of the industry follows the pattern of supply developed in the first thirty years of the century. The larger cities, notably Stockholm, produce and distribute their own electricity (as well as other public services), while others have banded together to form production companies. The best example of the latter development is Sydkraft, the second largest electricity producer, which was initially set up as a co-operative by a number of towns in the south of the country. It has retained this ownership structure, though a small number of shares are also owned by insurance companies and individual private investors (Sydkraft (1986, p. 45)).

There are also a number of production companies that have links with private industry. Many of these emerged out of the timber industries and mainly serve rural consumers as well as their own production needs. In addition there has been a substantial element of cogeneration by industry (where electricity is a by-product rather than a subsidiary activity) and such capacity is sold to the major suppliers or to local distributors. Nonetheless, the relative share of private production companies has diminished as Vattenfall has grown and as some

---

[7] For a contemporary review of the inter-war development of the Swedish industry, see Haldane (1938).
[8] Lucas (1985, pp. 108–9), Krangede Power Pool (1986), and Smith (1953, p. 336).

private operators have sold out to local authority companies. As in other countries, this shift has been accompanied by rationalization, with the number of utilities falling from 3,000 in the 1940s to 1,000 in the 1960s.[9]

While the transmission system is wholly owned by Vattenfall, the major producers have access to it. They effectively form a club within which long-term supplies of power are negotiated and short-run purchases and sales are facilitated. Members of the grid pay a membership fee and transactions are conducted on the basis of spot prices and longer-term contracts (Krangede Power Pool (1986)). Within this grouping, Vattenfall plays a dominant strategic role as a source of lower-cost power because it possesses much of the country's hydro and nuclear capacity.

Implications of Diverse Ownership Structures

There is no single prevailing structure of ownership for European ESIs, nor is there any clear evidence that any one structure stands out as superior from the variety of mixtures of the private and public sectors. This observation is supported by a study conducted by the International Energy Agency (1985) which found no evidence for believing that either privately or publicly owned ESIs were inherently more efficient. The structure, moreover, depends on the particular responses of countries to State intervention and the relationship between government and economy: it has been a response in part to technical change and rationalization—particularly regarding the perceived inefficiencies of very small operators—and in part to wider interventionist considerations, especially in the UK and France.

Given the existing structures, there has been a general move to greater involvement by public authorities in the ownership of the sector (on the grounds of perceived market failures and the achievement of other public objectives). There has also been a relative decline in the size of the private sector as a result of perceived under-investment.[10] This has in part been a result of uncertainties over future national energy policies (such as the recent Swedish decision to phase out nuclear power) and the relatively easier terms on which publicly owned utilities can borrow. In part it may also be explained by the differing incentives between private and public enterprises. If the public participants undertake to *guarantee* sufficient capacity to meet aggregate electricity demand, then they will typically create a planning margin for excess capacity. This will depress future expected prices and hence deter private sector profit-maximizing investment decisions.

[9] For example, last year, Sydkraft purchased part of Skandinaviska Elverk's distribution network—see 'Asea sells one-third share in electricity network', *Financial Times*, 21 April 1986—while Vattenfall has taken over a number of rural distributors as their concessions run out or as the costs of maintenance grow. On the concentration of the industry, see United Nations Economic Commission for Europe, Committee on Electric Power (1969, p. 111).

[10] This point was made in interviews with officials in Germany and Sweden and at the IEA. It is also incidentally a factor in the US; see Robinson and Sharpe (1986).

There are, however, signs that the trend towards the gradual extension of public ownership may be halted or reversed, at least at the national level, as privatization and liberalization policies seek to encourage new producers and greater private ownership of ESIs.[11] Nonetheless our first observation is that, overall, private ownership has been a declining phenomenon for much of the post-war period in the European ESIs considered.

Whether the system is private or mixed, it is typically characterized by the granting of local and regional concessions which mark out supply areas for production and distribution. Where natural monopoly exists, duplication is wasteful and entry is typically not a feasible option. Concessions, moreover, are hard to overturn, not merely because of the terms on which they are granted but also and more fundamentally because they have a capitalized value. Thus where natural monopoly arises in ESIs, government intervention has been an almost universal phenomenon. Ownership has not, however, always proved to be the chosen instrument. Indeed public ownership can sometimes lead to abuses of monopoly power. In many countries, there exists a perception that the utilities' revenues are used as an additional source of local revenue, i.e. as local taxes.[12] Likewise there are cases where ownership can be used to influence policies unrelated to the business of power supply (such as the choice of fuel, price, etc.). These are dealt with in our discussion of regulation below.

Our principal conclusions in this section are therefore, first, that out of a plurality of ownership structures in Europe, none is obviously superior; second, that both private and public ownership have given rise to distortions in allocative decisions; third, that the presence of a substantial public presence alters private sector incentives; and finally, that in the case of local government involvement, ownership may not safeguard against monopoly-rent-seeking behaviour. Nevertheless, to the extent that ESIs are characterized by considerable market failures which European governments have typically sought to offset, some role for government is inevitable.

---

[11] One section of the Swedish Opposition wants Vattenfall privatized while, as noted, the German government is selling off its remaining holdings in utilities. Some countries outside this survey are also contemplating privatization: the Spanish INI is selling off its holdings in regional utilities (albeit at the same time as partly nationalizing the national grid), while some Belgian distributors have sold to the private sector.

[12] Concrete information on this is hard to obtain, though it was readily admitted by officials in all countries that charges are used as a source of revenue. (This was certainly the case in the UK during the inter-war period; see Foster, Jackman, and Perlman (1980, pp. 138-42) and Hicks (1938, p. 114) on the controls over this.) A Dutch official put a figure of 500 million guilders on the revenue element from all utility provision (not counting cross-subsidies from it to employment etc.) and new policy is designed to prevent this. In Sweden, Stockholm Energi has been taken to court over such charges. In Germany it has been seen as a source of cross-subsidy, according to Parris, Pestieau, and Saynor (1987, p. 123). Officials argued that, with the squeeze on expenditure being focused on the local governments, they needed all the revenue they could get, a view echoed in Sharpe (1981) and Council of Europe (1985, pp. 61 ff.). Yet the latter report also indicates that some local authorities are unable to cover costs, let alone derive revenues (p. 18).

3. Organizational Structure

The trends of concentration and increasing public participation have run parallel to changes in the organization of European electricity supply. Since the war, no one form has dominated, but there has been a strong tendency towards horizontal and vertical integration. The move to greater public ownership, particularly at the local level, has encouraged horizontal integration of electricity distribution with the supply of other utility services. In some cases, concentration of ownership has taken the form of the vertical integration of production and distribution activities.

Since the way in which an industry is organized is a fundamental issue in the privatization debate, we next outline some of the costs and benefits of horizontal and vertical integration and then examine how these are managed in our three models.

Horizontal and Vertical Integration

Horizontal integration takes three forms: integration between electricity distributors; between electricity producers; and between different utilities (or mixed utilities). All three forms of integration are common in Europe.

Where services are provided by the same authority, certain economies of scale are possible over billing, metering, maintenance, cable-laying, marketing, and servicing. Against these benefits must be set the costs. These arise from the increased market power and differences in consequent regulatory methods. Electricity for domestic consumption is regulated to some extent in all the countries examined, whereas electricity heating and gas heating are typically not (since they are regarded as substitutable). Where energy utilities operate in both monopoly and competitive markets, they may have opportunities to use the surplus earned in the former to cross-subsidize the latter by transferring joint costs. This result is most likely to arise where there is only limited transparency in cost allocation. (See below on the complications posed for regulation, as well as Hammond, Helm, and Thompson (Chapter 15 in this volume).) Moreover, the information advantages that the joint utility possesses over the regulator make it harder to monitor effectively.

Vertical integration, where production, transmission, distribution, and other functions are performed by the same firm, is also prevalent throughout Europe. It also features prominently in the privatization debate in the UK. In theory, vertical integration allows for a more effective transfer of power from producer to distributor and minimizes the transactions costs (Williamson (1975)). It may even be seen as ensuring a more secure and guaranteed supply than a system where all functions are separate. However, as with horizontal integration, it gives rise to the possibility of cross-subsidy and price discrimination, which may be hard for regulators to detect. Thus the monopoly of distribution may allow a vertically integrated utility to transfer costs from more competitive sectors, such as production or appliance sales, to more monopolistic sectors, such as domestic supply. As will be seen below, this gives the utility flexibility when confronted by general regulatory rules.

Ownership and Organization

Responses to the problems of horizontal and vertical integration in Europe have included the separation of functions to different parts of companies or even different companies. In some cases (most notably the Netherlands) the problems that integration poses are being addressed in wider reforms of the industry.

Nationalized systems tend to separate out elements of the industry vertically. In the UK, Area Boards function as separate companies, but co-ordinate with the CEGB under the umbrella of the Electricity Council. In France, distribution is a separate section of the industry, though it is partially horizontally integrated with Gas de France (GdF).

In the *mixed system*, there is considerable horizontal integration of utilities at the distribution level: many towns in Germany and the Netherlands jointly administer distribution of utility services. In Germany, there is also some vertical integration between production and transmission for the largest companies (and in certain cases this may extend to distribution).

Central to the impact of integration on performance are the ownership and organization of the co-ordination function, and hence the main grid. The Dutch government is seeking a major reorganization to break the vertically integrated structure of the industry (where the producing companies own the transmission network SEP), thereby introducing more competition between producers and breaking the dependence of local distributors on a particular provincial producer.[13] In addition, responding to the problems of size and investment behaviour, the reforms aim to integrate the present structure of sixteen producers, to create utilities sufficiently large to finance the construction of new power-stations. The reforms are designed to make the SEP more independent of the producing companies: instead of merely transferring power through the SEP, the government's plan envisages the producer utilities selling to SEP, which will in turn average out the costs of production before selling to the distribution utilities. This separation will replicate a number of the features of the Swedish model outlined above.

Integration is also being pursued for the distribution system by horizontally integrating gas and electricity (and, where it is provided, district heat) distribution tasks through larger 'energy utilities' throughout the country.

In the *Swedish State incumbent model*, organization is broken up by the use

---

[13] The Dutch proposals had not been finalized at the time of writing, and this discussion is based on interviews. The process of creating energy utilities is going ahead. On this, see International Energy Agency (1986, p. 352).

of the power pool and also strict accounting conventions. The major companies, including Vattenfall and Sydkraft, keep their distribution functions in separate affiliates or subsidiaries from the business of production. The advantage of this system is that short-run price signals can be used to co-ordinate production decisions, whilst consumer prices can transmit longer-run decisions, relevant to the investment decision in durable consumption items.

Implications

The processes of horizontal and vertical integration have stemmed from the gradual development and concentration of utilities. The advantages of such integration have to date been generally perceived as outweighing the problems in all the countries included in this study. However, the innovations that some countries have developed in recent years suggest that some of the disadvantages are now being given greater consideration. The actual or potential problems with integration have been increasingly recognized by governments, as opportunities for cross-subsidy are exploited and regulation undermined. Integrated companies have been able to strengthen their market dominance at the expense of potential entrants. As we shall see in the next section, integration leads to important consequences for potential manœuvre under general regulatory rules.

4. Regulation and Competitive Entry

The problems that have been identified with ownership and organization arise out of the substantial market failures in electricity supply and must be considered jointly with the regulatory approach employed. This section addresses the pattern of regulation and competition within the countries examined, drawing particular attention to the plurality of regulatory forms and the interaction with ownership and organization. We note a spectrum of controls which ranges from direct intervention of the sort practised in the UK to much lighter forms which prevail in much of Europe, and draw a number of general lessons for the UK if more decentralization is to be introduced.

To the extent that all countries contain a degree of regulation, the second part of the section examines the modes of regulation, i.e. rate-of-return, cost, and price methods. We outline the arguments for and against the different approaches, and show that they are all practically based on cost factors. The problems of applying regulatory rules and of gathering information for the regulator are also examined, as is the way in which these problems are dealt with in each of our three stylized classificatory models of European ownership and organization. Finally, we outline the potential for competitive forces in our models, taking up the argument for competition in electricity supply. There are three ways in which this can occur: competition amongst the established suppliers, entry by small carriers, and competition from other countries. We look at how far these are present in the countries examined and consider whether regulation for competition at both the national and supra-national levels may be required.

The Scope of Regulation

At first sight there is little uniformity to the way in which European ESIs are regulated; instead, a wide range of regulatory rules and conditions exist. It is clear, however, that all-embracing general regulatory controls are not the norm. In recognizing the need for explicit and wide-ranging controls, the UK lies at one end of a spectrum of control, while most European countries tend toward much lighter methods.

For expositional purposes, we can identify certain broad characteristics of regulation that are present to varying extents in our case studies. The first is the statutory monopoly element present in the industry in all countries: the utility takes over or is granted a concession to supply in exchange for an obligation to supply and provide a safe and reliable service. This concession is the basis for another characteristic of regulation—the *de jure* or *de facto* exemption from the country's cartel law, an exemption tempered by other controls. In certain systems (notably West Germany's) there is a perception that the utility's behaviour is controlled by the presence of a competitive alternative to electricity for particular market segments. In others, the element of public ownership plays this role. Exemption from the cartel law may be accompanied by explicit controls for those consumers unable to take advantage of other sources of supply: some form of scrutiny over domestic tariffs is present in all the countries examined. Finally, there is a tendency for controls (especially price ones) to be used by regulators for broader macroeconomic or political objectives.

The Nationalized Systems

The nationalized systems assigned concession rights to the new State enterprises either directly or through an adaptation of the old procedures. In France most local authorities retain the right to grant distribution concessions but invariably assign them to EdF. Private production was not ruled out but, aside from own use, the electricity could be sold only to the ESI (though in the UK the 1983 Energy Act now permits the renting of the grid). Public supply is the monopoly of the nationalized company.

The granting of these rights typically confers an exemption from cartel law in nationalized systems, but it does not exclude the utilities from scrutiny or control, though these controls generally derive from the wider regulatory rules set for nationalized enterprises.[14] In the UK there is an obligation on the supplier not to show *undue preference* to a particular customer in negotiations. There are

---

[14] On the theory of public enterprise objectives, see Rees (1984a).

also controls on the formulation of the bulk supply tariff (BST) and of tariffs to the consumer. More recently, investigating the performance of the ESI has been the responsibility of the Monopolies and Mergers Commission as part of an efficiency audit of nationalized industries. In France, the Contrat du Plan imposes a range of obligations on the EdF in respect of investment, pricing, and general efficiency, and the Cour des Comptes audits the industry's performance.

The fact that the industry is publicly owned does not remove the need for regulation in such systems. These controls extend beyond the setting of tariffs for domestic consumers. In fact, such controls are often used for objectives beyond those of providing an efficient ESI. Tariff controls have been used to meet broad anti-inflationary targets, to the utility's detriment, as in the case of EdF during the 1970s when its finances were damaged by government controls on tariff increases. In the UK, the government has imposed external financing limits on the industry for many years.

Decentralized Mixed Systems

In mixed systems, the concession to supply power confers an effective exemption from close regulation. In Germany, the 1935 Energy Act recognized the need to assign monopoly rights in regions and granted local authorities the right to control utilities in their area (though these rights were counterbalanced by an obligation to supply all customers efficiently and reliably). Exemption from antitrust law was clarified by the 1957 Cartel Act, which, under Article 103, permits public utilities to engage in concerted practices, though this was circumscribed by controls on prices for domestic customers and an obligation not to price discriminatorily. In the Netherlands, the exemption is implicit in the concession.

The exemption from close regulation is partly explained by the ownership factor. To the extent that provision of services is a local responsibility (and this, we noted above, is generally the case at the level of distribution), it may become a more significant electoral consideration. Hence, it is argued, the local electorate provides a check on monopoly-rent-seeking activity.

Light regulation of industrial customers in these countries is also justified by the presence of competing sources of supply. The competition is generally not primarily from within the ESI (the potential for such competition is discussed below), but from other fuels (as measured by cross-elasticities). Gas is typically the prime source of competition, especially to industrial customers. Hence both gas and electricity supplies to industrial users are typically exempt from direct regulatory control (Commission of the EEC (1984b, p. 10)). Industrial users are left to negotiate their own contracts with the ESI, with the latter aware that they can substitute other fuels or even generate their own power. In the German case, there is scope for legal action by industrial customers who consider they are being subject to discriminatory pricing; the 'as-if' rule provides for some control over utilities abusing their position, by requiring that customers be charged on a similar basis to similar customers in other parts of the country.

The main focus of controls in mixed systems is the domestic customer: there is a general recognition that little potential exists for competitive substitutes in this market and that some regulation is required. In Germany, the rules on price control are contained within the Bundestariff Ordnung Elektrizat (Federal Directive on Electricity Tariffs) which sets conditions for the types of tariffs to be offered to domestic, agricultural, and small industrial consumers. Under this regulation, Tariffs Supervision Authorities examine and regulate the rates of increase at the Laender level. In the Netherlands, the Ministry of Economic Affairs can intervene if it considers the rise in prices too great.[15]

Regulation in decentralized systems tends to be weak as much because of the minimal federal involvement as in spite of it. As noted, the development of ownership on a decentralized basis (with considerable local or regional government participation) was accompanied by the allocation of most regulatory powers to regional authorities, as for example under the German 1935 Energy Act. With ownership and authority coinciding in many cases, the controls on planning and investment in the 1935 Act have not been widely used. By contrast, in a case like North Rhine Westphalia, where the Land is not a shareholder in the principal utility RWE (which is owned by private investors and smaller municipalities), the Land government is willing to exercise these powers. Even so, the resources it applies are fairly small (according to officials, only two full-time personnel were involved in scrutinizing tariff proposals).

As a response to weak and sometimes inconsistent regional and local regulations, the national governments in both the German and Dutch cases are becoming more involved in a quasi-regulatory function. Furthermore, national concern for overall energy policy requires at the minimum a co-ordinating role. But, in the absence of direct regulatory authority, broader policy instruments are sometimes also represented. For example, the Dutch use national general price control laws in their efforts to reduce inflation.

Dominant State Incumbent

As might be expected, the Swedish system of mixed ownership with a dominant State incumbent employs elements of both regulatory systems. Again, granting a concession effectively exempts the utility from cartel law; the main State involvement is an investigation of tariffs by the SPK (the State Cartel and Prices Authority). The SPK reviews the tariff proposals of the two largest utilities for the government. Low voltage customers are also able to complain to an energy prices arbitration group within the energy department if they consider that they are being unfairly charged or discriminated against.

The principal method by which the government regulates is indirect, through ownership of the largest utility. In particular, the tariffs set by Vattenfall (as lowest-cost producer) effectively set prices for the ESI as a whole. The

---

[15] See International Energy Agency (1985).

government also controls Vattenfall through ownership, though not as comprehensively as is the case in the nationalized model discussed above. Nonetheless, budgetary controls, the process of State auditing, and the possibility of stronger intervention by the government are powerful constraints on the activities of Vattenfall (and, through its key role in the sector, on the rest of the industry, public and private) (National Economic Development Office (1977) and Verney (1959)). A key element in the Swedish system, moreover, is the extent to which informal contacts among the various elements in the ESI, and between them and the government, obtain many of the objectives of a more rigid regulatory structure. In this, Vattenfall plays a key role, backed up by the threat of more explicit government action against the utilities.

To summarize, regulation varies from system to system in its intensity and the means of application. All systems accept that ESIs contain natural monopoly characteristics, and have historically granted monopoly concessions and therefore effective exemption from cartel law. The cartel law exemption is justified either because ownership supplants it for control, or because competition from other utilities or fuels limits monopoly exploitation. The nationalized systems in practice impose the greatest regulatory obligations on their industries, closely followed by State incumbent systems. In mixed systems, where the structure is perhaps closest to a private industry, regulatory control is lightest.

Overall, however, there is a core element of regulation in all systems, focusing on domestic prices. Given, therefore, that regulation exists to some degree in all systems, what form should it take?

Modes of Regulation

Traditionally regulation has focused on the rate of return (as in the US). This method has been subject to much theoretical and empirical criticism (Averch and Johnson (1962) and Joskow and Schmalensee (1983)) and its shortcomings have prompted the development of other systems of regulation, primarily focusing on price. In this section, we first outline the respective merits of these systems and demonstrate that each requires, both in theory and in practice, a common basis in costs. The principle of long-run marginal cost and its practical analogues therefore lies behind both methods of regulation. Despite the apparent radical differences, price and rate-of-return regulation have much in common when applied to European ESIs.

Having demonstrated the common basis of these systems, we explore the practical problems of deriving reliable information on costs. We discuss the concept of 'cross-regulation', and the extent to which independent or 'yardstick' information can be obtained. Finally, we relate these practical difficulties in implementation to our three models of ownership and organization.

Rate-of-return versus Price Regulation

The dominant mode of regulation in the US and many European countries is the rate-of-return method whereby the regulator sets prices according to a maximum permitted rate of return on capital employed. By limiting the profits that utilities make, it is argued, price increases are constrained and monopoly incentives overcome. As has been noted in both the theoretical literature and the empirical evidence, however, such a system contains little incentive towards reducing costs. The Averch–Johnson theorem demonstrates that rate-of-return regulation can tend towards capital bias.

One alternative to the rate-of-return method is to base regulation on price. In both the UK and France, new price-based regulatory rules have set the allowed rates of increase below the rate of inflation, thereby taking into account the cost differences that the industry faces compared with the general increase in prices. This system of price indexation has been made explicit in the 'RPI $-X$' regulatory rule for UK privatized industries. This rule may well be applied to a privatized ESI.[16]

However, there are a number of drawbacks with the system as it is currently implemented. The first is the lack of relevance of the RPI to cost structures of the ESI. While labour and certain other components may be significant, the fact that the ESI is capital- and fuel-intensive suggests that a different index may be of greater relevance. The X factor should adjust for this in theory; but the design of it so far seems to replicate some of the shortcomings of rate-of-return regulation, and the method may fall into many of the traps of regulatory capture and capital bias symptomatic of rate-of-return regulation: the requirements for information and for preventing unnecessary investment remain (Helm (1987a)). Nonetheless, a better designed X (in particular taking into account the performance of other ESIs) could act as a powerful basis for regulation. In this sense, the French use of price indexation as one of a number of regulatory measures may be more effective.

Both rate-of-return and price regulation ultimately derive their rationale from a notion of long-run marginal cost (LRMC) pricing. Though this principle has been seen as primarily a creature of publicly owned utilities, it prevails throughout the European ESIs. Determining the rate of return and determining X in RPI $-X$ regulation requires detailed information and judgements about costs. In assessing costs, the time period is crucial, and countries adopt various measures of the long run (and even within countries there is considerable debate on how it should be determined). It is seen as providing a predictive element by which to commit investments for future capacity.

---

[16] On the design of RPI $-X$, see Littlechild (1981) and Helm (1987a).

Deriving Cost Information

Having established that both price and rate-of-return regulation are dependent on cost information, we now look at the regulatory problems associated with deriving this information, prior to examining how our three models cope with these requirements.

The first problem we can identify is that of *cross-regulation*. Cross-regulation can arise in circumstances in which joint costs are allocated between two activities affected either by a general regulatory rule (such as $RPI - X$) or by differing regulatory rules. The firm can adjust prices to some extent independently of the relative costs of each activity, and thus cross-subsidize between monopoly and competitive sectors. The problem for the regulator in deriving information is to assess how far costs are correctly attributed to the different elements of the industry. For example, where a distribution utility is engaged in the supply of more than one product, one of which is regulated and one of which is relatively unregulated, the regulator may have problems in assessing whether the allocation of joint costs is fair or whether it masks cross-subsidies. Thus the British gas industry is regulated in its relatively monopolistic domestic market but not in the more competitive industrial sector. There is scope for the British Gas Corporation to pass on joint costs to its domestic sector, giving itself a competitive advantage against rival fuels in the commercial market. A similar situation prevails in many European ESIs.

This problem is closely related to that of *regulatory capture*. Regulatory capture arises when the target has a monopoly or quasi-monopoly control over the information required by the regulatory agency to carry out its function. The possibility for capture may decline as competitive sources of information become available, facilitating a degree of 'yardstick competition' (Vickers and Yarrow (1985) and Shleifer (1985)), though this may be offset by the degree of government ownership or the existence of formal or informal mechanisms of self-regulation or collusion.

Finally, problems arise with the difficulties of forecasting and the extent to which non-cost constraints (giving second-best results) distort implementation of regulation based on cost data. To the extent that LRMCs are based on future projections of demand and supply, they are prone to forecasting errors. These errors will tend to be biased towards excess capacity where the industry is publicly owned, reflecting a desire for security of supply and the managerial objectives of nationalized industries. Indeed, excess capacity as a consequence of over-optimistic growth forecasts is the norm throughout Europe, and this can act (as we noted above) as a serious disincentive to private sector investment. Extraneous (often political) factors impinging on the ESI may dictate different principles for the setting of input prices (e.g. coal subsidies) and these may further undermine the usefulness of LRMC.

Regulatory Focus and Ownership

These general problems are reflected rather differently in the case studies, since ownership and organization directly affect the type and status of information available to the regulator. We now look at each of our three models in turn.

In *nationalized systems*, there has been a move towards the use of price-based systems of regulation: in France the price indexation formula has been employed, while in the UK it has been used in privatized utilities. Deriving the cost information is facilitated by the accounting separation of different parts of the ESI, as well as through scrutiny by institutional mechanisms—Monopolies and Mergers Commission reports in the UK and through the efficiency audits in the French system. The selection of the appropriate measure of LRMC, however, is less clear-cut, to the extent that considerable distortions are imposed by financial obligations in both (either in the use of external financing limits in the UK or the massive nuclear investment programme in France). In addition, both systems have a tendency to over-capacity based on poor forecasting records. Such factors have undermined the practical usefulness of LRMC pricing, even though the nationalized industries are theoretically the best placed to carry it out effectively.

In the *mixed systems*, especially in Germany, rate-of-return regulation has been the principal method employed. However, the effectiveness of such regulation has been questioned, given the poor resources devoted to regulation by Laender governments and despite their efforts to develop cost-reducing incentives.

The decentralized nature of these systems yields considerable scope for some yardstick competition. However, the paucity of regulatory resources must work against this. Such regulation may also be weakened by the reliance of regulators on utility associations such as the VDEW, which acts as a forum for the German ESI and in many respects performs a self-regulatory role. Informationally it is at a considerable advantage *vis-à-vis* individual Laender regulators. In the Netherlands, the pattern of self-regulation and of information provision is even tighter, with producers, industry associations, and the transmission organization closely knit. Until recently, they have been left to determine most aspects of policy. To some extent the government's efforts to reform the organization of the industry, noted above, are motivated by the need to overcome these obstacles to effective regulation. LRMC is employed to a much lesser extent in these countries, especially where private sector firms concentrate on average rather than marginal project returns. Additional second-best distortions arise in both countries from the use of utility revenue for wider expenditure, and in Germany from the so-called 'coal penny', a tax on electricity to subsidize the use of domestic coal in power-stations (though arguably the fact that this is explicitly done is superior to the UK system of obliging the CEGB to purchase UK coal at higher than market prices).[17]

In the *Swedish model*, there are certain advantages in information gathering being performed by the government (as owner of Vattenfall) and the SPK. The extent of regulatory capture may still be significant, however, as a result of the 'Swedish system'. On the one hand, it may hinder the process of gathering information, through the network of formal and informal contacts and the 'search for consensus'; on the other, it is also designed to gain a regulatory result, backed up by threat of more direct action. On the more precise issue of pricing, the country has made efforts to move away from direct LRMC towards the use of short-run marginal cost (SRMC), in the price-based co-ordination of its production, through the Nordic grid power pooling system. Price information partially replaces the need for cost data. This important regulatory advantage will be returned to below.

Whatever the merits or drawbacks of different modes of regulation, all require cost information. We have identified some of the problems of deriving accurate information and have indicated the importance of comparative or competing sources of information. In this respect, if no other, competitive mechanisms within the industry are vital. To these we now turn.

Competition

The presence of potential and actual competition is a determining feature of the impact of regulation. There are two considerations here—the impact of regulation *given* the degree of competitive pressure, and the scope for intensifying competition. In this section we concentrate on the latter. We outline the major sources of potential and actual competition, focusing primarily on production. These are analysed at three levels—competition between existing large-scale producers, competition from new entrants, and competition between countries at the European level.

So far we have noted the impact of other sources of fuel supply, rather than direct competition within the electricity sector. In recent years the latter has become an issue in its own right (particularly with increased interest in the potential of cogeneration in many countries). The reason for this is partly disillusionment with the performance of existing public sector firms and partly the result of further changes in technology. We therefore briefly examine existing competition, before turning to potential entry at the national and international levels.

Competition amongst Established Suppliers

In practice, competition amongst established suppliers is rare (though it may manifest itself in the extent to which large customers choose their location on the basis of the cost of power). Such competition is, of course, absent in wholly nationalized systems. But also it does not play a large part in most decentralized

---

[17] On the Kohlpfennig, see James (1982, pp. 192-4).

and mixed regimes, thanks to the principle of regional monopoly: although the systems as a whole are decentralized, distributors are generally constrained to purchase from the regional supplier. The proposed reform of the Dutch system by breaking the regional link may provide for such competition amongst suppliers.

The most extensive use of competitive mechanisms is found in the State incumbent model. The Swedish power pool allows the largest public and private producers to determine production through a quasi-spot market in power transfers, based on the lowest available costs. It replaces the need for a planned merit order system (as used in the UK) by a market-orientated price system. It is not, however, a pure market one, being confined to established large producers and hence oligopolistic. Entrants on a smaller basis have to sell to their local concession holder (as in most of the rest of Europe).

Competition and Entry

New producers may be encouraged by the changes in entry terms. In this respect, regulation can enhance the potential for competitive pressure. There are two stages to such regulation. The first is to gain access to the grid or to a particular distributor. Common-carrier provisions are, however, quite limited in a number of European countries, most notably in Germany. The second is to have terms of entry that make competition worth while. In practice, this has proved the major stumbling-block to wider competition.

Most entry potential lies in cogeneration. In many areas, power produced as a by-product was the initial source of supply. More recently, aside from meeting the needs of the producer, industrial cogeneration has made a significant contribution to public supply (as much as 5 or 10 per cent). Since the 1970s, however, it has suffered partly as a result of the oil crises (with hydrocarbon-based plant suffering major fuel cost increases) and partly as a result of increased cheap power production from the ESI itself.[18] The decline of cogeneration has, however, given rise to concern over the terms on which such power is purchased. Moreover, concerns over mix of energy supply, particularly in the light of the decisions of certain governments to phase out or suspend nuclear power, have increased interest in cogeneration and alternative sources of energy, both of which could be expected to be provided by small producers either as by-products or as their main activity.[19]

Nationalized systems have normally taken power on the basis of the fuel cost to the supplier, which for most potential producers has been uncompetitive with the price at which they can purchase power from the system. The UK has tried to

---

[18] Hyden (1985) discusses the role of industrial cogeneration in Sweden. The Swedish government's energy policy aims to encourage such power generation in the wake of the decision to phase out nuclear power by 2010.

[19] The Swedish Energy Administration is reviewing entry conditions, while the VDEW recently reviewed its agreement with small producers, and, as noted, the Dutch reform proposals are designed to encourage such small-scale generation (see Commission of the EEC (1984a, p. 144)).

encourage more private production by improving the terms of entry, and it may be that EdF will be required to move in the same direction by the present French government. The results of the UK experiment have been meagre, however, partly because the terms of entry are still set by the incumbent supplier (Hammond, Helm, and Thompson (Chapter 8, this volume)).

A similar pattern emerges in the *mixed systems*, where the industry has been obliged to establish rates of purchase for cogenerators under threat of government action. Such government pressure is typically indirect, through utility legislation, licences, and negotiation with regulatory authorities. The Dutch reforms are designed to allow greater small-scale production, particularly by the distribution companies themselves. An additional obstacle to entry in Germany is the 1935 Act, which gives the Laender the power to prohibit new capacity construction. This power has on occasion been used against proposals for new CHP schemes, normally where the Land itself is the main supplier of power to the region.

In *Sweden*, cogeneration is well established and is encouraged by the new non-nuclear energy policy. There are still complaints about the terms of purchase and the government energy administration department is currently investigating these.

International Trade in Electricity

To the extent that reforms of entry conditions are being considered within the EEC, the possibility of trans-border purchases arises. Due to the differentials in energy costs across borders, industrial customers may consider the import of cheaper foreign power. International transfers of electricity occur between utilities regularly for purposes of peak load management and in most cases broadly even out. The nature of the trade is changing, however, as surpluses and short-falls in European electricity supply have emerged. The existence of substantial over-capacity in the French nuclear programme has rendered the country a substantial net exporter of electricity to its neighbours.[20] However, attempts to sell power direct to customers rather than to utilities have largely failed, as EdF and local utilities have colluded to prevent sales, based on the principle of reciprocity. In particular, a number of large German companies have attempted to buy French power direct but have been obstructed both by the refusal of utilities to transmit it and by the government's refusal to allow new lines to be constructed.

If, however, there is a trend towards comparative advantages and surplus capacities amongst the European ESIs, and if there is a trend towards growing competition within national systems, especially in energy-intensive industries, there is the basis for a common European power market. The market will

[20] On trade prospects from France to the UK, see Dowlatabadi and Evans (1986). See also the Annual Report of Nordel for details of trade in Scandinavia. On the European trade position, see United Nations Economic Commission for Europe, Committee on Electric Power, *Electricity Statistics*.

develop gradually from the bilateral contracts between EdF and neighbouring utilities. Arbitrage between these contracts and competition for industrial customers bypassing the local utilities will intensify the development of the European electricity market. Furthermore, this market will be based on price rather than quantity or merit order planning, and hence replicate many of the features of the Nordic grid. The parallel is marked also to the extent that the role played by EdF will be similar to that of Vattenfall.

But because the new market will be price-based, it does not follow that regulation is redundant. Rather, since the natural monopoly element has moved to the European rather than national level, regulation will be required to coincide with it. Without co-ordinated regulation at the European level, competing national regulatory agencies may encourage utility migration, and thus force the weakest regime to dominate. Experience in Germany, noted above, with competing Laender regulators, may be repeated at the European level. A pro-nuclear policy in France will encourage utilities to enter joint ventures with EdF, rather than build better-located plant in Switzerland or certain West German Laender.

One possible starting-point is the Treaty of Rome. Its provisions on competition policy could be used to prohibit the abuse of market position by the established utilities. Controls on public enterprises could be adapted to require transparency in costs and prices.

5. Conclusions

This paper has provided a partial survey of a number of European ESIs. We have looked at the importance of ownership, the impact of organizational structure, the plurality of regulatory approaches, and the scope for and impact of competitive entry. In this final section we present a number of tentative conclusions and policy implications for the UK relevant to the privatization debate.

The first conclusion relates to ownership. Despite the current vogue for privatization in the UK (and to a lesser extent in a number of other European countries), it is hard to avoid concluding that private ownership *per se* makes very little direct difference to performance. Though much more quantitative research on the relative performance of utilities in the two sectors is required, qualitative evidence gathered suggested that the private sector tended to underinvest, and that technical performance (particularly in small-scale distribution) tended to be lower. Certainly there is little evidence to suggest technical inferiority for the UK and French nationalized systems.

However, if ownership *per se* has little direct effect, the interaction of ownership on the one hand, and particular organizational structures and regulation on the other, is of great importance. In the paper we have identified three stylized types of ownership and organizational structure, and demonstrated the impact of these on performance. We focused attention at the organizational level on vertical and horizontal integration. Vertical integration between production and distribution gives rise to cross-regulatory difficulties and reduces the amount of information available to regulators. Horizontal integration between different industries yields billing and metering economies (though there is no reason why specialist firms should not take on these roles), but reduces information and yields cross-regulatory incentives. We therefore conclude that vertical and horizontal disintegration increases the potential for efficiency incentives, improves regulatory implementation and monitoring, and enhances potential competitive entry. There is thus evidence to support the division of the ESI in the UK prior to privatization and explicitly legislating against joint activities between electricity and gas.

Regulation in Europe was identified as pluralist. Nevertheless we have shown that both price and rate-of-return regulation display the common characteristic of ultimate dependence on cost information. The similarity of regulatory problems across the European ESIs is explained by the informational problems of deriving accurate cost data. We have identified a number of these and shown how each of the different stylized types copes with them. The implications of this section of the study are that the debate for the UK should focus less on the relative merits of $RPI - X$ and rate-of-return regulation, and more directly on potential sources of independent cost data. Dissecting the industry to avoid vertical and horizontal integration separates out cost and profit centres, and hence tends to generate more data and encourage the development of yardstick competition.

Finally, the study looked at the scope for enhancing competition. This we identified at three levels. First, competition between producers could be increased by the development of transparency in contracts. The most obvious way of doing so follows the Swedish example of introducing spot prices. Second, new producers could be encouraged to enter the system by explicit tariffs and regulation of the terms of entry. The UK Energy Act, despite its problems, forms one of the better European examples, inasmuch as it sets out to establish published private purchase tariffs (PPTs) and common-carrier facilities. Another aspect of competition emerges at the international level, where the possibility of trade is increasing. The development of French exports is likely to be the principal source of inter-European trade in electricity. This trend reflects the gradual technical movement of the natural monopoly element from the national to the European supergrid level. As arbitrage between bilateral sales develops and as the EdF in particular attempts to sell directly to companies rather than to utilities only, a European market will develop. Changes in the structure of the UK ESI should be designed to take this into account.

Part IV

Gas

14 Gas Privatization: Effects on Pricing Policy*

CATHERINE PRICE

1. Introduction

In November 1986 the government sold off the largest nationalized industry so far included in its privatization programme—British Gas. This paper addresses the issues of how privatization will affect the pricing decisions of the industry, taking into account the change of ownership, the different objective policies which this implies, and the market and regulatory environment within which it will operate.

To assess the likely effects of privatization, we must examine as a base-line the policies that the industry has followed while nationalized. However, this is not quite so straightforward as it may seem. Relevant factors include the policies that the government has recommended for nationalized industries, the extent to which the gas industry has followed them, and the effect of changes in both the objectives of the industry and its constraints when privatized.

Slater (Chapter 7, this volume) examines at some length the rationale for marginal cost pricing, and he considers as an example of some of the difficulties involved the case of setting a price for North Sea gas. He shows that there is in general a prima-facie case for marginal cost pricing in nationalized industries, and that governments have, on the whole, recommended such a policy (though not always unambiguously).

However, the difficulties of defining long-run marginal costs, and in the nature of the concept itself, have given rise to uncertainty about their magnitude. This has given the nationalized industries some discretion in applying the policy, so that there is an opportunity for them to follow their own objectives. Rees (Chapter 5, this volume) provides both an intuitive explanation of what these may be and a model to explain how a nationalized industry may follow its own objectives subject to government constraint. The management of the gas industry, and particularly its chairman Sir Denis Rooke, have been a powerful lobby vis-à-vis the sponsoring Department of Energy. There is evidence that the industry has pursued a policy of long-term sales maximization (as Rees's model would suggest, though not in the way that it predicts). Thus some markets have been protected by charging low prices, particularly the so-called 'premium' markets (domestic and firm industrial). This is in contrast to Rees's prediction that prices would be raised more above costs in those markets that had inelastic demand, but can be explained by the desire to capture and protect these markets in the long run. Here the dual nature of fuel prices, as an indication to

---

* Catherine Price is a Lecturer in Economics at the University of Leicester.

consumers both of appropriate investment in fuel-burning appliances and of the intensity of their use once purchased, is important. In persuading the premium markets, which cannot easily switch to alternative fuels, to invest in gas appliances, the industry assures itself of a tied market as the price of gas rises in the future.

The likely effect of privatization on these policies depends both on how far the shareholders can enforce a new profit-maximizing objective (rather than the long-term sales maximization policy of the management) and on how effective the constraint on average revenue will be. Rees has extended his model to look at likely effects of privatization, and this paper applies the analysis to the specific influence of the average revenue constraint on price discrimination and differentiation in different markets and activities of gas supply. How far these various tendencies will be realized depends on their relative strength, as well as on institutional factors, such as the influence of the Office of Gas Supply (OFGAS).

2. History of Gas Pricing in the Public Sector

The body that was privatized as British Gas PLC grew out of an industry nationalized and rationalized in 1949. The British Gas Corporation (BGC) was incorporated in the Gas Act of 1972, which vested in the new Corporation the interests of the twelve Area Boards and the Gas Council which had been set up in 1949. The 1972 centralization arose as the industry's activities changed from those of a production and distribution business (whose activities could sensibly be divided regionally) to that of a centralized distributor of North Sea gas, which relied on a national transmission network. The Area Boards retained functions as local cost and administrative headquarters (for example the issuing of bills), but the basic organization was unified, and decisions on policy (including pricing) were undertaken centrally. (This is in contrast to the electricity industry, which maintains a regional structure similar to that of the gas industry before 1972, including autonomy in pricing decisions for each area.)

The BGC inherited obligations from the previous Area Boards which were little changed from those laid down by the nationalization Act (Gas Act, 1948), since the reorganization was undertaken primarily for administrative purposes. Thus the clauses giving guidance about pricing remained as they had been, with their attendant ambiguities. There were two such obligations in the Act—that the industry should cover costs taking one year with another, and that it should show no undue discrimination between consumers. Both these clauses, and the general pricing policy of nationalized industries, had been subject to debate and reinterpretation in the twenty-three years between 1949 and 1972, but the basic requirements on the gas industry were little changed or elucidated by the 1972 Act itself.

The formal obligations of the industry gave little specific guidance about prices. More informally the government had come to recommend a basic policy of marginal cost pricing in nationalized industries (see Slater (Chapter 7) for discussion both of the rationale of such a policy and its application to the fuel industries). The government seemed to recommend such a pricing policy, particularly in the 1967 White Paper (HM Treasury (1967)), though the industries themselves showed some reluctance to accept it. And the guidelines remained unclear on some important aspects. In particular, the issue of marginal cost pricing where this led to losses because of decreasing average costs was never satisfactorily resolved. This was especially relevant for gas, where both economies and diseconomies of scale operate. In the local distribution sector, gas is a classic 'natural monopoly' with average costs decreasing over the entire range of market demand; thus marginal cost pricing for this sector of operations would mean an inevitable loss. But in gas production a combination of historical accident and the exploitation of less accessible and more expensive reserves as demand increases raises marginal costs above average costs; thus marginal cost pricing in this sector would generate profits. These factors are discussed in some detail in both Newbery's and Slater's chapters (2 and 7 respectively). The exact balance between the economies of scale in distribution and the diseconomies of scale in production is difficult to determine, but in recent years diseconomies of scale have predominated.

How far the industry has itself pursued marginal cost pricing is not clear. The fact that it has generated large profits over the past decade suggests that it may indeed have done so. It has been subject to considerable political influence, both overt and covert—in the early 1970s it was required, with other nationalized industries, to keep prices down in the interests of counter-inflationary policy. A decade later another Conservative government required that prices be raised in order to reduce the public sector borrowing requirement and so fight inflation. More specifically, the gas industry was subject to detailed instructions on its domestic gas prices—first when Mrs Thatcher vetoed an increase when she took office in the summer of 1979, and six months later when the Secretary of State for Energy announced a 30 per cent real increase over the next three years (Howell (1980)).

The industry's profitability has similarly been viewed with mixed feelings over the past ten years. When it first became apparent, current cost accounting was introduced with much greater enthusiasm than in other nationalized industries, presumably to obscure the large profits, often seen by the public as evidence of exploitation. Various other accounting alterations were made over the years, which progressively diminished the apparent size of the profits. (This trend was reversed for the 1985–6 financial year, when privatization was expected, so that declared profits in that year were rather larger than they would have appeared under the previous system.) The government itself felt some threat from the industry's profitability; its ability to finance internally all investment in recent years has deprived the Treasury of its classic 'hold' over nationalized industries, namely refusal to sanction the investment programme. One government response was the introduction of a gas levy, a tax on the rent which the BGC obtained from fields signed under old agreements at very low prices and not subject to petroleum revenue tax. This was supplemented by a negative external financing limit, effectively forcing the industry to lend its profits to the National Loans Fund. All of this suggests that the government was trying to make acceptable the profits that resulted from marginal cost pricing. Indeed, in the House of Commons debates in January 1980, the Secretary of State for Energy justified the price increases largely in terms of the cost of marginal supplies from the North Sea (Howell (1980)); further price increases in 1984 were also defended by the Department of Energy on marginal cost grounds (Energy Committee (1984)).

Some aspects of marginal cost pricing have proved convenient to both industry and government. The consequent profitability gave the industry independence and an obvious political advantage with the government, and the chairman, Sir Denis Rooke, has never been slow to point out the contribution that the BGC has made to the exchequer. But the main objective of the Corporation during the late 1970s seemed to be more closely related to sales maximization and the long-term security of particular markets, than to resource allocation. Evidence on this can only be 'circumstantial', but is suggestive nevertheless. For a number of years the BGC refused to acknowledge that the gas market would eventually become more expensive to supply, and that demand would consequently decline. Indeed, forecasts of demand seem to have been made at some times totally independent of any assumptions about relative prices, suggesting either that price elasticity of demand is zero or that prices were fixed in order to achieve a given level of demand (rather than prices (and costs) determining appropriate demand).

Such protection of markets raises the issue of charging differential prices to different sectors of the market. We have so far discussed marginal cost pricing in the context of the general *level* of prices, and the argument that this policy maximizes welfare. However, in practice, industries have operated within overall financial constraints of various kinds, particularly the financial target and capital rationing (imposed by external financing limits) which Rees analyses. In such a case, the optimal pricing from the welfare point of view will be Ramsey pricing. Rees's model suggests that the pursuit of other objectives means that the firm is unlikely to adhere to Ramsey pricing. In any case, it is clear that blanket control on the industry, whether a financial target or (as under the privatized regime) a maximum average revenue constraint, gives the industry discretion to alter the balance of prices (and in particular their relation to costs) so as to achieve its own objectives within the control that the system exercises on the industry as a whole.

In the gas industry, such discretion can be exercised in a variety of contexts, most obviously in the different net prices charged to premium and non-premium markets. Costs vary between markets because of different average consumption per consumer and the varying proportion of demand that occurs at peak; peak demand determines the capacity of the system, and so demand that consists of a high proportion at peak is more costly to supply. The different prices charged to each market arise largely because the pricing system operated by the gas industry does not separately identify individual characteristics of demand, such as peak consumption. But the differential profitability of the markets shows how the industry can vary prices within an overall constraint on level of prices or profitability.

Rees's model on the objectives of nationalized industries suggests that a nationalized industry will set prices according to what the market will bear, i.e. prices will exceed marginal costs more in relatively price-inelastic markets. However, in the BGC's case the dominant objective seems to have been to maximize long-term sales revenue by *protecting* the premium markets. This suggests a recognition that they would be especially valuable to the industry as gas costs rise in the future and more fickle markets switch to the newly competitive alternative fuels. Such potential loyalty typifies both the domestic market (where much equipment lasts twenty or more years, and so movement between fuels is correspondingly slow) and the firm industrial market, where consumers do not have a readily available alternative fuel-burning capacity. Rather than raising prices in these comparatively price-inelastic markets, the BGC pursued a policy of lower prices and maximum market capture. By the end of the 1970s, there was a considerable discrepancy between price and 'beach-head cost plus distribution costs' in the domestic and industrial markets, with consequent differences in net profitability. The Corporation seemed to be pursuing long-term market maximization by favouring the least price-elastic markets (contrary to the short-term profit-maximizing behaviour of a discriminating monopolist or of a nationalized industry maximizing short-term turnover).

However, the BGC has eschewed a more direct form of price differentiation between regions. Each region costs a different amount to supply because of its distance from the beach-head(s), and regional differentiation of gas prices would be comparatively easy and cheap to implement through the existing regional structure within the company. As in the case of different markets, failure to implement regionally differentiated prices means that gas supply to each region yields different profitability. The motive for uniform prices is not obvious; the BGC has maintained that a system of differentiated prices would yield insignificant benefits and be unpopular with consumers (Energy Committee (1985)). While the benefits are not enormous, the costs of implementation would also be low, and any unpopularity with consumers who paid more would presumably be matched by the gratitude of those (near beach-heads) whose gas would become cheaper. Other fuels (including electricity and oil, gas's main competitors) are priced differently in different parts of the country, without attracting the odium from consumers that the gas industry fears. Perhaps the most realistic explanation is inertia, not surprising in such a large and profitable institution.

This failure to reflect fully the costs of supply to different groups constitutes what Phlips describes as the classic form of price discrimination in spatial economics. 'Sales at uniform delivered prices—where all consumers pay the same price regardless of their location—are treated as examples of indisputable discrimination that favours more distant customers "against" those who are located close to the seller's plant' (Phlips (1983)). This statement refers to the spatial case (as in regional pricing), but the model can be extended to other characteristics of consumption with analogous results. The lack of an explicit peak-load pricing system also constitutes such discrimination, and the inequality of net profitability between markets (which arises largely because of different costs from varying peak demands) is an implicit expression of it. The size of the standing charge may also discriminate against some consumers.

Here, some expansion of the marginal cost pricing principles discussed earlier is relevant to determine the optimal behaviour of a firm operating under minimum profit constraints. Nationalized industries have been effectively operating under such constraints imposed by financial targets and the government's reluctance to accept marginal cost pricing where it results in financial losses for the industry concerned. In such a case, the optimal prices to charge (i.e. those that maximize welfare) are Ramsey prices, which can be characterized as prices that diverge from marginal costs in each market in inverse proportion to the price elasticity of demand in that market. Another property of such prices is that marginal profitability is proportional to the difference between price and marginal cost in each market. This would have been the optimal pricing policy for the BGC if it were aiming to maximize welfare in the markets that it supplies. The lack of price differentiation between regions and time periods suggests that it did not pursue such a policy.

The BGC has for some years been defensive about pricing, showing little enthusiasm for open debate. In particular, it has dismissed out of hand some suggested policies for increased price differentiation. When the Energy Committee broached the question of charging different prices for gas consumed at peak and off-peak periods, the Corporation's reply was revealing. It maintained that it could not justify charging more for gas at a time (the winter) when people most needed it (Energy Committee (1985)). Yet this is precisely when a higher price would be justified by the additional utility that the gas yields to the consumers and by the higher cost of provision. A peak-load pricing system would be expensive to implement, and these administrative costs might counteract the benefits that it would yield. But the argument that gas should not be more expensive when most valued suggests the same sort of misunderstanding of the pricing mechanism and its role in resource allocation as does the premium markets argument.

One aspect of pricing which has attracted particular public attention is that of standing charges. This is the part of the tariff that consumers pay as a fixed cost, regardless of how much gas they use, and it has a long history of public misunderstanding. Because there are some costs that a consumer incurs, regardless of how much gas is used, it is appropriate to recoup such costs separately from the charge made per unit of gas used. The BGC had gradually increased the proportion of its revenue received in this way, so that by 1982 the standing charge to domestic consumers reflected reasonably accurately these fixed costs of being a consumer. However, the government introduced a rebate scheme (Howell (1982)) whereby consumers using only a small amount of gas would not be charged more in their standing charge than they had to pay as commodity charge for the gas used.

This scheme had popular appeal, but the rebate made little economic sense, since the costs properly reflected in the standing charge are independent of the quantity consumed. In effect, the scheme introduced discrimination in favour of users of small amounts of gas, at the expense of large consumers, though the cost of the scheme was not large. The rebate scheme was more or less imposed on both the gas and electricity industries, and the costs of meeting it were borne by the industries concerned. Even claims that the rebate scheme was beneficial on income distribution grounds (because users of small amounts of fuel would be poorer) were not justified by the evidence on fuel consumption. A case is sometimes made that fuel has particularly meritorious characteristics in consumption, because it forms a high proportion of expenditure of poor households (see Dilnot and Helm (Chapter 3, this volume)). The rebate scheme was designed to help such households, but analysis of the beneficiaries suggests that it was both wasteful and mean, i.e. it failed to help a large proportion of the target group and did benefit many who were not in this group (Gibson and Price (1986)). The standing charge rebate scheme was phased out after three years.

3. Efficiency of BGC Pricing Policy—Summary

The overall objectives of any nationalized industry are difficult to assess, particularly in the light of the confused framework and advice under which they have operated during the past forty years. However, the tensions between government and industries suggest that there has been some divergence in the objectives of the two bodies. Thus even if the *government* had pursued a policy of welfare-maximizing prices (generally marginal cost prices in the fuel industries), the BGC has had its own objectives, not necessarily mutually consistent.

In the last few years, marginal cost pricing has been a policy broadly favoured by government. While the BGC may have been happy to accept this policy for the general level of prices, the Corporation has tried to protect at least those markets that are likely to be loyal to it in the long term, at the expense of profits if necessary. (This is not surprising, given that it is the government, and not the BGC, that has received the profits, though the BGC has not hesitated to make political capital out of its contribution to the exchequer, where appropriate for its own purposes.) It is difficult to formalize the objective of long-term sales maximization (and eventual market protection) that the BGC has been following, partly because it depends so heavily on the gas industry's own perceptions of the future. However, from the point of view of economic efficiency it has clearly led to some distortion from the optimum in prices charged to different groups of consumers.

4. Pricing Efficiency in the Private Sector

Since privatization at the end of 1986, British Gas has operated subject both to different objectives and to different controls. The new owners are likely to be profit maximizers and to encourage the industry in this direction. The industry itself is largely unchanged as a management unit from that of the British Gas Corporation, and so the management's own objectives may be expected to continue, including that of long-term market protection. But since the management also own shares in the industry, they may revise their own objectives in the direction of profit maximization. How far this occurs depends on how many shares the management hold, and how large a proportion of income the dividends form.

As far as the level of prices is concerned, profit *maximization* would suggest that marginal costs should equal marginal revenue. The changes necessary to achieve this can be appreciated from the present size of the discrepancy: in the domestic market, most estimates of elasticity lie between 0 and -1, i.e. the industry is producing beyond the profit-maximizing sector of the demand curve (since marginal revenue is negative). In the industrial market, long-term elasticities are generally estimated at about -1.5, which yields a marginal revenue of 6 in the interruptible industrial market. Even the most conservative estimates of marginal costs in this market would not put these much under 15 pence per therm, so the gap is considerable. It would, of course, diminish on two counts as sales decreased—the price (and marginal revenue) would rise and marginal costs would fall. Thus unconstrained profit maximization would push prices up and output down in both the industrial and the domestic markets.

Because unconstrained profit maximization in monopoly markets is not publicly acceptable, British Gas operates under regulatory constraints in the tariff market, i.e. that market in which annual consumption per consumer is less than 25,000 therms. This constitutes mainly domestic consumers, for whom gas supplies about 85 per cent of the available market, and its monopoly power is obvious. The contract market, where prices are individually and confidentially negotiated, is considered too competitive to need constraints. Thus the formal structure of British Gas is as a private monopolist subject to constraint in a sector of its market that presently accounts for about two-thirds of the industry's turnover.

How far it will pursue maximum profits in either market is a moot question, and depends on the balance of the inherited management objective of long-term sales maximization, pressure for profit maximization from the new owners, and the regulatory control.

5. Regulatory Constraints

The form of regulation is an average revenue constraint, rather than rate-of-return regulation which has been more usual in the United States. The latter had proved difficult on a number of counts; in particular it encouraged over-investment in capital equipment, in order to raise the base on which profits could be earned. Thus the restriction on average revenue was seen as a more direct and satisfactory constraint on monopoly exploitation. It was particularly appealing in view of the government's argument that privatization would encourage managerial efficiency, since with controlled prices the enterprise could increase profits by reducing costs. Indeed the drive for managerial efficiency that such a constraint imposes, particularly if the regulatory lag is sufficiently long to avoid cost improvements being recouped immediately through a tighter constraint, is one of the strongest arguments in its favour. The constraint itself applies to the tariff market. Average revenue in this market must rise by no more than 2 percentage points less than the rate of inflation in each year. However, this constraint applies only to the non-gas costs (i.e. costs of transmission, distribution, consumer service, and administration) and not to the costs of purchasing gas, which can be passed on in full. This constitutes the 'RPI $-X+Y$' formula, reproduced in the Appendix.

However, the regulation does have disadvantages. One of these, analogous to the effect of a financial target, is that any constraint that applies to the aggregate revenue will introduce distortions into the way a profit-maximizing company sets its price in each market. Any firm maximizing profit will discriminate between markets, but not by charging uniform prices. Profits are maximized if half the difference in costs is passed on to consumers (assuming linear demand curves and constant marginal costs). The main incentive to introduce price differentiation is from the profit motive itself, and complications arise from restrictions on average revenue. The constraint merely affects the nature of that differentiation. The Energy Committee (1985) suggested that both peak-load pricing and regional differentiation of tariffs should receive further consideration. The pressure to maximize profits might provide British Gas with just the incentive to introduce price differentiation which it would not undertake on resource allocation grounds as a public enterprise.

Since revenue is limited, profits can be raised only by lowering costs (one of the justifications for this form of regulation), and this can be achieved by expanding low-cost markets and contracting those that are expensive. Profitability can then be raised by reducing costs, without increasing average revenue (though this might eventually incur a long-term cost by triggering an increase in the value of X in the formula).

The effect of the constraint can be analysed in a number of different cases.

These can by typified as markets with similar demands but different costs (the regional case), different demands but similar costs, and the *hybrid* case of different demands and different costs (the 'peak-load problem').

For markets with similar demand patterns and different costs, all prices would be higher under an average revenue constraint than they would be under Ramsey pricing that yields the same average revenue. In markets with similar costs but different demands, prices under an average revenue constraint would be higher than Ramsey prices in the market with inelastic demand, and lower in that with price-elastic demand. In the 'mixed' case, like that of peak-load pricing, the results are unambiguous in the high-cost, high-demand market, where the price will be higher than Ramsey prices. However, in the off-peak period, the result is unclear, and price may be either higher or lower than Ramsey pricing, depending on the tightness of the constraint and the differences in costs and demand (Bradley and Price (1987)).

The average revenue constraint could be chosen to bring prices as close as possible to the social optimum, though even this would not achieve the optimum itself (marginal cost pricing in each market). However, there is disappointingly little evidence that the government has considered such factors in choosing the average revenue constraint. In particular the average revenue achieved with marginal cost pricing is not the optimal one to use as an average revenue constraint. Yet the average revenue is based on that inherited from a nationalized industry operating with marginal cost pricing, at least as the *nominal* policy. We can see the probable effect of this constraint on price differentiation in two of the cases discussed earlier.

How is the constraint likely to affect regional gas tariffs? Here the optimum solution is for the costs of each region to be reflected in the prices charged. The existence of an average cost constraint will cause the producer to increase profits by restricting demand in expensive (and higher-revenue) markets, and expanding output in the cheaper, lower-revenue markets. But compared with the existing price discrimination inherent in uniform pricing, there may be some improvement. The differences in price would not be large—a maximum increase or decrease of about 10 per cent of revenue in the domestic and firm industrial markets if present price discrimination through average pricing over regions were to be abolished entirely. But the profit-maximizing position would be for British Gas to continue to absorb half of the regional cost differences—so changes in price might be closer to 5 per cent. Since the industry retains much of its regionalized structure from its pre-1972 days, it would be comparatively easy and cheap to differentiate regional tariffs; bills are generally already sent out separately by each region. The industry's costs would fall (and profits rise) by about £½ million, and the resultant change in demand might be worth about £6 million. This is not a large amount by gas industry turnover standards, but it constitutes a comparatively cheap way to raise profits while obeying the tariff limitations, and a significant resource saving, nonetheless.

Consider next peak-load pricing. There are some difficulties associated with measuring peak consumption (more so for gas, whose meters generally do not have an electricity supply, than for electricity), but the technology of metering is changing so rapidly that some sort of feasibility study for peak-load pricing would certainly seem justified. In the mean time it would be possible to charge consumers according to their appliance ownership—this would at least give an (admittedly rough) idea of the relative expense of different uses of gas according to its time profile during the year—i.e. space-heating is more expensive *on average* per therm consumed than cooking, because a higher proportion is taken during the winter peak. Both costs of gas and the size of the distribution system are determined by peak demand. At present, tariff consumers receive no message about the higher cost of winter supplies (though in the contract market some indication is given through the availability of cheaper interruptible supplies).

If British Gas were to introduce some system of peak-load pricing, gas would become cheaper in summer and more expensive in winter. If demand is at all responsive to price, winter demand would decrease and summer demand increase—though this might take some time. Response would probably occur mainly in customers' choices of new appliances (not, as the BGC suggested, by everyone deciding to turn off their heating in cold spells). Thus the peak-load pricing system would be effective mostly through consumers' investment, rather than consumption, decisions. Such a scheme would reduce BGC's average costs, since the balance of expensive and cheap gas would change in favour of the latter. But since the restriction it faces is on average *revenue*, this would be a legitimate way of raising profits. Because little work has been done on time-of-use studies for domestic gas, it is difficult to quantify the implications for either demand or profits; but savings may well exceed the costs of such a scheme, and so justify its introduction.

6. Standing Charges

Immediately prior to privatization, the BGC announced a reduction of 20 per cent in the level of standing charges. This was no doubt largely to allay public fears, which had been mirrored in the requirement of the industry's post-privatization licence that it should not increase standing charges more quickly than the rate of inflation. In the light of this restriction, it is somewhat surprising that the industry lowered the base on which it would apply. From the point of view of economic efficiency, it is a retrograde step to move charges in any category away from the level that best reflects costs (and evidence suggests that standing charges were previously not above the optimal level). Such a change inevitably introduces distortions in consumption, and will result in some consumers using gas which yields them less benefit than it costs society to provide. Costs are likely to be significant both in terms of the industry's finances and in terms of resource efficiency; in the domestic market alone, the financial cost will be about £100 million per year.

The main reason for lowering standing charges was presumably to reduce consumer fears at a particularly vulnerable moment for the industry and the government. If public opinion is so strong as to cause the industry to behave in a manner apparently against its own interests, such pressure may continue to deter British Gas from outright profit maximization after privatization; though now that the 1986 Act is passed and the industry sold, it may be less receptive to such pressures.

The constraint is less binding on standing charges than on non-gas costs as a whole, i.e. those costs that arise from distributing the gas rather than from buying it. The tariff constraint formula allows for the costs of purchasing gas to be passed directly to consumers, while the $RPI - X$ restraint applies to non-gas costs only. The non-gas element of costs would then fall in real terms. Since standing charges should reflect only non-gas costs (gas costs should all be recovered through the running rate, which is a reflection of the costs per unit commodity supplied), there is no reason why the control on standing charges should be different from that for non-gas costs as a whole.

However, the net effect of the tariff formula is likely to be that gas tariffs *overall* will rise faster than the rate of inflation. The costs of gas purchase from the North Sea have been rising at more than double the rate of inflation (see Newbery (Chapter 2)). Since gas costs now constitute over half the total cost of gas supplies, their higher rate of increase will thus outweigh the real reduction in the value of non-gas costs that can be passed on to tariff consumers. Real average tariff revenue is likely to rise, and the clause on standing charges is more restrictive than on average tariff revenue, though less so than on non-gas costs alone.

7. Storage Costs

The exemption of gas purchase costs from the tariff formula seems reasonable, but it may have some rather odd (and quite unintended) implications for British Gas decisions. The most obvious potential distortion is where the gas/non-gas costs are substitutes for each other, i.e. in providing for the winter peak. Because more gas is demanded in winter than in summer, this is provided partly by buying more from the oil companies during the high-demand winter period, and partly by providing storage for gas bought in the summer but used in the winter. Both these methods have their costs—the oil companies that sell gas to BGC charge a higher average price per therm if there is heavy discrepancy between winter and summer supply (since it costs them more to supply gas unevenly); and if these costs are saved by having a more regular supply of gas from the North Sea, additional storage capacity must be provided, along with the running costs of putting gas into, and taking it out of, store. In the past, British Gas has presumably minimized costs by balancing these two methods of meeting peak demand so that the marginal costs of providing an extra unit of storage by either means are identical.

However, the regulatory formula invites British Gas to distort this balance by providing 'too much' of its peak capacity through negotiating uneven supply contracts. The additional cost of this is an 'allowable gas cost' and so is exempt from tariff control; whereas storage is a 'non-gas cost', and so the amount that the company will be able to pass on will be regulated by the 'RPI $-X$' clause. Pressures to maximize profit will lead British Gas to try to raise its price above the current level, and the company will be anxious to find ways to allow it to do so in the tariff market, despite the regulatory controls. By inflating the provision for peak under the 'gas costs' heading, the company will be able to raise average revenue more than if this gas were made available through storage. Thus the formula encourages British Gas to provide more peak gas through variable supply loads, and less through the provision of storage, than would be the minimum cost mix.

8. Regulated and Unregulated Markets

If the industry aims at sales rather than profit maximization, it will set marginal revenue equal to zero in each market. This would mean lower prices and higher output than in the profit-maximizing case, and might coincide with the social optimum. But since the overwhelming pressure on the new industry will be to maximize profits, the sales maximization objective is likely merely to modify these slightly. Since the profit-maximizing price is higher than the social optimum, such pressure is likely to be socially beneficial.

In the unregulated contract markets, competitive differences are likely to lead to sharp variations in prices, depending on relative elasticities of demand. (Even if the general case for the competitiveness of this market is accepted, there are clearly pockets of monopoly power, particularly in the short-term firm industrial market.) Indeed falling oil prices in 1986 have seen the BGC respond by significantly lowering prices to those customers who have dual fuel-burning capacity, while leaving the price unchanged for those not able to switch away from gas. (In November 1986, prices of 14.5 to 26 pence per therm for gas delivered under similar contracts were reported.) Thus in this market there is likely to be sharp price discrimination between different markets, with price differences based on elasticities of demand rather than costs of supply.

Profit maximization in either market depends on the constraints imposed by the competitiveness of the market or the regulations imposed by the licence under which British Gas will operate. The licence itself provides only for regulation of the tariff market, since the government argues that there is sufficient competition in the contract market (generally larger industrial consumers) for this market to need no outside control. Not everyone agrees with this analysis: the Energy Committee (1986a) in its report on gas regulation was doubtful, as were consumer organizations and some of British Gas's competitors in this market. Since gas supplies 36 per cent of the industrial market, it is difficult to be altogether convinced by the Secretary of State's assurances (and those of Denis Rooke) that the industry will be controlled by competitive forces. Own-price elasticity for industrial gas is usually estimated as fairly low, rarely numerically greater than -2, which is a far cry from the infinitely elastic (or at least very highly elastic) demand curve expected in a competitive market. There would certainly seem to be enough monopolistic power for the gas industry to raise prices significantly above their marginal costs. How great this gap will be is difficult to estimate but an increase of at least a few pence is likely, even in the present depressed oil price situation.

The existence of a regulated and an unregulated market may itself distort gas industry behaviour. If, as most independent observers consider, there are elements of monopoly in the unregulated contract market, this will be much more profitable than the tariff market, where average revenue is kept down. Pressure on profits will make this unregulated market much more attractive than the regulated one, though British Gas management, anxious to build loyalty with 'longer-term' domestic consumers, may consider the tariff market has other advantages. But in terms of profitability, the industry is likely to concentrate on the more attractive contract market. Since the industry cannot discourage the sales in the tariff market through price rises (because of the average revenue restrictions), it may try to discourage them through other means—perhaps rationing—to increase gas available for the more profitable outlets. The balance of advertising may also change once the industry is privatized; the above analysis suggests that domestic advertising would become less intensive, and contract market sales promotion more so. There will also be an important role for the Office of Gas Supply in apportioning costs to the different markets, particularly if a dramatic alteration in the balance of markets supplied results in significant cost changes.

9. Implications for Regulation

The regulatory structure for British Gas has now been fixed for the twenty-five year duration of the initial licence. In the regulated tariff market the combination of profit maximization under constrained average revenue will encourage distortions between different categories of consumer. This does not necessarily occur with the tariff basket constraint on prices applied to British Telecom, which gives the industry an incentive to move toward Ramsey pricing (Bradley and Price (1988)). However, it is clear that the average revenue constraint does introduce its own potential for distortion.

Its role in encouraging productive efficiency may be more justifiable. Because British Gas will be able to keep the profit from any cost reductions, at least until the average revenue constraint is revised in five years' time, it may be encouraged to make economies that it was disinclined to do when the government appropriated the profits. However, the early decision to double the salaries of top management is not an encouraging start to a cost-cutting exercise, unless one seriously believes that management's marginal productivity will be greater in a private firm. How far other cost reductions will be made remains to be seen, and will be difficult to judge without a substantial efficiency study.

10. Conclusions

Because of its uniquely powerful position and the decision to privatize it as a whole, the gas industry retains considerable monopoly power which will influence its behaviour once privatized. It is also subject to average revenue limitations under its licence—though these will of course only be as effective as the enforcing agency, OFGAS. This is likely to suffer an information shortage even more severe than that recently experienced by the BGC's overseers—the Department of Energy—who were themselves handicapped by lack of information. There is also the danger, well known from United States experience, of capture of the regulatory body either by the industry or by the government, which might then be able to impose the sort of unhelpful interference that has led to so much criticism of its handling of public enterprises. Some would say that sponsoring departments have already shown some signs of capture by their sponsored nationalized industries. But even if the controls work as intended, they incorporate some odd incentives to behaviour, particularly when coupled with a shareholder's objective to maximize profit and a management objective to maximize long-term sales. The result will be some inevitable distortions in resource allocation—e.g. in the unregulated contract market, in the structure of tariffs (because of the restriction and now the lower base of standing charges), and in the relative profitability of the two markets. But ironically profit maximization may lead to better price differentiation than during nationalization, particularly in the tariff market where restrictions on average revenue mean that profits can be raised only by reducing costs.

Appendix*

Condition 3: Restriction of Gas Prices to Tariff Customers

1. The Supplier shall—
   (a) exercise its power to fix tariffs under section 14 of the Act so that no increase in the tariffs last fixed by the Supplier under section 25(3) of the 1972 Act has effect before 1st April 1987; and

*This Appendix is an extract from Department of Energy (1986) and is reproduced here with the permission of the Controller of Her Majesty's Stationery Office.

(b) in setting its prices for tariff customers having effect on or after 1st April 1987 take all reasonable steps, having particular regard to the interests of those customers, to secure that in each Relevant Year its Average Price per therm shall not exceed the Maximum Average Price per therm calculated in accordance with the following formula—

$$M_t = \left(1 + \frac{\text{RPI}_t - 2}{100}\right) P_{t-1} + Y_t - K_t$$

where
M_t = Maximum Average Price per therm in Relevant Year t;
RPI_t = the percentage change (whether of a positive or negative value) in the Retail Price Index between that published or determined with respect to October in Relevant Year t and that published or determined with respect to the immediately preceding October;

$$P_{t-1} = P_{t-2}\left(1 + \frac{\text{RPI}_{t-1} - 2}{100}\right)$$

but, in relation to the first Relevant Year, P_{t-1} (and, accordingly, in relation to the second Relevant Year, P_{t-2}) shall have a value equal to the Average Price per therm in the financial year commencing on 1st April 1986 less the Allowable Gas Cost per therm in that year all calculated as if the financial year commencing on 1st April 1986 were a Relevant Year;
Y_t = Allowable Gas Cost per therm in Relevant Year t;
K_t = the correction per therm (whether of a positive or negative value) to be made in Relevant Year t (other than the first Relevant Year) which is derived from the following formula—

$$K_t = \frac{T_{t-1} - (Q_{t-1}M_{t-1})}{Q_t}\left(1 + \frac{I_t}{100}\right)$$

in which

T_{t-1} = Tariff Revenue from Tariff Quantity in Relevant Year $t-1$;
Q_{t-1} = Tariff Quantity in Relevant Year $t-1$;
Q_t = Tariff Quantity in Relevant Year t;
M_{t-1} = Maximum Average Price per therm in Relevant Year $t-1$;
I_t = the percentage interest rate in Relevant Year t which is equal to, where K_t (taking no account of I_t for this purpose) has a positive value, the average Specified Rate plus three or, where K_t (taking no account of I_t for this purpose) has a negative value, the average Specified Rate.

The subscript t represents the Relevant Year.

15 Regulation of the Gas Industry*

ELIZABETH HAMMOND, DIETER HELM AND
DAVID THOMPSON

1. Introduction

In this memorandum we consider the requirements for the regulation of the British Gas Corporation (BGC) in the light of the Gas Bill and Proposed Authorisation.

The regulation of BGC should be directed toward the following objectives:

(a) promotion of efficient pricing policies both in regard to the overall *level* of gas prices and in regard to the *structure* of charges;
(b) promotion of the productive efficiency of BGC's operations;
(c) promotion of competition; in particular providing conditions for the entry into the industry of competing gas suppliers and for 'fair' competition with substitute fuels;
(d) specification of standards, or duties, for activities (in particular in relation to safety) where there may be a conflict between commercial and wider social objectives.

The design of a regulatory structure to achieve these objectives needs to make a distinction between activities where competition already exists or is feasible (e.g. appliance sales, appliance maintenance) and activities where competition is unlikely to be strong and where BGC will continue to enjoy monopoly power (e.g. supply of gas to domestic consumers). In the former case, regulatory policy should be directed toward ensuring that fair competition is able to take place and that BGC's size, and monopoly position in other markets, do not enable it to distort the terms of competition. In the latter case, regulatory policy should aim to provide the type of incentives to efficiency that would be provided by a competitive market.

In the three sections that follow, we assess how far the provisions in the Gas Bill and the Proposed Authorisation will achieve the first three objectives (that is, those relating to the efficiency of gas supply) and we make a number of recommendations which will, we believe, make the proposed regulatory framework more effective.

---

*This article is an extract from Memorandum 27 in Energy Committee (1986a).

At the time of writing this paper, Elizabeth Hammond was a Research Officer at the Institute for Fiscal Studies. Dieter Helm is a Fellow in Economics at Lady Margaret Hall, Oxford and a Research Associate at the Institute for Fiscal Studies. At the time of writing this paper, David Thompson was a Programme Director at the Institute for Fiscal Studies.

2. Regulation of Gas Prices

The basic principle that should govern the regulation of gas prices is that prices should be set so as to achieve efficiency in resource allocation in the gas sector and in related energy sectors. This requires that prices in different markets are set in relation to the marginal costs of supply (in particular where substitute fuels such as electricity follow this principle). There is evidence to suggest that both the present *level* of gas charges and the *structure* of charges are not efficient in this sense.

First, the level of prices; a number of studies have concluded that at present, gas prices are set substantially below the estimated long-run marginal cost of supply.[1] Essentially this has arisen because the costs of supplies from the Southern Basin are much cheaper than those of the more recent Northern Basin supplies; BGC's prices are set at a level which covers the average costs of supply but which is substantially below the cost of supply from the most recently developed fields. The effect of this is that gas stocks are being depleted too rapidly and customers are being encouraged to invest (through the installation of equipment such as central heating boilers etc.) in too high a level of gas consumption. In the short term, consumers benefit from lower gas prices but this works to their disadvantage in the longer term as the exhaustion of lower-cost sources of supply forces gas prices to rise whilst their investment in equipment such as central heating inhibits a switch to other fuels. The first requirement for the regulation of gas prices is therefore to allow prices to rise to levels that are more closely related to the marginal cost of supply and thereby to ensure an efficient rate of depletion of gas resources. It would be undesirable, however, to implement such an increase in a single year. As already noted, gas consumers typically invest in ancillary equipment and there is therefore limited flexibility in switching to alternative fuel sources in the short term. Regulatory policy should be directed toward realigning prices over a longer period (say five to ten years) and toward providing clear signals to potential consumers of the targeted level to which prices will rise. This will enable consumers to make a more efficient choice in deciding between investment in alternative fuels.

An increase in prices will clearly raise the potential profits of BGC. This raises the central regulatory issue of how this benefit, which effectively reflects the economic rents associated with the ownership of contracts for the low-cost sources of gas supply, can be retained by the government rather than the shareholders in BGC. To the extent that rising profits are anticipated in the bids made for BGC's shares on flotation, this benefit might be capitalized in the proceeds obtained from the sale of BGC. However, the problems encountered in setting the flotation price in previous asset sales indicate that this is only likely to happen, at best, imperfectly.[2] An alternative approach would be to modify

---

[1] See Hammond, Helm, and Thompson (1985) for a review.
[2] See Mayer and Meadowcroft (1985).

taxation through the Gas Levy, either by providing for an overall higher rate than at present or by providing for a disaggregated levy which would apply at a separate rate to each of BGC's existing contracts. The separate rates would be specified at levels which, in each case, raised the costs of supply closer to estimated long-run marginal costs. Both methods will raise revenue, but the second approach is likely to be more effective.

Much of the debate on the regulation of the level of prices charged by BGC (Condition 3 in the Authorisation) has so far centred narrowly on methods of restraining increases in charges. Whilst important, we suggest that this preoccupation is essentially misplaced. The issue that is most central in the regulation of prices but is also the most difficult is, we suggest, ensuring that prices are allowed to rise to a level (related to marginal costs) that will encourage efficient levels of consumption and production both in the gas sector and in related energy sectors. Complementary taxation policies are required (as we note in the previous paragraph) to ensure that the benefits of higher charges are enjoyed by the general taxpayer rather than BGC's shareholders.

Once prices have risen to an efficient (marginal cost related) level, we consider that regulatory control of the type proposed in Condition 3 (subject to the suggestions made in the following section) is required to prevent any further increases in charges which would exploit BGC's monopoly of supply. In particular, competition from other fuels is unlikely to prove sufficiently powerful to restrain BGC's pricing freedom.[3] We are therefore concerned that the pricing control (Condition 3) may be discarded too readily (under the circumstances outlined in pages 17 and 18 of the Authorisation) and suggest that the control should be retained unless the Monopolies and Mergers Commission (MMC) concludes that this would be against the public interest.

Turning to the structure of gas charges, there is evidence to suggest that the present structure of gas charges is inefficient. To achieve efficient resource use, charges in different markets should be related to their respective costs of supply. Where costs differ, then so too should charges. The present uniform price structure fails to reflect the significantly different costs of supplying gas to different areas and at different time periods. In consequence, too high a level of consumption is being encouraged in areas where gas is comparatively expensive to supply and vice versa; and this is likely to impact adversely upon industrial consumers' choice of location.

The second requirement for the regulation of prices is therefore that prices in different markets should be set in relation to their respective costs of supply. 'No undue preference', in Clause 14 of the Bill, should be interpreted so as to mean that prices in different markets should bear a consistent relationship to their respective costs of supply. In practice, this means that prices in different Regions should be set in relation to the costs of supply in each Region (we discuss further how this might be implemented below). This principle also

---

[3] See Hammond, Helm, and Thompson (1985) for a review.

suggests that prices should be higher in periods of peak demand (during winter months) and lower in off-peak periods.

3. Promoting Productive Efficiency

The incentive to achieve productive efficiency provided by the proposed regulatory framework lies in the restriction placed on price increases to tariff customers. Within the restricted level of prices, any increase (or reduction) in BGC's costs will be directly reflected in its profitability. The incentive provided to improve efficiency will, however, only be effective if the regulatory authority (either the Director or the MMC) is able to reach an independent view on the level of costs, and hence the level of prices, that could be achieved by a fully efficient enterprise. If on the other hand the regulatory authority is unable to form an independent view on the scope for cost reductions, and has to place substantial reliance on the forecasts and assessments made by BGC, then the pricing formula will in practice become very similar to a 'cost-plus' or rate-of-return pricing formula. As we noted in our earlier memorandum (Memorandum 14 in Energy Committee (1986a)), experience overseas (in particular in the US) has shown that rate-of-return regulation has often been associated with inefficient performance by the regulated enterprise.

To reach an independent view on the scope for cost savings, the regulatory authority requires competing sources of information. One possibility would be to review the costs and efficiency of gas suppliers in other countries. A more direct approach would be to compare the costs and efficiency of BGC's different Regions. The 'British Gas Efficiency Study' carried out by Deloitte, Haskins, and Sells (1983) indicated significant differences in the cost efficiency of BGC's various Regions. Establishing each of BGC's twelve Regions as a separate cost and profit centre would provide comparative information on the costs of distribution and supply. Deloitte's analysis suggests that this approach would be feasible, based upon analysis already prepared internally in BGC, and that for this reason it would not be expensive.

We recommended in the previous section that prices in each Region should be set in relation to the individual Region's costs of supply so as to reflect the economic characteristics of each Region (in particular their accessibility to supplies from the North Sea). To the extent that differences in Regions' costs also reflected the different levels of efficiency achieved by each Regional organization, this would provide more direct consumer pressure to improve performance in the less efficient Regions.

The gas price formula specified in the Proposed Authorisation provides for an increase in the costs that BGC pays for gas to be passed directly on to consumers. This reflects a situation where gas supply costs are likely to increase (as the costs of more recently developed sources will typically be above the present average cost of supply) but where the likely increase in costs is difficult to forecast. However, the approach proposed means that BGC is able to recoup directly costs associated with newly contracted sources of supply. This places BGC in a very strong bargaining position in relation to newly developed sources. Effectively whatever price BGC offers for supplies from a newly developed field, it will be enabled to recoup this from consumers through an increase in the permitted maximum average price. This has implications for the emergence of competing suppliers (an issue we turn to in the next section). A more effective approach may be for the appropriate allowance to be specified in advance on the basis of negotiation between the Director and BGC. To the extent that newly negotiated contract prices were above (or below) the pre-specified level, BGC would incur a loss (or gain) in profits. But the distortion in bargaining between BGC and gas producers inherent in the proposed system would be reduced.

4. Promoting Competition

The promotion of competition is not included as an explicit duty of the Director, or Secretary of State, in Clause 4 of the Gas Bill. We regard this omission as both surprising and of serious detriment. Recent experience with the deregulation of sectors as diverse as express coaching and electricity supply[4] demonstrates the substantial competitive advantages enjoyed by incumbents with a long experience of statutory protection from competition, and the scope that exists for such incumbents to prevent competitive entry (in some cases perhaps unintentionally). This experience indicates that the removal of statutory restrictions on competition may not, by itself, be sufficient. Accompanying safeguards that reduce the scope for incumbents to deter market entry by competitors are also required. The omission of a duty to foster competition is particularly important because the criteria that the MMC is required to consider in deciding upon a proposal for a modification of the Authorisation are precisely the duties set out in Clause 4 of the Bill.

The provisions in the Bill will not allow effective competition to emerge for gas supply to domestic customers. Under the proposed legislation, competition will be relevant for the supply of gas to the larger industrial customers and for activities such as appliance sales.

In the case of activities where competition already exists—the sale of appliances, installation, and maintenance—the regulatory requirements should be limited to ensuring that BGC is not able to compete unfairly by cross-subsidizing such activities from the far larger gas supply business. There are two requirements here. First, these activities should be identified as separate audited profit centres in BGC's accounts (in practice these activities might be privatized under separate ownership). Second, the Director should be enabled to seek a modification to Condition 3 (the price regulatory formula) of the Authorisation

---

[4] See Davis (1984) and Hammond, Helm, and Thompson (Chapter 8, this volume).

if he considers that such cross-subsidy is taking place.

In the case of activities where competition is feasible but is not taking place (that is, the supply of gas to customers using more than 25,000 therms a year), the position is more complex. We noted in our previous memorandum (Memorandum 14 in Energy Committee (1986a)) that the scope for fair competition is significantly restricted by BGC's ownership of contracts for the supply of gas at prices substantially below the costs incurred in more recently developed fields. Whilst BGC's prices continue to be set at levels which cover only average costs, rather than marginal costs, then competitive entry is unlikely to be feasible. This provides a second powerful reason for regulatory policy (and associated taxation arrangements) to be directed toward raising the level of gas prices to reflect more closely the marginal costs of supply. In the absence of such a policy, competition is unlikely to emerge until low-cost supplies of gas are exhausted.

Even in this event there are two further requirements for ensuring that fair competition is able to take place. First, the potential supplier requires access to the use of BGC's transmission network, and Condition 9 of the Authorisation provides for the specification by BGC of 'typical' charges. To ensure that such charges are specified at levels that are 'fair', the operation of BGC's pipe-lines should be established as a separate, audited cost and profit centre in BGC's accounts with internal transfer charges for the conveyance of BGC's own gas supplies. The Director should be enabled and required to establish on this basis that the 'typical' charges computed by BGC are set at reasonable levels.

Second, the provisions in the Gas Bill (Clause 14) relating to the publication of tariffs and the duty not to show undue preference do not (necessarily) apply to the supplies to customers who take over 25,000 therms a year. The scope therefore exists for BGC to set tariffs for such customers at anti-competitive levels and to cross-subsidize from other activities. In principle, any such anti-competitive behaviour might be dealt with within the provisions of general competition policy. We believe that there is a strong case, however, for following a more positive policy and requiring that the terms of contracts are published and that the Director is enabled to audit the terms of a sample of contracts on a periodic basis to ensure that they are set on a basis which represents fair competition with other potential suppliers.

Although competition to BGC may not emerge immediately (and, as we noted above, how quickly competition emerges will depend on the policies adopted on the regulation of gas prices), we believe that it is important that the regulatory framework is established in a form that enables competitive suppliers to enter the market (in particular by adopting the two proposals set out above).

5. Summary

In summary we suggest that the regulation of BGC could be made more effective than at present proposed. There are three principal requirements:

(1) The regulation of BGC's prices to domestic consumers should ensure that prices are set in relation to the costs of supply, in order that efficiency in resource allocation is achieved. Complementary taxation policies (in relation to the Gas Levy) are required to ensure that the benefits of increased charges are returned to the government.
(2) Each of BGC's Regions should be established as a separate cost and profit centre and prices in each Region should be set in relation to their respective costs.
(3) The promotion of competition should be an explicit duty of the Secretary of State and the Director. To facilitate the development of competition in the supply of gas to industrial customers, BGC's transmission network should be established as a separate cost and profit centre and the Director should be enabled to audit periodically a sample of contracts between BGC and industrial users to establish that the terms of these are not anti-competitive.

Part V

Coal

16 Liberalizing the British Coal Industry*

COLIN ROBINSON AND EILEEN MARSHALL

1. The Aftermath of the Strike

There is a widespread perception that the British coal industry's fortunes have improved significantly since the end of the 1984–5 strike. That strike is now seen as a major turning-point in an industry which, in terms of output and employment, has been in decline for most of the last seventy years and which has also suffered from poor industrial relations.

Productivity has increased sharply in recent months. According to Press reports, output per man-shift achieved a record in early December 1985, exceeding 3 tonnes.[1] In some Areas—such as South Yorkshire, South Wales, and Kent—the increase has evidently been such as to surprise the National Coal Board (NCB). Pit closures have reached a scale (24 in the second half of 1985)[2] which would almost certainly have provoked industrial action in the past, and yet there has been virtually no opposition. It appears that only a few collieries, such as Bates and Horden in the north-east, remain on the Board's early closure list; they are now being considered under the new pit review procedure. At the same time that closures of high-cost pits are proceeding, the NCB is pressing ahead with its plan for new 'superpits' such as the proposal to mine between Coventry and Kenilworth.

Given rising productivity at remaining pits, the closure of so many high-cost mines, and restocking by the Electricity Boards and other big consumers, it is hardly surprising that the Board's financial position has improved considerably, though whether it continues to do so will depend partly on oil price trends. No figures appear to have been published but it has been claimed the NCB may break even '. . . substantially before the target date of the start of the 1987–88 financial year'.[3]

Another change is that the Coal Board's sales forecasts, which we have criticized for many years as being far above what could reasonably be expected,[4] have been considerably reduced. It is reported that the Board now anticipates UK coal consumption in 1990 will be no more than 110 million tonnes.[5] That is a

---

* This paper appeared as a submission to the Energy Committee (1986b).
   Colin Robinson is Professor of Economics at the University of Surrey. Eileen Marshall is Lecturer in Industrial Economics at the University of Birmingham.
[1] For example, 'Coal industry productivity hits post-nationalisation record', *Financial Times*, 24 December 1985; 'South Yorkshire tops coal output list', *Financial Times*, 19 December 1985; 'British Coal's goal for 1986—3.5 tonnes a man shift', *Coal News*, January 1986.
[2] 'South Wales hit hardest by pit closures since July', *Financial Times*, 4 January 1986.
[3] 'Coal industry productivity hits post-nationalisation record', *Financial Times*, 24 December 1985.
[4] For example, in Robinson (1979) and Robinson and Marshall (1981).
[5] 'The NCB's unequivocal commitment to the market', *Financial Times*, 16 October 1985.

far cry from the forecasts of 135 million tonnes in 1985 and 170 million tonnes in 2000 which were put forward in the industry's plans of the mid-1970s and were still being used in the early 1980s. Our own forecasts[6] made from 1979 onwards were of a coal market of 75 to 110 million tonnes by the end of the century, assuming no change in government policy towards coal; the Coal Board's latest figures seem consistent with that view.

Important alterations in union structure have also occurred since the strike, for the first time since the National Union of Mineworkers (NUM) was formed in 1944 from the earlier Miners' Federation of Great Britain. The new Union of Democratic Mineworkers (UDM) is reported to have at least 30,000 members, primarily in Nottinghamshire, south Derbyshire, and the north-east, and to be attracting members in some other areas.[7] A further significant event in labour relations came when the NUM agreed, after some resistance, to enter negotiations towards a productivity-based wage agreement;[8] the UDM had earlier accepted an incentive-related pay increase.

The emergence, post-strike, of an industry with less high-cost capacity, increased productivity, improved finances, a more realistic view of sales prospects, and a less monopolized work-force is generally seen as a considerable advance as compared with the state of the coal industry in the 1970s and early 1980s. It would, however, be unwise to assume that the upturn in the fortunes of British coal will persist without any further action being taken. Before suggesting what form such action might take, it is helpful to look at the industry in longer-term perspective.

2. A Brief Historical Perspective

Figure 16.1 illustrates the extent of the decline in British coal since the beginning of this century; in 1983, the last 'normal' year, production of deep-mined and opencast coal of 117 million tonnes was only 40 per cent of the 1913 peak of 292 million tonnes.

During the inter-war years coal, like other 'staple' British industries, declined primarily because export markets were lost; home consumption of coal remained about constant. The industry was run down during the Second World War so that by the mid-1940s it was producing only some 190 million tonnes a year. By that time, nationalization (long advocated by the miners) was seen as a means of reviving the industry and supplying the anticipated rising demand from British consumers. Nationalization of coal was not particularly controversial in party political terms and it is hardly surprising, given the sorry state of the industry, that it appeared to offer a way forward.

---

[6] Robinson (1979) and Robinson and Marshall (1981).
[7] 'Fight is on for the allegiance of Britain's miners', *Financial Times*, 6 December 1985; 'Scargill claims drift back to NUM', *Financial Times*, 31 December 1985.
[8] 'Pit pay talks to resume after NUM concession', *Financial Times*, 9 December 1985.

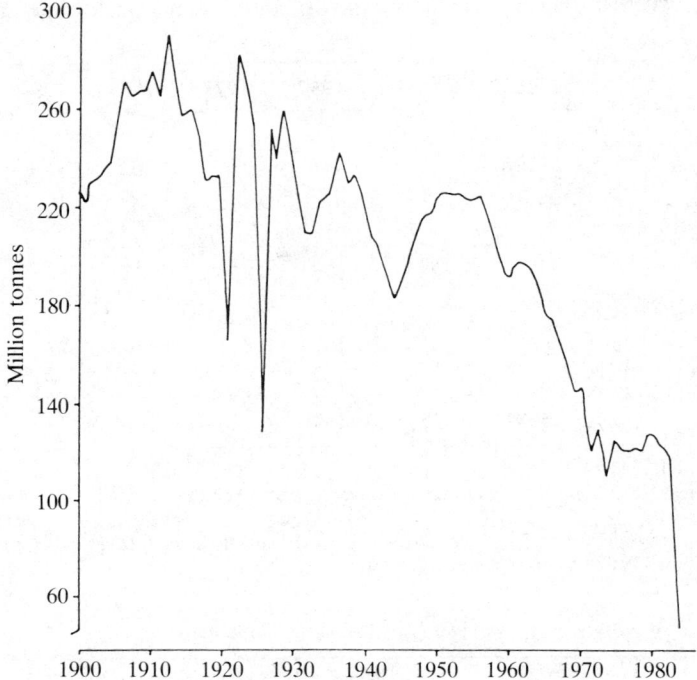

Figure 16.1 UK coal production, 1900–84

In the event, such hopes were disappointed. After a temporary increase in demand, by the late 1950s decline had resumed—this time as home consumption fell because of competition from low-priced oil and later natural gas (Table 16.1).

From the early 1970s onwards, as Figure 16.1 shows, output fell more slowly. But there was no sign in Britain of the revival there was in some other coal industries. Table 16.2 illustrates the changed trend of world coal consumption after the first oil 'shock' of 1973–4. In the previous ten years, coal's share of world energy had dropped considerably from 40 to 28 per cent, but between 1974 and 1984 there was an increase to 30 per cent. In Britain, however, coal's market share continued to fall: between 1973 and 1983 (the last strike-free year) coal's share declined from 38 to 35 per cent.

The failure of British coal to revive after the oil shocks is puzzling in the sense that, until the fall in oil prices early in 1986, the price of industrial coal averaged only half to two-thirds the price of fuel oil (measured in pence per therm) in the years following 1973. The long time-lags, which are always present in the response of fuel consumption to price changes because of the inherent inertia of energy markets, are, no doubt, one explanation.

Table 16.1. British coal mining: supply and demand since nationalization (million tonnes)

	1947	1957	1967	1977	1983	1984[a]	1985[a]
SUPPLY[b]							
Deep-mined	190	213	168	107	102	35	75
Opencast	10	14	7	14	15	14	16
Imports	1	3	—	2	4	9	13
	201	230	175	123	121	58	104
DEMAND							
Home	188	216	167	124	111	77	105
Exports and bunkers	5	8	2	2	7	2	2
	193	224	169	126	118	79	107

[a]Strike-affected.
[b]Including licensed mines, but excluding slurry recovered other than by the NCB from dumps, ponds, and rivers.
Sources: Ministry of Power, *Statistical Digests*; Department of Energy, *Digests of Energy Statistics*; and Department of Energy, *Energy Trends*.

Table 16.2. World primary energy consumption, 1964–84

	1964		1974		1984	
	Mtoe	Percentage of total	Mtoe	Percentage of total	Mtoe	Percentage of total
Coal	1,505	40	1,691	28	2,180	30
Oil	1,420	38	2,760	47	2,844	39
Natural gas	604	16	1,088	18	1,410	20
Hydro	221	6	344	6	485	7
Nuclear	4	—	62	1	282	4
TOTAL	3,754	100	5,945	100	7,201	100

Note: Mtoe = million tonnes oil equivalent.
Source: British Petroleum, *Statistical Review of World Energy, 1985* and *Statistical Reviews of the World Oil Industry*.

But in Robinson and Marshall (1985) we have attributed coal's continued decline in such favourable competitive circumstances to two additional factors. First, given the monopolistic position of the NCB in coal (though not in the British energy market as a whole), consumers probably doubted whether in the long term coal prices would stay depressed relative to oil prices. Second, after the disruptions to coal supplies in 1971–2 and 1973–4 and the subsequent threats of similar union action, it is likely that consumers feared further serious interruptions.

Forty years on from the nationalization Act, the underlying problems of the coal industry remain. In addition to its monopoly of '. . . working and getting the coal in Great Britain . . .', it has had considerable protection from the competition of other fuels such as government pressure on the Electricity Boards to buy coal, limits on coal imports, a tax on fuel oil, and government grants to firms that convert to coal from oil and gas. Furthermore, there has been very substantial investment in British coal, averaging about £700 million a year in the early 1980s (two to three times that in the West German industry). Yet consumption has fallen and the NCB has made large losses in recent years (Table 16.3).

Table 16.3. National Coal Board financial results, 1980–5 (years ending March; £ million)

	1980	1981	1982	1983	1984	1985
OPERATING PROFITS/(LOSSES)						
Collieries	(122)	(107)	(226)	(317)	(595)	(1,773)
Opencast	110	157	157	192	211	140
Other mining activities[a]	6	12	21	23	27	9
Non-mining activities[b]	26	17	5	3	(1)	(43)
Total operating profits/(losses)	20	79	(43)	(99)	(358)	(1,667)
Social costs less grants[c]	(17)	(29)	(61)	(49)	(74)	(53)
Interest	(184)	(256)	(344)	(364)	(467)	(520)
Other items	22	(1)	20	27	24	15
Deficit grant	159	149	428	374	875	2,225
SURPLUS/DEFICIT AFTER DEFICIT GRANT	—	(58)	—	(111)	—	—

[a]Rents, shipping terminals, etc.
[b]Manufacture of coke and smokeless fuel, chemicals, distribution of fuel and appliances, estates and land, engineering, computer services, and income from related companies and partnerships.
[c]Costs incurred as a result of closing uneconomic capacity and the transfer or redundancy of employees which are met wholly or partly from government grants. In 1984 such government grants amounted to £270 million.
Source: National Coal Board Annual Reports and Accounts.

It seems very unlikely that the coal industry's problems have disappeared as a consequence of the 1984–5 strike and its aftermath. Most likely there is a considerable temporary element in coal's recovery as major customers (especially power-stations) restock, workers try to restore earnings, and attempts are made to repair the damage done to management–labour relations during the strike. Without wishing to belittle the considerable efforts made by coal industry management and workers to improve the condition of the industry, it seems to us that now is an appropriate time to consider fundamental restructuring measures. In Robinson and Marshall (1985) we have discussed such measures, the benefits they might bring, and means of dealing with consequent problems. Briefly, the idea is to liberalize the British coal market, freeing overseas trade in coal, introducing private capital, and encouraging competition among several British sources of supply. Our conclusions are summarized in the rest of this memorandum after a brief description of the industry's present structure.

3. The Present Structure of the British Coal Market

Demand

Since the post-war peak of coal consumption in the mid-1950s, sales to most markets have declined (Table 16.4), rapidly up to the mid-1970s and more slowly subsequently. Some markets, such as the railways and gasworks, have virtually disappeared.

Table 16.4. Coal markets in Britain, 1957 and 1983

	1957		1983	
	Million tonnes	Percentage of total	Million tonnes	Percentage of total
Power-stations	47	22	82	74
Coke ovens	31	14	10	9
Gasworks	27	13	—	—
Industry	38	18	7	6
Domestic	36	17	8	7
Railways	12	5	—	—
Other	25	11	4	4
	216	100	111	100

Source: Department of Energy, *Digests of UK Energy Statistics*.

Electricity generation is now far and away the largest market for coal, accounting for nearly three-quarters of coal consumed in Britain, although power generation sales have fallen somewhat from their peak of about 90 million tonnes in 1980. Coal, in turn, has recently supplied about three-quarters of the fuel used by the electricity supply industry in a strike-free year. The close relationship between the coal and electricity supply industries is an important feature of the market to be taken into account in any restructuring proposals.

Supply

Although coal supply to the British market is dominated by the NCB, there are some imports and a small private sector. In 1983, the last full year before the recent strike, deep-mined coal production was 102 million tonnes, opencast (strip-mined) output was 15 million tonnes, and imports were 4 million tonnes (Table 16.1).

Net foreign trade in coal has generally been small relative to home supplies, except during the 1984-5 strike and its aftermath (Table 16.1). Imports have periodically been subject to government control, usually with the aim of protecting the Coal Board from foreign competition.

The private sector is a minor supplier to the British market, but some explanation is needed of how it came into being and how it coexists with a nationalized supplier. On vesting day under the 1946 nationalization Act (1 January 1947) the NCB took over from the Coal Commission (appointed under the 1938 Coal Act) coal deposits in the ground in Great Britain. There were over 1,400 collieries, but the NCB granted licences to about 480 of the smallest of these (each employing no more than 30 workers underground) to continue in private ownership as 'licensed mines'. There are about 160 such licensed deep-mines today producing about 0.75 million tonnes a year (about 0.5 per cent of total British deep-mined coal output) compared with 1.4 million tonnes a year (0.7 per cent of the total) twenty years ago. The 30-man limit still applies, though in 1981 the Coal Board agreed with the government that it would '. . . exercise the greatest possible flexibility on the maximum size of licensed deep-mines, giving sympathetic consideration to applications for mines employing over 30 persons underground'. Mine-owners at present pay to the Coal Board a royalty of 50 pence per tonne which the NCB proposes to raise in stages to £2 in 1988.[9] Although some of the mine-owners are very critical of the royalty they have to pay, one can at least assume that the small mines that continue in business are profitable rather than making losses on the scale of the NCB's £6.50 per tonne (on average) in deep-mining activities.

All opencast coal is produced by private companies, but on larger sites they act only as contractors to the NCB's Opencast Executive (or, in Scotland, to the Area that controls both deep-mined and opencast production). There are, however, about 60 relatively small sites that the NCB licenses to private companies to produce and sell the coal. Annual production varies around 0.7 million tonnes (about the same as the licensed deep-mines). Until recently, under the provisions of the relevant legislation,[10] the NCB would grant such licences only if, in its opinion, the workable coal reserves were likely to be no

---

[9] 'NCB seeks fourfold rise in mine licence royalties', *Financial Times*, 16 July 1985.
[10] The Coal Industry Nationalisation Act 1946, the Opencast Coal Act 1958, and the Coal Industry Act 1977.

more than 25,000 tonnes. However, in 1981 the government agreed with the Coal Board that the limit should be raised to 35,000 tonnes and that adjacent sites with reserves totalling 50,000 tonnes could be licensed for private operation.[11] The Monopolies and Mergers Commission in 1983 recommended that legislation should be passed to raise the workable reserves limit to 100,000 tonnes on the grounds that licensed opencast contractors provided '. . . a small but valuable element of competition to and comparison with the operations of the Opencast Executive' (Monopolies and Mergers Commission (1983a, para. 11-95)). That recommendation has not so far been implemented.

Licensed opencast operators have to pay a royalty of £16 per tonne to the NCB (compared with the Board's profit on opencast of less than £15 per tonne in 1983-4) although in some cases they deliver coal to the NCB's Opencast Executive at a specified price instead of paying the royalty and selling the coal themselves. The size of royalty payable has been a matter of dispute in the past. Under the 1981 agreement the Coal Board agreed to set royalties 'at levels which permit efficiently-managed operators to develop their licences profitably' and to reduce royalties for new licences where accounting evidence is produced to show that 'profit expectations would otherwise be cut to unreasonably low levels' (Monopolies and Mergers Commission (1983a, para. 11-82)). Nevertheless, the opencast operators complained to the Monopolies and Mergers Commission that royalty rates were too high and that the NCB licensing policy was generally too restrictive.

Although licensed mines and opencast sites produce only about $1\frac{1}{2}$ million tonnes per year in total,[12] the private sector has a significance out of proportion to its size. Its presence in a market dominated by a State monopoly indicates that it is possible to operate profitably on a small scale in British deep-mining and opencast production despite both the arbitrary limits on the size of private activities and the setting of royalties above the NCB's average profits on the relevant activity.

4. Privatization and Liberalization

In previous memoranda to the Committee (Energy Committee (1985, memoranda 13 and 28)), we have distinguished between privatization and liberalization, criticizing the British Gas privatization plans on the grounds that they do very little to liberalize the market. There is no need to repeat these arguments except to point out that, in the case of gas, there are some elements of natural monopoly which, therefore, justify regulating *some* parts of the British Gas Corporation's present business on privatization.

---

[11] See Monopolies and Mergers Commission (1983a, para. 11-82).
[12] There is, in addition, 2-3 million tonnes a year of coal recovered by private firms from dumps, ponds, and rivers.

These elements are absent in British coal. Deposits of coal are naturally dispersed around the country. There will be economies of scale in production from individual deposits, and managerial, marketing, and distribution economies may be achievable by grouping a number of pits under one management. However, it is extremely improbable that, to exhaust economies of scale, the *whole* British industry must be placed under one central management. The more likely outcome of such centralization is managerial 'diseconomies' as the organization becomes larger than can be efficiently managed and costs consequently increase. Thus the disadvantages of monopoly—the ability to manipulate prices to consumers and to pressure government in protective measures—are unlikely to be offset by cost reductions associated with larger sizes.

The coal monopoly in Britain is, in our view, a political creation: it is difficult to find any justification on economic efficiency grounds for nationalization. Moreover, the monopoly has had to be politically maintained by increasing protection. Since the industry is not naturally monopolistic, there is no case for privatizing the NCB as a whole. To do so would merely transform an unnatural public monopoly into an unnatural private monopoly. Our proposals are to liberalize the coal market, including competition-increasing privatization measures. We first discuss ways and means of liberalization, then turn to the advantages of so doing and the problems that would arise.

5. Ways and Means of Liberalization

Freeing Imports

An essential step in a liberalization programme would be to remove the limitations on coal imports which have been used on and off for about thirty years to protect British coal from overseas competition. Governments have relied on arm-twisting of big consumers, especially the Central Electricity Generating Board (CEGB) and the British Steel Corporation (BSC), rather than explicit controls which would have run counter to international trading obligations.

The CEGB has been particularly affected by these back-door restrictions; at present, in the aftermath of the miners' strike, it appears to be the only organization that is still being induced by government to keep its imports substantially below what it would wish. Present controls on CEGB imports date back to early 1981 when, after the threatened miners' strike had been called off, the government significantly increased the protection it gives to British Coal. At that time the CEGB was pressed into limiting imports to only about $\frac{3}{4}$ million tonnes of coal a year. Total coal imports into Britain, which had risen to $7\frac{1}{2}$ million tonnes in 1980, then fell back to around 4 million tonnes a year until the most recent miners' strike when they increased to 9 million tonnes in 1984 and about $12\frac{1}{2}$ million tonnes in 1985.

It appears that the CEGB, which is potentially Britain's largest coal importer, was allowed as a temporary stockbuilding measure to bring in $2\frac{1}{2}$ million tonnes of overseas coal in 1985,[13] but that is probably very much less than it would like. Before the 1981 restrictions, the CEGB had planned to build one or more terminals to increase its annual coal import capacity to about 15 million tonnes. It still appears keen to import significant quantities so as to keep competitive pressure on the Coal Board. Lord Marshall is reported to have said in December 1985 that the Generating Board wanted imports '. . . to keep some involvement in world markets and to remind the NCB of world prices'.[14]

A very imperfect attempt to provide the CEGB with the effects of import competition without having the imports is the Joint Understanding on coal supplies and prices between the NCB and the CEGB. This restricts imports by its provision that the CEGB will take 95 per cent of its coal supplies from the NCB, provided the price of those supplies rises no more than the rate of retail price inflation. Indirect import competition is introduced by an arrangement whereby 8 per cent of CEGB coal supplies (some 6 million tonnes annually or enough to supply the Thames-side power-stations) are paid for at a price intended to be competitive with imports; the other 87 per cent of NCB-supplied coal (about 65 million tonnes) is at a substantially higher price. The two prices were reported early in 1985 to be £36 and £44 per tonne respectively plus delivery charges; the lower price is reported to have been reduced since the fall in crude oil prices.

In the principal coal-exporting countries, such as Australia, the United States, and South Africa, the cost of coal at the mine averages only around one-third of the average pit-head cost of British coal. Transport costs, however, approximately double the cost of coal from such countries by the time it reaches western Europe and there are trans-shipment costs to Britain because of the absence at present of deep-water coal terminals.[15] Imported coal is therefore unlikely to undercut British coal in the main central coalfields—which have considerable transport cost protection—unless the pound appreciates considerably or there is a further decline in already depressed world coal prices.

In our view, if there were freedom to import in the next ten years or so, coal imports would probably not exceed 20 million tonnes a year. Nevertheless, they would be important in giving British fuel consumers—and especially the CEGB which suffers most from import restrictions—a choice of supplier. A free trade policy in coal, with imports always a possibility, would stimulate efficiency in the production and sale of British coal. Even if no other change were made to the structure of the coal market, it would put competitive pressure on the NCB and give the Board an incentive to reduce its costs. Moreover, once consumers found

---

[13] 'CEGB chief in warning to coal board on imports', *Financial Times*, 12 December 1985.
[14] See footnote 13.
[15] It has, however, been reported that the BSC deep-water terminals at Hunterston and Redcar have, on occasions, been used to import steam coal. See *FT European Energy Report*, 15 November 1985, pp. 14-15, and 'South Africa faces mounting pressure against coal sales', *Financial Times*, 29 November 1985.

it worth while to expand coal-importing facilities (and possibly to build more coastal power-stations) so that substantial imports became an established feature of the British market, the diversification of supply sources would enhance the security of coal supplies as compared with the present dependence on the NCB.

Privatizing Coal Production

For the foreseeable future, the bulk of coal demand in this country is likely to be supplied from British reserves rather than from abroad. As well as lifting restrictions on imports, liberalizing the coal market therefore implies competition in the supply of British coal and competition for the use of the relevant resources, both labour and capital. In contemplating privatization we first need to consider the ownership of coal reserves.

Ownership of Coal Reserves

The NCB has the duty under the 1946 Coal Industry Nationalisation Act of '. . . working and getting the coal in Great Britain, to the exclusion (save as in this Act provided) of any other person'. Thus, as explained above, private deep-mines and opencast sites have to be licensed by the NCB and pay royalties to the Board.

There has been controversy over the apparently high royalties set by the Board, but, regardless of royalty rates, licensing of private operations and royalty collection seem to us to be functions of the State, not of a corporate body. The idea that a nationalized corporation should so act seems to be based on the assumption that because the NCB is a 'public' corporation, it will pursue the 'public interest' and can therefore be entrusted with holding reserves and receiving royalties on behalf of the nation. However, the Coal Board clearly has every incentive to avoid any significant expansion of private sector activities, whatever the 'national interest' might demand. Furthermore, if one accepts that rent is available from the exploitation of a natural resource and that it should accrue to the community as a whole, then it should be channelled directly to the State (not via an intermediary with its own corporate objectives), with some mechanism for distributing it to society. The present system gives the NCB a dominant market position and allows it to maintain that position through a regulatory role which permits it to minimize private sector entry to the market and to collect substantial royalties from those small private operators it allows to exist. Another important failing in the present system is that the community as a whole derives no natural resource rent whatsoever from the mining activities of the NCB itself.

It would be more satisfactory to remove from the NCB ownership of unworked coal and the right to receive royalties, placing property rights to coal either in the hands of private landowners or in a revived Coal Commission[16] or in the Crown. The last is the system that operates for oil and gas reserves; they are owned by the State, which lays down terms and conditions (including taxes, royalties, working obligations, and environmental constraints) on which private firms or public corporations may exploit them. In the British North Sea, for example, rights to explore for oil and gas reserves and to produce from any that are found are allocated primarily under a 'discretionary' regime (under which Civil Servants decide which applicants are worthy to receive such rights), though with occasional limited cash auctions. A discretionary system is used also for awarding licences for onshore oil exploration and production.

We think auctioning licences would be preferable.[17] However, we would regard even a discretionary award system for coal exploration rights as an improvement on a system in which one corporation, which happens to be nationalized, pays no royalties on its own activities but vets applications from private companies to work coal and, if it grants permission, collects royalties from those companies.

Opencast Coal

Opencast coal-mining seems an obvious candidate for privatization in the interests of liberalization. Private companies already extract opencast coal, carry out associated civil engineering works, transport coal, and restore the sites. In the words of the Monopolies and Mergers Commission (1983a, para. 11-7), the Opencast Executive is '. . . essentially a development and management organisation, its tasks being to find and evaluate workable sites, plan their development and secure the necessary statutory approvals, let contracts, and supervise the subsequent opencast mining on behalf of the NCB'.

The size of the private sector could be substantially increased if the upper limit on reserves that can be worked privately were to be raised. The limit was originally established in the Opencast Coal Act of 1958 at 25,000 tonnes, and then raised in 1981 to 35,000 tonnes or 50,000 tonnes for adjacent sites. So far the government has shown no signs of accepting the Monopolies and Mergers Commission recommendation that the limit should be raised to 100,000 tonnes. Legislation to allow such an increase would permit more companies to extract and sell opencast coal themselves rather than working under the NCB's supervision. However, the royalty of £16 per tonne now payable to the Board on private opencast operations probably needs to be reduced if private activities are to be expanded significantly.

A more radical move would be to place all opencast operations in the private sector, taking away from the NCB the regulatory and rent collection roles which we have already argued are more properly the functions of government. The

---

[16] The 1938 Coal Act established a Coal Commission which acted as owner of coal resources and collector of royalties, but not as a producer.
[17] For reasons explained in Dam (1976) and Robinson (1981).

Board would no longer be the judge of whether private companies should be allowed to develop smaller opencast sites as licensed operators, paying the Board a royalty. Nor would it act as an intermediary, supervising companies working as contractors on the larger sites. It is difficult to believe that experienced civil engineering companies really need such supervision. Companies wishing to work opencast sites would apply to the Department of Energy under a procedure which would need to be laid down (as it already is for oil and gas) and which would incorporate environmental protection safeguards.

The present government has already moved some small way towards equalizing the terms of competition between the NCB and private opencast operators, and towards freeing opencast production of some of the constraints imposed on it in the past, by announcing in a circular[18] its intention to bring planning procedures for NCB opencast coal applications into line with those for licensed (private) opencast working and for other minerals. Under 'transitional arrangements' (apparently still in force), from 1 March 1984 the Board has had to apply to local planning authorities for planning permission and, should permission be granted, to the Environment Secretary for authorization to work the site (under the 1958 Opencast Coal Act). The transitional arrangements operate until the promised repeal of the relevant provisions of the 1958 Opencast Coal Act.

Before the 1983 White Paper (Department of Energy (1983a)) which preceded the circular, the government had endorsed a target of about 15 million tonnes a year for opencast output. As Table 16.1 shows, output has been around that rate in recent years. The Commission on Energy and the Environment in 1981 had also recommended that, on environmental grounds, opencast output should not exceed 15 million tonnes a year. However, the present government

. . . see[s] no case for continuing to endorse a target for opencast output. Each project should therefore be considered in terms of the market requirement for its planned output . . . The overall level of opencast output will in practice be determined by the market subject to the acceptability of individual projects as determined by the planning system.[19]

We support this move away from a centrally determined opencast output target to a more decentralized system. Relatively low-cost, high-quality opencast coal production should not be held down, as it evidently has been on occasion, in order to avoid cuts in deep-mined output. In a liberalized market, opencast coal would be allowed to compete with deep-mined coal, though in a densely populated country such as Britain the environmental costs of both forms of production need to be estimated and imposed on the producing companies; privatization would make essentially no difference to the environmental problem since there is no reason to believe nationalized industries are better guardians of the environment than private companies.

---

[18] Joint Circular from the Department of the Environment (3/84) and the Welsh Office (13/84) to local authority and local Planning Board Chief Executives, 27 February 1984.
[19] Para. 15 of Joint Circular, referred to in footnote 18.

Subject to such environmental regulations (covering, for example, noise, dumping of waste, and land restoration) there seems no reason why opencast coal production should not be fully privatized.

Deep-mined Coal

Privatizing deep-mines raises more difficult issues. Many pits at present make losses (Table 16.3), even without taking into account interest payments; some losses are so large that keeping the pits in question open is more a means of disguising unemployment than of producing coal for the consumer. The probable opposition of the National Union of Mineworkers (NUM) and other mining unions would also reduce the attractiveness of mines to potential private owners. Despite such problems, in our view privatization offers benefits not only to the community at large but specifically to miners. In the following section we set out some of the ways in which more competition could be introduced into the supply of British deep-mined coal.

Relaxing Existing Restrictions on Licensed Deep-mining

In the same way that restrictions on licensed opencast mining could be relaxed, so they could be on licensed deep-mining. The present limit of 30 men working underground in a licensed deep-mine was set in 1946, since when there have been great advances in the technologies of mining coal and of controlling and monitoring underground operations. Even though the limit is interpreted with more flexibility by the Coal Board, it is still a serious impediment to private deep-mining activity. If it were raised, there would be some expansion of licensed mining which would provide more competition for the NCB. But a continuation of a system which allows the Board to vet potential competitors and to collect royalties from them is undesirable. Of course, it would be possible for applications to be considered and royalties to be collected by the Department of Energy and the Treasury, but that leads on to more fundamental proposals for changing the structure of coal supply in Britain.

Joint Ventures for New Pits

One way of introducing more private capital into British deep-mining which falls short of complete privatization would be for private companies to invest in the 'superpits' which are likely to be developed in the next twenty years. Despite early technical problems, the Board's big new mining complex at Selby (10 million tonnes a year) is already in production, and Asfordby in Leicestershire (about 2 million tonnes a year) is due to begin producing coal in the mid-1990s. But several other big prospects have already been identified—for example, Witham in Lincolnshire and South Warwickshire (near Coventry). Such projects might well be attractive to private investors and would bring wider advantages. In the first place, some new ideas about production, distribution, and marketing would penetrate the British market from companies with mining experience abroad. Secondly, the presence of private shareholders would help to keep down costs. Thirdly, keener competition would be introduced into the market. Some of these benefits could be obtained from a less thoroughgoing privatization scheme under which the Coal Board and private investors undertook joint ventures in new pits. However, to ensure that the benefits were passed on to consumers, the private companies would have to market their share of output separately without collusion with the Board. If the provision of private capital for joint ventures resulted in all the coal being sold by an unreconstructed NCB, the benefits would mainly accrue to the new private owners in the form of monopoly profits. Joint ventures might be a starting-point for a new pit privatization scheme, but we would not envisage them as more than a transitional solution. Private investment in British mining is likely to be inhibited by the continued existence of a nationalized organization which would be seen both as a possible subsidized competitor and as a joint-venture partner subject to government interference.

Selling Existing Pits

A more fundamental approach to coal privatization would be to offer existing as well as future pits to private investors with the aim of establishing a number of suppliers of British coal, competing both with each other and with imported coal. Private investors could include existing companies with expertise in mining, groups of Coal Board managers interested in 'management buy-outs', or miners wishing to run pits as worker co-operatives.

Superficially, much of the British deep-mining industry as it stands is an unappealing prospect to potential investors. It is a heavy loss-maker (Table 16.3) and many of its pits are very old. However, the state of the industry as a whole conceals the presence of numerous pits in the central coalfields that are profitable at present and that are likely to remain so even if imports are freely allowed, because of their transport cost advantages. It is also significant that the very small private sector which has been tolerated by the Coal Board seems to have been able to operate profitably despite the extremely restrictive conditions imposed on it. Quite possibly the NCB's concentration on 'superpits' has led it to neglect smaller mines[20] which could be more profitable if they were run by private companies prepared to invest in technologies appropriate to such mines.

More generally, potential investors might well calculate that the existing state of the industry gives little indication of its future potential under private ownership. After all, it has been run for forty years by a centralized State corporation and subjected to constant government interference and the attentions of a powerful union. Centralization would disappear and

---

[20] See Burns, Newby, and Winterton (1985).

government interference would be much diminished with a liberalized coal market, but the union problem might remain since the British coal industry has suffered from monopoly in the labour market as well as in the product market. Even with existing union arrangements, especially given the recent break-away tendencies in the NUM, we think there would be reasonable bids for some existing pits. However, a further diminution in union monopoly power would probably be necessary before the industry became a really attractive proposition to private investors.

Selling Existing Pits Jointly with Power-stations

The most far-reaching and probably most desirable proposal would be to offer 'packages' of existing pits and power-stations to private investors. The close relationship between the coal and electricity supply industries means there is considerable logic in such an idea, and its implementation would avoid an undesirable increase in the buying power of the CEGB (below). Unlike coal, there are some natural monopoly elements in electricity supply (for example, the bulk transmission grid and local distribution of electricity to households). However, power generation itself is not a natural monopoly activity; privately owned power-stations[21] could supply electricity to a State-owned (or private but State-regulated) national grid which would in turn supply distributors.

The offer of pits and nearby power-stations in the central coalfield areas should be very attractive to private investors, subject to the qualifications mentioned above. It has the advantage of privatizing two nationalized industries in one operation, introducing private ownership into those parts of the electricity supply industry where it is most appropriate. Moreover, it would result in a very substantial reduction in monopoly power in two fuel industries. At present, as we have seen, there is bilateral monopoly bargaining between the CEGB and the NCB over coal supplies for power-stations. If the NCB were split into several supply companies with no change in the position of the CEGB, the latter would stand in a very powerful monopsonistic situation as the main customer for the competing suppliers of British coal. To avoid this accretion of power to the electricity supply industry, it seems to us desirable that there should be a move to private ownership of power-stations—in some cases, with associated pits—at the same time that coal-mining is privatized.

Conclusions on the Forms of Liberalization

Some forms of liberalization would clearly be more straightforward than others. Lifting import restrictions and full privatization of opencast coal-mining would probably be the easiest first steps, though both moves would no

---

[21] Private power generation has been made somewhat easier by the provisions of the Energy Act 1983.

doubt be opposed by the NCB and unions because of possible encroachment on the market for deep-mined British coal.

Privatization of deep-mining raises more complex issues, partly because more radical changes to the structure of the coal-mining industry (and, ideally, of the electricity supply industry) would be needed. Furthermore, given the unfortunate history of British coal-mining under private ownership, strong emotions are attached to 'public' ownership of the coal-mines, even though the present form of ownership gives no transferable property rights to the public whilst imposing on taxpayers a heavy burden of support to offset mining losses and on consumers unnecessarily high prices. Both management and unions in the coal industry may well perceive their interests to be to hold on to such market power as they now have. Monopolies, however, have a habit of promoting their own demise. Although the NCB has a virtual monopoly of coal-mining in Britain and the NUM contains the majority of miners, coal now supplies only about one-third of UK energy consumption, so that there are other fuels to which consumers can turn. Coal's market power has already waned significantly, as the most recent strike showed. It is a serious indictment of the NCB and the NUM that, even in times such as the 1970s and early 1980s when oil prices were soaring, they were unable to take advantage of their apparently much-improved competitive situation versus other fuels and to promote expansion in the industry.

6. Liberalizing or Reform from Within?

In our view, there is a clear case for reduced government control of British coal. Nationalization was intended to rescue the industry from the serious condition into which it had fallen between the two World Wars. The rescue act did not work. Even to the miners, who for so long had wanted nationalization, it must seem to have failed. Moreover, the monopoly created by politicians has had to be politically maintained because it is a monopoly of British coal alone, which leaves the NCB exposed to competition from other fuels and from overseas coal producers. Increasingly severe protective measures to shelter British coal from home and overseas competition have therefore been imposed by governments. The coal industries of some countries have expanded in the wake of the two oil 'shocks' of the 1970s, but the British industry has not. It is still in decline, as it has been for most of the last seventy years. Prospects for the next few decades are poor too, unless positive actions are taken to restructure the industry.

In saying that the coal market should be liberalized we are not, of course, suggesting that there should be no government intervention. Governments would have to ensure free entry to the industry in the interests of avoiding the formation of monopolies. Moreover, private markets generally 'fail' in various ways, in the sense that they under-produce or over-produce certain goods and possibly do not produce some goods at all. In the case of coal, there are various external costs and benefits which markets will probably not adequately take into account. For example, the production, transportation, and consumption of coal all have impacts on the environment which ideally should be 'internalized' in the costs of the producers, transporters, and consumers so that their actions reflect the full costs to society, rather than just private costs. Moreover, there are costs to individuals and society from the run-down of an old-established, regionally concentrated industry such as coal-mining which need to be incorporated in pit closure decisions. We would regard the operation of a strong competition policy and the internalization of social costs and benefits as essential features of a properly functioning liberalized coal market.

It might be argued that a safer course of action than liberalization is simply to proceed with the present reorganization of the NCB so that the industry becomes more efficient, bringing benefits to fuel consumers (in the form of lower fuel prices than they would otherwise have paid) and to taxpayers (in the form of reduced subsidies for the industry). After all, the government has said that the Board must break even by 1987-8, which might be regarded as putting on sufficient pressure. An attempt to privatize might seem unduly provocative and a source of unnecessary upheaval in the coal industry.

However, we have doubts whether reorganization from within, supplemented by government pressure from without, will be sufficient to give the British coal industry the kind of future to which it ought to be able to look forward. Government injunctions to break even at some time in the future have been given to the NCB so many times in the post-war period—most recently in the Coal Industry Act of 1980 which specified 1983-4 as the date by which deficit grants would disappear—that one can only be sceptical about the worth of such political pronouncements. Break-even, which is a rather arbitrary efficiency target, has not been achieved as a consequence of such orders; indeed, subsidies to the coal industry were greatly *increased* from 1980 to 1984.[22] Even as the government now says it wants break-even to be achieved by 1987-8, it is planning to give the NCB about £2.65 billion of taxpayers' money in the next two financial years. The British coal industry seems to us to need much less political interference and much more influence from the market. Yet so long as it remains in the hands of a centralized, State-owned corporation, for whose activities Ministers are answerable to Parliament, it seems very unlikely that governments will be able to overcome their urge to meddle in its affairs, allowing the industry to respond to market pressures. Nor will the NCB's management be able to resist asking the government for more time and money.

7. Gains from Liberalization

Reduced Monopoly Power of Labour

In British coal there has been a monopoly not only in the product market but also in the labour market; one gain from liberalizing the coal market which

---

[22] See Robinson and Marshall (1983).

could not so readily be achieved by reform from within is that the monopoly power of labour would most likely be reduced, with consequent benefits not only to fuel consumers and taxpayers but in the long term to miners as well.

The NUM, which is the most powerful of the mining unions, came together in 1944 from the earlier Miners' Federation of Great Britain (MFGB), partly because of the belief that, with nationalization approaching, there should be a centralized trade union to face and bargain with a centralized corporation. The evidence of recent years suggests that the monopoly power of the NUM, displayed by frequent threats of strikes and other disruptive actions, has seriously damaged coal's market prospects. With many union–management disputes (under governments of both major parties), including national strikes in 1972, 1973–4, and 1984–5, British coal has a far worse record of supply interruptions than the Arab oil producers. Not surprisingly, therefore, potential consumers are reluctant to convert to coal, despite subsidies of a quarter of conversion costs and low-interest loans from the EEC. The monopoly position of the NUM may also have made consumers doubtful whether, in the long run, British coal prices would remain significantly below oil prices (see Section 2 above).

Even though the NUM is now a divided union, losing members to the Union of Democratic Mineworkers, its power to interrupt supplies and to raise wages and prices is probably still seen as considerable. Demand for British coal will continue to be depressed because of fuel consumers' fears of supply interruptions and price increases. Yet so long as the coal industry is managed by a single large corporation, there will be some logic in the argument that a countervailing force of one large union is required. One of the advantages of having a number of coal supply companies is that it would weaken the case for having one union and there would probably be a natural tendency for bargaining to become more decentralized.

We are not suggesting that liberalization implies large numbers of coal suppliers in the British market, as there were just before nationalization. In present circumstances, there would probably be a relatively small number of competing suppliers, of sufficient size to take advantage of technical economies of scale, carry out research and development, and advance the technologies of mining and mining management. Nor are we suggesting moves to break the power of labour in British mining, but only to concentrate bargaining at plant or company level, and hence focus on efficiency improvements and enhancing the industry's market prospects. With the ending of product and labour monopolies, there would be ample scope for decentralized bargaining over terms and conditions of employment.

Wider differences between areas than now in wages and other conditions would almost certainly emerge. But such differences are an essential feature of a changing, competitive industry in which conditions of production vary across the country. Attempts to minimize wage differentials in the past have reduced incentives in the most productive areas (despite the NCB's productivity scheme) and have stopped workers pricing themselves into jobs in other areas. Quite possibly, some mines deemed 'uneconomic' under nationally negotiated wage agreements would become economic with wages determined locally and with the development of technology more suited to smaller pits.

From the miners' point of view, the principal advantages of more decentralized bargaining would be to focus on direct benefits for miners (rather than on political or quasi-political ends) and the real prospects it would offer of stability (and later expansion) in British coal as consumers' fears of monopoly action diminished and their confidence returned. Otherwise, a continued decline in production and the introduction of labour-saving technology, partly in an attempt to reduce the monopoly power of labour, will rapidly reduce the labour-force in mining. The belief that a 'strong' centralized union, acting as the NUM has recently acted, will be able to preserve miners' jobs is in our view false: in the end, miners' jobs depend on the decisions of fuel consumers who have every incentive to keep themselves beyond the grasp of monopolies.

Less Politicization of Decision-making

In all organizations, resources tend to flow naturally into uses where expected returns are highest. In nationalized industries—where political and bureaucratic interference in pricing, output, investment, and other vital decisions is endemic—lobbying of Ministers and key Civil Servants and other similar activities are potentially high-return activities which consequently flourish, occupying management time and crowding out the search for efficiency. Lunching the right people at the right time is likely to be far more rewarding than a more diligent search for ways of reducing costs.

We have little doubt that substitution of the incentives of the economic market-place for those of the political and bureaucratic market-places would unleash very large efficiency gains in British coal-mining. Such gains must have been suppressed for many years, not because of inherently inefficient management and workers, but because in the nationalization scheme of things they were low-priority activities which had to take their place below dealings with the Civil Servants and politicians.

Competition on the supply side of the British coal market would reduce lobbying and political manipulation to a much greater extent than would lesser degrees of liberalization. If, for example, liberalization took the form only of allowing imports, the NCB and the mining unions between them would continue to constitute a powerful force which would constantly be lobbying for reintroduction of restrictions, either directly or by subsidization of home-produced coal. Consumers would fear success of the coal lobby, or an interruption of imports by the unions, which would diminish benefits of free trade. On the other hand, competing British coal suppliers and local bargaining with workers would present a less cohesive force and would have less incentive to try to capture bureaucrats and politicians by lobbying.

Efficiency Gains

Liberalization of the market would bring positive benefits, not only from the reduced politicization of decision-making, but also from competition among home and overseas suppliers of the British coal market which would exert downward pressure on costs and prices. After forty years without such competitive pressures, it would be surprising if there were not large efficiency gains to be realized. Full liberalization, including the privatization of British coal supply, would also result in efficiency pressures from the capital market as shareholders influenced management to keep costs down.

A less obvious efficiency gain is that the insurance costs of countering the monopoly power of coal-miners and coal management would no longer be incurred in a fully liberalized market. That monopoly power has not only depressed the market for British coal; consumers who have continued to burn coal, sometimes under duress from government, have felt obliged to hold excess coal stocks and to take other costly actions (such as having stand-by fuel sources) to carry them through supply disruptions. The CEGB in particular, which has been forced by government to take more British coal than it would have wished, has kept excess stocks of coal, oil, chemicals, and other materials; it has held high-fuel-cost oil-fired stations in readiness; and it has incurred enormous costs when strikes have come by using those oil-fired stations.

Between the end of December 1980 and December 1983, power-station coal stocks were increased from 18.6 million tonnes to 31.9 million tonnes, principally because of coal strike fears. Then when the 1984-5 strike came, the CEGB switched 39 million tonnes coal equivalent of its fuel purchases away from coal (mainly to oil, supplemented by additional imports of electricity from the Scottish Boards and some small extra use of natural gas). In the process it incurred extra costs estimated at about £2 billion.[23] Adding the costs of building extra stocks of coal before the strike in order to ride out its effects, the full costs to the CEGB (and thus to the fuel consumer and taxpayer) must run to several billion pounds. There have been additional longer-term insurance costs because the British nuclear power programme has probably been larger than it would have been if indigenous coal had seemed to be a secure source of fuel supply for power-stations.

8. The Problem of 'Uneconomic' Pits

At present, the majority of the pits in the British coalfields do not even cover their operating costs before payment of interest. Some make huge losses. Although some of these loss-making pits might well become profitable under private ownership, not all would do so. The problem of 'uneconomic pits',

---

[23] Central Electricity Generating Board, *Annual Report and Accounts*, 1984-5, paras. 15-18.

which has been at the centre of recent disputes between the Board and the mining unions, including the 1984–5 strike, would undoubtedly become a major issue in a privatization programme. However, the problem would in some ways become easier to handle.

One of the principal sources of dispute in the most recent strike was over whether or not there should be some 'independent' element in the pit review procedure—that is, whether some independent party should have a say, or even the final decision, in cases where the Board proposed to close mines. Judgements on whether or not pits are 'uneconomic' are not straightforward, even if social costs and benefits are neglected. Conventional accounting statements are not particularly helpful since they are backward-looking and are a mixture of avoidable and unavoidable costs. The information required for pit closure decisions, however, relates to *future avoidable* costs and revenues—that is, what costs would not be incurred and what revenues would be lost if a pit were to be closed. There is a considerable element of judgement in such decisions; for instance, estimates would have to be made of the effects on revenues and costs of increased investment in the pit, assuming that it was not too close to exhaustion. An advantage of allowing bids for existing mines is that it would, in effect, be an independent review procedure. Potential investors (including Coal Board employees) could make up their minds whether or not pits could be run profitably and support their judgement with their money. Thus a much wider group would be involved in decisions on whether or not pits were 'economic'.

Inevitably, there would be no bids or only very low bids for some mines—those where expected avoidable costs exceeded avoidable revenues. Thus, the truly uneconomic pits (as perceived by investors) would be isolated. That, in itself, would be an advantage since, for the first time, there would be a definition of such pits that did not come from the Coal Board. The pits in question would not necessarily have to be closed. It would depend on the estimated costs and benefits to society of keeping them open. Several economists have pointed to the social costs of pit closures,[24] and some have argued that practically all existing pits should be kept open on the grounds that, at a time of high unemployment, the 'shadow wages' of mining labour are close to zero. Their case is that since miners put out of work by pit closures would either not find work or would displace people who would otherwise have been employed, such closures would result in a net increase in unemployment. On this argument, it is misleading to classify pits as 'uneconomic' on the basis of the Coal Board's accounting methods. Instead, miners' wages should be expressed as zero or some low figure in the accounts, on the grounds that if they were not producing coal they would be producing nothing; most pits would then appear 'economic' on this form of social accounting.

As they stand, such analyses are far too partial.[25] The only social cost or

---

[24] For instance, Glyn (1984).
[25] See Robinson (1985).

benefit they include is the unemployment effect. They ignore the monopoly-enhancing effect which is costly for society and which occurs when the employees of a nationalized industry that is already being heavily subsidized are, in effect, given guaranteed employment. A proper cost–benefit analysis, given the present structure of the industry, would have to include these social costs and other external effects of coal-mining such as environmental costs. It would also need to allow for possible spill-over effects on other industries whose work-forces might equally well claim that they should be protected from unemployment.

With privatization, however, the monopoly problem would be much diminished and it would be sensible to carry out cost–benefit analyses for those pits that could not be sold, to determine whether or not they should be kept open. If a case could be made, on a pit-by-pit basis, for continued operation, the best method of subsidy would probably be a once-for-all capital payment to those who would manage the mine (for instance, a workers' co-operative). It would then be up to the new owners to operate without further subsidy (and without government interference). Where cases could not be made for keeping pits open, it would seem reasonable to pay once-for-all subsidies to offset the social costs of closure—both to the employees who lost their jobs and to local authorities who could attempt to attract new industries with better long-run prospects than ageing pits.

9. Conclusions

It seems to us that, under an organizational form that essentially involves a large corporation being confronted by a large union, the British coal industry in the post-war period has demonstrated inflexibility and an incapacity to respond to market circumstances. The result has been a waste of the nation's resources and damage to the long-run interests of British coal itself.

It is difficult to imagine a period better designed for a revival of British coal than the 1970s and early 1980s. Oil prices soared and coal's principal competitor in new power-stations, nuclear power, lacked credibility as in virtually all industrial countries except France, nuclear building programmes were set back by technical problems and lack of public acceptability. Ministers, Civil Servants, the Coal Board, and the mining unions all professed to believe that British coal sales would increase. They did not. Output and employment in the industry continued to decline, even though it received increasing support from the rest of society. Customers were evidently unwilling to turn to a fuel source controlled by a State monopolist engaged in a running battle (interspersed with occasional periods of collusion) with a powerful union. They were fearful of frequent supply interruptions and long-run price increases.

Recent acceptance by the Coal Board and the government that 100 to 110 million tonnes a year is the likely market for British coal shows a sense of realism, assuming the industry is not restructured. But that should not be regarded as the upper limit on coal consumption in Britain. Our previous market estimates (see Section 1 above) were based on the assumption that the structure of the coal market and government policy towards the industry would remain unchanged. Structural changes of the sort proposed in this paper would, in our view, significantly improve the outlook for British coal.

In a liberalized market there would be a number of rival home and foreign suppliers to whom existing and potential consumers of British coal could turn. We would expect that kind of market to accelerate the breakdown of the labour monopoly which is already occurring, with bargaining much more locally-based than now; technological progress should be rapid and relevant to different scales of operation (as it has been in the supply of British North Sea oil, where remarkable advances have occurred not only in the development of big fields but also in the ability to exploit small and previously inaccessible deposits); the diversification of supply sources would enhance security of supply, thus quelling the fears of coal consumers after their experience of recent strikes; decisions would be much less politicized; and there would be sustained pressure to reduce costs, keep prices down, and generally to raise efficiency. The government's role in this kind of market would be to maintain competition and to internalize external costs, not to intervene constantly in the running of the industry.

Despite the post-strike revival in British coal, the history of the industry suggests that continued monopolization of British coal supply and of the supply of British coal-miners is a recipe for continued decline. A liberalized market, however, offers the prospect of a more efficient coal industry, responsive to the needs of consumers and gradually reversing the long-run decline of the industry.

17 The Economics of Coal*

BILL ROBINSON

1. Introduction and Summary

For most of the post-war period, the coal industry has been run by planners. Prices have been fixed by long-term agreement. Production has been determined by volume targets. In consequence any unexpected variations in costs have been reflected in the profitability of the industry. For much of the period, the industry has run at a loss, reflecting a persistent failure to close unprofitable pits as fast as was necessary.

However, the problems of coal cannot be blamed entirely on bad planning. The industry was operating in a difficult environment, in which its market share was being continuously threatened, first by oil and then by gas. A market response to this threat would have been to price coal at a competitive level to conserve share, and hold costs down to that price—by closures if necessary.

The actual response was to keep prices at the highest level that the industry could get away with (helped by official arm-twisting of the electricity industry, which was the major customer) in order to minimize the losses that were an inevitable consequence of hitting the planned output targets. The result was a steady erosion of market share as customers switched to alternative fuels.

The tenfold increase in oil prices in the 1970s presented the industry with an opportunity to improve its competitive position. From 1974 onwards, coal was priced in its non-CEGB (Central Electricity Generating Board) market at a level which enabled it to conserve its market share. But the industry continued to make losses as demand fell consistently below official projections. The reason was that although coal was holding its share of the energy market, that market was itself falling, because of the slowdown in growth of gross domestic product (GDP) and the switch to less energy-intensive forms of production.

It took some time for this to be recognized. In the mid-1970s the industry was geared up to expand to meet the ambitious production targets of the 1974 Plan for Coal and its successors. In this climate of optimism, the industry acceded to a series of wage demands which carried the miners to the top of the national pay league. Investment was stepped up, but the productivity of the industry as a whole did not respond. In the coal industry there are always pits that are at the

---

* This paper is an edited version of three articles originally published in the London Business School's *Economic Outlook* in 1984 and 1986.

At the time of writing this article, Bill Robinson was Senior Research Fellow at the London Business School; he is currently Director of the Institute for Fiscal Studies.

The author is especially indebted to Louis Turner who guided his reading and made many valuable comments. Responsibility for errors is the author's alone.

margin of viability, as seams of coal run out or become too expensive to exploit. To keep costs down, it is important to keep closing such pits. But in the expansionary climate of the 1970s these closures were exceedingly difficult to implement. As a consequence, the industry acquired a 'tail' of uneconomic pits in which the avoidable costs of production were well above the selling price of the coal produced.

The failure to close uneconomic pits in the 1970s was at the heart of the difficulties faced by the industry in the early 1980s. The difficulties were compounded by the massive—though not permanent—rise in the foreign exchange value of the pound, which held the price of imported coal well below the average price of domestic production. This raised a fundamental and recurring question: should the electricity industry be allowed to use more imported coal? The argument in favour was simple: such coal was cheaper. The argument against was that cheap imported coal would hasten the decline of UK productive capacity, and leave the country vulnerable to any future increase in the cost of imported coal. It would also increase unemployment and worsen the balance of payments.

The balance of payments and unemployment arguments can, in theory, be dismissed in a world of mobile resources and flexible exchange rates. But even in such a world it may be worth subsidizing high-cost domestic production in order to assure security of supply. Whether this is worth doing depends on the relative costs and benefits of such a policy. The cost is the subsidy required to keep marginal pits in production. The benefits are the future savings obtained when the costs of alternative (foreign) supplies have risen above domestic costs in those subsidized marginal pits. The costs to the industry and its customers are immediate and certain. The benefits are distant and uncertain.

Put in those terms it seems unlikely that a pit producing coal at over £100 per tonne (as some were in 1981) could ever be worth subsidizing (the price of imported coal over the period 1980–5 fluctuated between £30 and £38 per tonne). However, domestic production at £38 per tonne might be worth subsidizing, even though today's price of imported coal is £25 per tonne, since history suggests it may not always be so cheap. This example shows that any decision about pit closures must in the end depend heavily on an essentially commercial judgement about future cost and price movements. Since the cost of production in each pit, as well as the cost of imports, fluctuates greatly from year to year, it does not always make sense to close a high-cost pit, even from a narrow commercial point of view.

The opponents of pit closures take the argument further, and claim that the true costs of keeping a high-cost pit in production are lower than the commercial cost, since society as a whole has to bear the cost of the unemployment that results when pits are closed. These arguments can be used to justify a much higher rate of subsidy, since the exchequer cost of throwing a miner out of work, in terms of unemployment and other benefits paid, and taxes and National Insurance contributions forgone, can be of the order of £6,000–£7,000 per annum, depending on family circumstances. However, it is clear that accepting this argument would set a dangerous precedent, since any loss-making activity could claim a subsidy on the same grounds. Pit closures, like bankruptcies, are a harsh but necessary part of the process whereby resources are continually redeployed to meet the changing demands of society. Obviously, however, they are more acceptable when there is a realistic chance that the labour released will in fact find other employment, and less acceptable when the general level of unemployment is high.

However, even if the argument about closures is conducted rationally, by weighing up the costs and benefits of subsidizing marginal pits, there are no clear-cut answers. The cost of keeping unprofitable pits open depends on the wages and productivity of coal-miners (and on the tax and benefit system). The benefits depend on the expected future price of alternative sources of fuel. The appropriate size of the industry depends crucially on the ratio of costs to prices. Given the enormous uncertainties that surround the key parameters, the academic analyst can only examine illustrative simulations. The results of these simulations suggest that, on most reasonable assumptions about future cost and price movements, the coal industry will continue to shrink. But the existence of a highly profitable core of 'superpits' means that it is unlikely to disappear altogether.

2. Historical Background

The National Coal Board (NCB) was set up by the Coal Industry Nationalisation Act 1946. The Act gave the Coal Board exclusive rights to 'working and getting the coal in Great Britain' (Monopolies and Mergers Commission (1983a, para. 2.4)). In return for this monopoly, the Coal Board was required to 'secure the efficient development of the coal mining industry'. It was required to supply coal 'at such prices as may seem to it best calculated to further the public interest in all respects'. The Act also requires the Coal Board to operate 'so that its revenues shall not be less than sufficient to meet its outgoings properly chargeable to revenue account (including interest) on an average of good and bad years'. The terms of the Act thus established the Coal Board as a monopoly supplier, and the break-even requirement for the *industry* meant that some loss-making *pits* would always be kept in operation. With hindsight it is easy to see that a statutory requirement to 'secure the efficient development of the industry' provides no strong inducement to close a loss-making pit. If each pit had been required to break even, taking one year with another, the history of the industry would have been very different.

The existence of unprofitable pits has long been a source of annoyance to the Central Electricity Generating Board (CEGB), which believes that electricity consumers subsidize uneconomic pits. This is a view that is also widely held among UK industrialists who do not always appreciate that the same happens in other European countries. 'The CEGB would like to see the NCB become a streamlined producer of economically priced coal, with no cross-subsidisation of uneconomic pits by profitable ones . . . A price structure based on cost of production at pits . . . would give the correct signals for generating electricity using the cheapest source of fuel' (Monopolies and Mergers Commission (1981)).

As we shall see in more detail below, the existence of inefficient pits does not in itself mean that coal prices are too high—the cross-subsidy in the industry could in principle keep the price down to market levels. The best guarantee of this would be that the CEGB had the right to buy in the cheapest market—including imports. However, 'Government wishes, overt or covert, have had a strong influence on the quantity of foreign coal burnt at power stations, and *when political constraints permit* [the CEGB] buys modest quantities of coal at prices which are significantly lower than those of comparable NCB coal' (Monopolies and Mergers Commission (1981), italics added). Without these constraints, the CEGB would import more cheap coal and lever the domestic price downwards.

Electricity generation is now the main use of coal. The price at which coal is sold to the CEGB is a crucial determinant of the financial position of the coal industry. That price is the outcome of negotiations between two State-owned near-monopolies. In such circumstances it is hardly surprising that the price is determined in no small measure by political factors.

The close relationship between coal and electricity is not a new phenomenon, but their degree of interdependence has grown strikingly since the war. In 1947 (when steam trains were still significant users of coal) only 14 per cent of coal output was sold to power-stations. Today the figure is closer to 70 per cent, as Table 17.1 shows. Conversely, over 80 per cent of the fuel used for electricity generation is coal. In these circumstances, 'the' price for coal in the UK is, to a large extent, what the CEGB can be made to pay for it. Although imports are always a constraint on the domestic coal price, the large transport cost for coal gives the domestic producer a degree of local monopoly. The power-stations have been built, by and large, around the coalfields (see map, reproduced from

Table 17.1. Who buys British coal? (million tonnes)

	1982	1983	1985–6
Electricity supply industry	80	82	86
of which: CEGB	78	76	79
Other	2	6	7
Gas, coke, etc.	13	12	14
Industry	7	7	8
Domestic and other	11	10	10
Total	111	111	118

Source: *Digest of UK Energy Statistics.*

Monopolies and Mergers Commission (1981)), and that enables the Coal Board to charge prices significantly above what would be charged if the power-stations were closer to the ports.

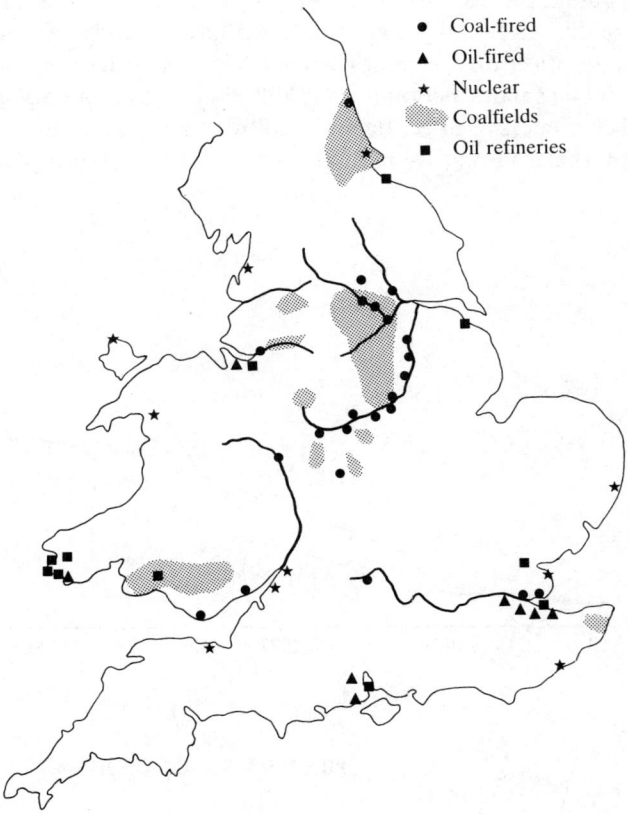

Major power-stations showing relationship to fuel sources

3. The Price of Coal and the Size of the Coal Market

Between 1973 and 1984 the dollar price of oil rose tenfold—a fivefold increase in real terms. Coal competes with oil (notably in electricity generation), so a rise in oil prices increases the demand for coal. In theory therefore the oil price shock should have raised coal prices and profits, and led subsequently to an expansion of the industry as new higher-cost sources of supply were brought on stream. What we observe in the UK is in stark contrast to this theoretical prediction: a declining industry which continues to make substantial losses. How has this state of affairs come about?

The short answer, which again we know from elementary theory, is that either the price of coal has not increased as rapidly as the oil price and/or the cost of producing coal has increased more rapidly. In fact both of these things have happened.

Before we pursue the discussion of the price of coal however, it is important to understand the institutional background. We have already seen that. Coal is sold at different prices on different markets. Most coal these days is sold to the CEGB, and only a small fraction of the Coal Board's total output goes directly to industrial customers at prices determined by demand and supply. In most of the discussion that follows, we use these prices, as representing 'the' price of

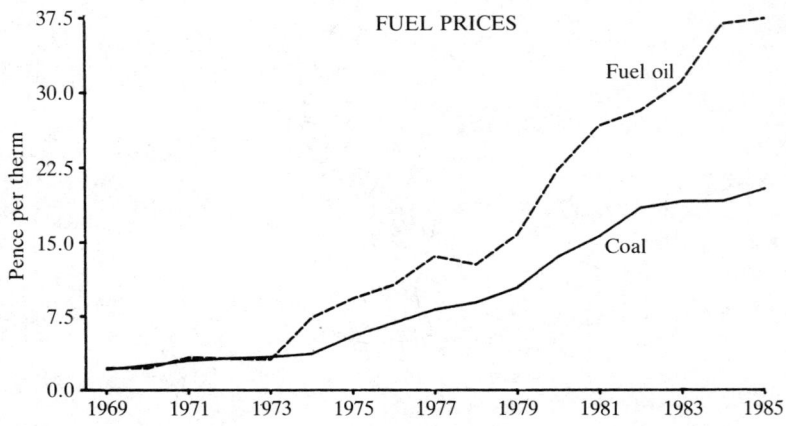

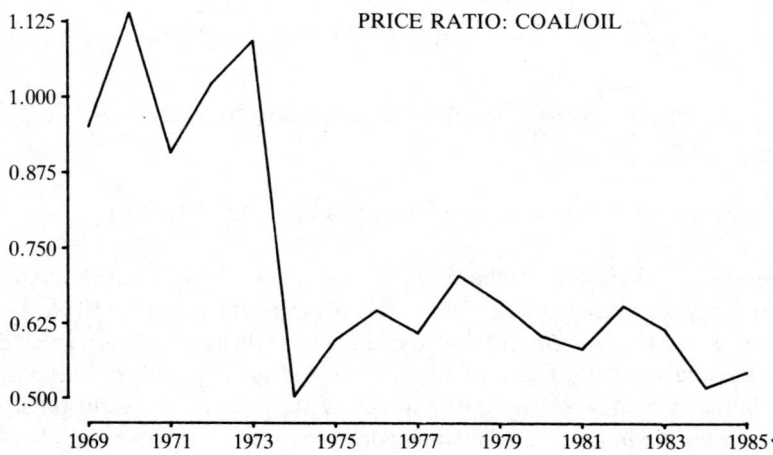

Figure 17.1 Fuel prices and price ratio of coal to oil

coal, but it must be understood that the bulk of coal output is actually sold on long-term contract to the CEGB under a three-tier price structure. These negotiated prices do respond to market conditions, but much more slowly.

Figure 17.1 shows the behaviour of oil and coal prices delivered to large industrial consumers since 1969. Before the oil price shocks of the early 1970s, the two prices moved fairly closely in line. Since then there has been a tendency for coal prices to move up with oil prices, but over the past ten years coal has been some 30 to 40 per cent cheaper than oil.

The fact that such large variations in relative prices have been possible illustrates a very important point about the energy industry. The response to any change in relative prices is slow. The CEGB is constrained politically. And the large industrial consumers who are free to choose their energy source on economic grounds alone are constrained technically. This is because consumption of energy generally requires capital equipment which cannot be changed straightaway. If the price of coffee rises, consumers can switch to tea straightaway. But if oil goes up in price, those with oil-fired central heating (or power-stations) cannot in general switch to gas or coal so quickly. We should not infer from this, however, that the *long-run* price elasticity of demand for different forms of energy is low. If the relative price difference is expected to persist, users will gradually switch in large numbers.

The long-term substitutability of different fuels is illustrated by the post-war history of the energy industry.[1] At the end of the war, coal accounted for over 90 per cent of total energy use in the UK. By 1972 that share had declined to only 36

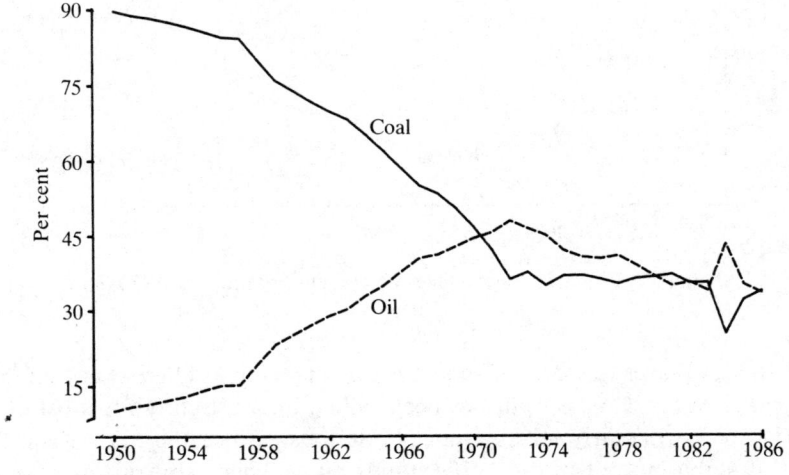

Figure 17.2 Percentage shares of UK energy market, 1950–86

---

[1] See estimates of own- and cross-price elasticities in the Department of Energy paper (Chapter 4, this volume).

per cent. As long as the prices of coal and oil were closely matched, coal lost market share at a rapid and accelerating rate (Figure 17.2). This is hardly surprising. Compared with oil, coal is bulky, difficult to transport, and poses severe waste disposal problems. At the same price per calorie, any user would prefer oil. Over time, many did in fact switch. And coal's problems were compounded from the late 1960s onwards by competition from North Sea gas (supplied, at least in some periods, at below marginal cost), which offered many of the advantages of oil plus greater security of supply.

The steady conversion to oil came to an end with the first oil crisis. The sharp rise in oil prices permitted the coal industry to find a price at which (mainly non-CEGB) users are broadly indifferent between coal and oil. With a price advantage of around 30 to 40 per cent, the coal industry has arrested the catastrophic fall in its market share. However, it has not *increased* its share since 1974. The markets lost by oil have been replaced by gas (mainly) and nuclear energy (Figure 17.3). Clearly the oil price rise has not permitted a previously healthy coal industry to make supernormal profits. It has rescued a chronically uncompetitive industry from a state of terminal decline.

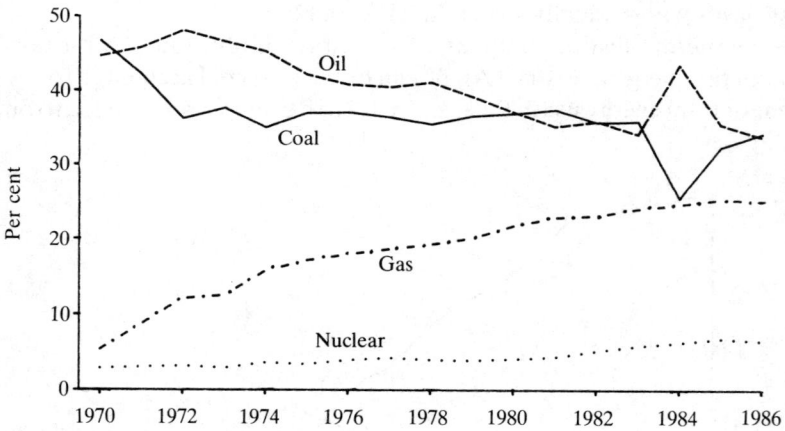

Figure 17.3 Percentage shares of UK energy market, 1970–86

Figure 17.4 places this decline in historical perspective. There is nothing new about pit closures. Coal output has been falling intermittently for most of the century and particularly steeply and continuously since the mid-1950s. The period of stable output in the 1970s stands out in sharp contrast against this historical background. But, as Figure 17.5 shows, there is an even sharper contrast between the post-war decline in coal demand and the projections of future demand on which official policy has been based since the mid-1970s. What was the basis for such optimism?

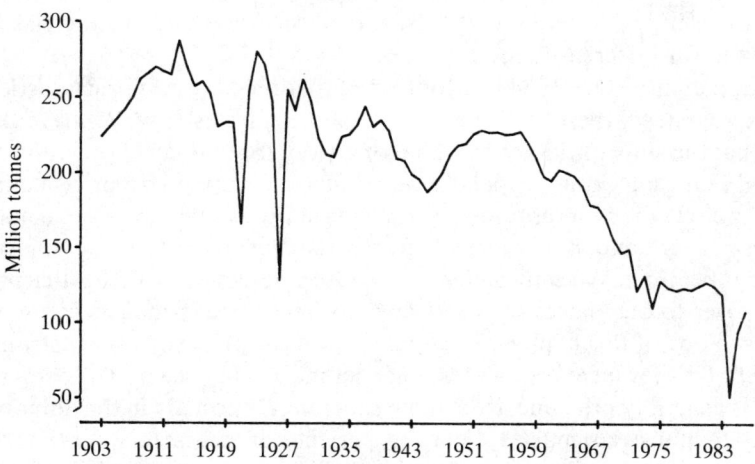

Figure 17.4 UK coal production since 1900

Demand for coal depends on total demand for energy and the share of that market taken by coal, which depends in turn on its relative price. Because of the long lead times which are a feature of the industry—it takes many years to prospect for oil, dig a new coal-mine, or construct a power-station—energy experts spend much time endeavouring to forecast energy demand ten to fifteen years ahead. As we have seen in public inquiries (Vale of Belvoir, Sizewell), the case for a new coal-mine or a new nuclear power-station stands or falls on the

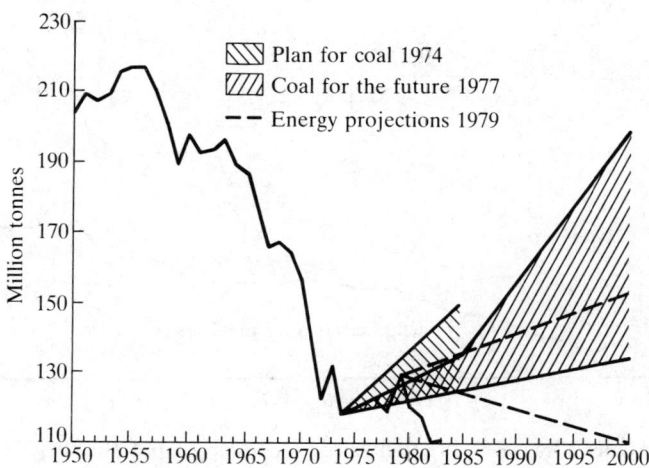

Figure 17.5 Coal output since 1950 with projections to 2000

size of the energy market in the 1990s, and on the price- and income-elasticities of demand for different fuels.[2]

The experience since 1974 has confirmed that the *long-run* price elasticity of demand for energy is much higher than the short-run elasticity. We have already noted that the substitution of one kind of energy for another is generally a long process, requiring capital expenditure, so that the response to price changes is slow. Precisely the same arguments apply to energy as a whole. We can make do with less energy and still enjoy a high standard of living, but the adjustment process takes time. We can insulate our houses, design more fuel-efficient cars, move closer to our places of work, give up 'industrial' products in favour of services. None of this happens overnight, but if the price signals are strong and persistent (as they have been), these changes inexorably occur. The demand for energy is generally price-inelastic in the short run, especially in the domestic (as opposed to industrial) market, but the experience of the last ten years has shown that the long-run price elasticity is nevertheless quite high. Capital equipment reaching the end of its productive life is being replaced today with more energy-efficient equipment in response to price changes that occurred ten years ago; and many of the effects of the second oil price shock have still to be felt.

The point is illustrated by Figure 17.6, which shows the behaviour of the energy ratio in the UK since 1950. The energy ratio relates total consumption of energy to gross domestic product (GDP), and this ratio has been falling steadily over the post-war period. Up to 1973 this simply meant that energy demand, though rising, grew less rapidly than GDP. Since then, output growth has been slower and the decline in the energy ratio more rapid. Consequently we have seen an absolute fall in total energy consumption in this country, as Figure 17.7 shows.

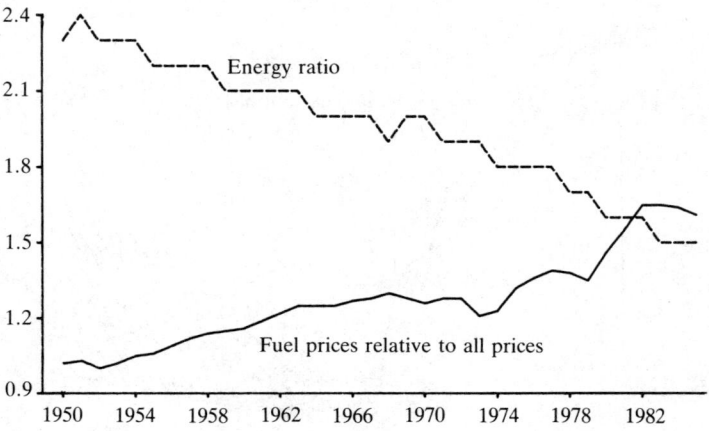

Figure 17.6 Relative fuel prices and energy ratio, 1950–84

[2] See Department of Energy paper (Chapter 4, this volume).

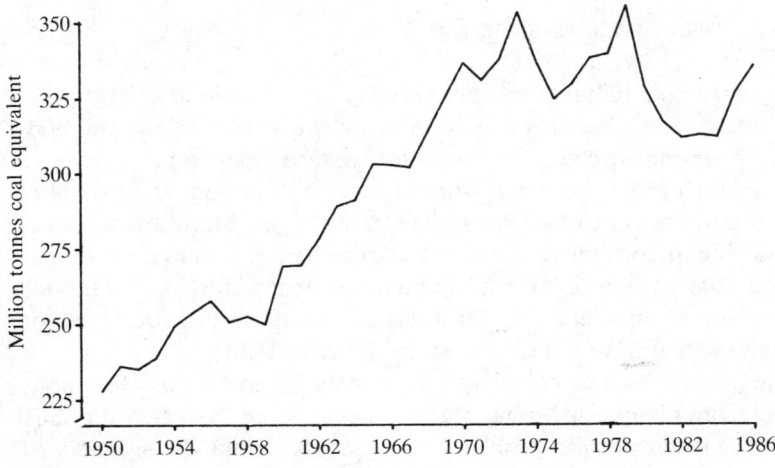

Figure 17.7 UK energy consumption, 1950–86

This brief excursion through the recent history of the energy industry—and the position of coal within it—enables us to answer the question raised at the outset; namely, why is coal not a profitable and expanding industry? Part of the answer is that although there has been a considerable improvement in coal's competitiveness *vis-à-vis* oil since the first oil crisis, the starting-point in the early 1970s was so uncompetitive that the improvement was insufficient to carry coal into an era of expansion. The other half of the answer lies in the demand for energy as a whole, which has proved much lower than was anticipated. The oil price shock of 1973 did not prevent the exercise of consumer sovereignty. There is a normal downward-sloping demand curve for energy. A monopoly supplier can fix the price at any chosen level, but he does not thereby repeal the laws of supply and demand. The consumer will determine the quantity sold at any given price. When the price of energy went up, demand fell.

The adjustment to higher energy prices is slow and probably far from complete. Even though energy prices are now falling again, the fuel-saving habits inculcated over a decade of high prices will persist. The resumption of growth in the economy since the 1980–1 recession has barely stabilized the demand for energy, which fell sharply during the recession. There is little reason to expect a substantial increase in energy demand in future, though the recent revival in manufacturing industry will, if it persists, arrest or even reverse the downward trend in energy use established since 1973. If total energy demand is stable, there is little prospect that the coal industry, which at present relative prices cannot increase its market share, will see a growing demand for its products. With hindsight we can see that the plans to expand the coal industry in the aftermath of the energy price shocks of the 1970s—including the Plan for Coal—were mistaken.

4. The Costs of Producing Coal

Although the coal industry was not able to match the rise in the oil price in the 1970s, coal prices nevertheless rose sharply. Between 1972-3 and 1981-2 the price of deep-mined coal rose from just under £7 per tonne to over £35 per tonne. Allowing for the rapid inflation over that period, this represented an increase in real terms of 54 per cent. However, even with this increase in prices, the Coal Board continued to make a loss on its deep-mining operations. The reason is that costs also rose substantially in real terms. And although costs increased less rapidly than prices, the improvement was insufficient to eliminate the large losses that were being made in the early 1970s.

Table 17.2, which is calculated from data given in the Monopolies and Mergers Commission (MMC) report on the National Coal Board published in June 1983, illustrates the problem. In the 1970s, miners' wages moved from below the average manufacturing wage to substantially above it (Figure 17.8), taking them from twelfth position in the earnings league to first or second place. Although mining productivity also grew faster than manufacturing between 1972-3 and 1980-1, real unit costs in mining increased by 49 per cent compared with only 2 per cent in manufacturing. Had mining wages grown in line with manufacturing wages over that period, the industry would have been in surplus by 1981-2.

Table 17.2. NCB deep-mines: unit operating costs and revenues (£/tonne, 1980 prices)

	1972-3	1981-2	% change
Revenue	20.3	31.2	54
Costs			
Labour	11.8	17.6	49
Depreciation	1.3	1.9	44
Other[a]	9.1	15.1	65
Total	22.3	34.6	55

[a]Overheads, materials, repairs, power, heat, and light.
Note: *Excludes* interest charges.
Source: MMC (1983a, Appendix 3.3, vol. 2, p. 25).

However, Table 17.2 shows that wages were not the only problem—not even the major problem. Other costs rose even faster. This highlights a particular problem of the coal industry, which is that costs have a built-in tendency to rise. The older the mine, the further is the seam of coal from the pit-head and the greater are the geological difficulties and the costs of extraction. For any particular mine, costs are rising all the time. So to maintain profitability, the industry must continually close down high-cost capacity and open new lower-cost mines.

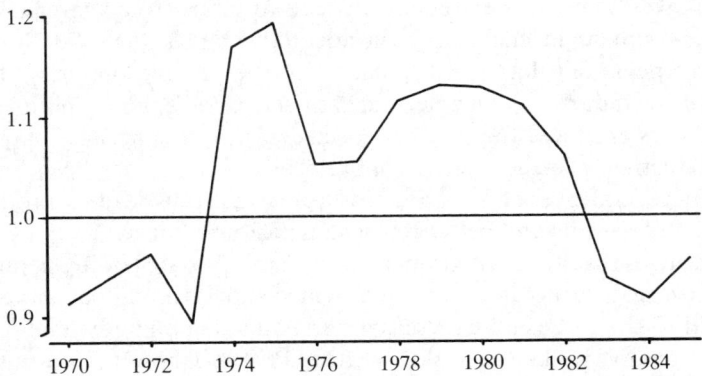

Figure 17.8 Earnings in the coal industry relative to manufacturing

One problem in the 1970s was that there had been inadequate investment in the 1960s, so an insufficient number of new coal faces were brought into operation. But it also seems probable that an insufficient number of old pits were closed down. Pit closures present many problems. One little recognized technical difficulty is that the costs of extracting coal in a particular pit can vary sharply over time—an annual variance of 15 per cent is the norm for the industry. Under these circumstances it would clearly be wrong to close a pit just because it makes a loss in one particular year. Against a background of over-optimistic demand forecasts, the temptation to give a loss-making pit another chance must have been strong—especially as closure always imposes severe disruption for the workers and families involved. But whatever the reason, it is clear (with hindsight) that the failure to close uneconomic pits sufficiently quickly has been an important factor in the industry's chronic problems.

To keep the industry competitive, it is necessary not just to close old pits, but also to invest in existing and new pits. If management blames the unions for resisting closures, the miners in their turn blame management for inadequate or

Table 17.3. Investment per employee (£ per man, 1980 prices)

	Coal industry	Manufacturing industry
1974–5	1,158	597
1975–6	1,744	618
1976–7	1,901	666
1977–8	2,087	747
1978–9	2,608	811
1979–80	3,203	856
1980–1	3,387	857
1981–2	3,311	932

Source: MMC (1983a, Appendix 3.14, p. 64) and National Income 'Blue Book'.

ill-chosen investment. It is extremely difficult to judge from aggregate figures whether investment in coal has been adequate, but Table 17.3 shows that investment per man has been greater in the mining industry than in manufacturing industry as a whole, and has risen more rapidly. This investment has not always produced the returns expected of it, partly because output (and hence productivity) in the new and more efficient mines has been held back because of general over-supply. Pit closures would thus lead to productivity gains in the newer pits and better returns on past investment.

The problems that are created when uneconomic pits are left in operation too long are revealed starkly in Figure 17.9, which shows the supply curve for the industry in 1981-2. The curve is constructed from data on individual pits from the MMC report, ranked in order of cost. It shows how the output of the industry can be increased by moving up the supply curve, bringing into production successively higher-cost pits to the point at which the (marginal avoidable) cost of the marginal pit is just equal to the price of its output. As the figure shows, there was by 1981 a large tail of pits where costs were far in excess of any likely return.

5. The Role of Imports

In Figure 17.9 there is a horizontal line which represents the price at which coal can be sold. Why a horizontal line rather than a downward-sloping demand curve? Because the UK coal industry can be considered as one relatively small supplier of coal to the world coal market. Under the classical assumption of

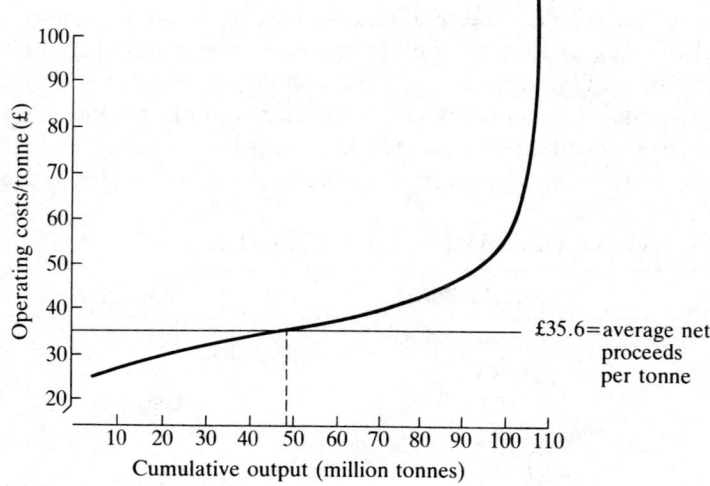

Source: MMC from NCB information.
Figure 17.9 Deep-mined production: unit operating costs against cumulative output, 1981-2

perfect competition, the market price cannot be affected by the action of a small producer, who simply takes the price as given and regulates his output accordingly.

This model does not really fit the UK facts because of the pressures on the CEGB to buy British coal at cost-plus, whatever the price of imports. In consequence, the world price for coal does not directly determine the UK price. Even so, this framework is instructive, because it suggests that the expansionist plans of the 1970s, drawn up in the wake of the oil price hikes, contained not one but two fundamental flaws: not only has the demand for energy in general and coal in particular proved more price-sensitive than was assumed, but also the benefit to *UK* producers is far less than was assumed. The rise in price has not only reduced total demand; it has also increased competing supplies. As the price of coal has risen, the industry has expanded—but it is the low-cost producers in the United States, Australia, and South Africa who have benefited, not the high-cost marginal British producers.

This description is an over-simplification, because of yet another important special characteristic of the coal industry: transport costs—especially over land—are a substantial proportion of the total cost of a tonne of coal delivered to its point of use. This makes it difficult to judge in the abstract whether or not imported coal is cheaper than domestically produced coal. It may be cheaper for a coastal power-station to buy imported coal, but not cost-effective for an inland power-station—especially one located near a coal-mine. Since over three-quarters of total coal output is used to generate electricity, the competitiveness of imports over the medium term depends crucially on where new power-stations are located.

Although high import costs effectively give UK producers a local monopoly of supply in some parts of the country, the potential for coal imports does loosely link UK coal prices to the world price. For Thames-side power-stations, imports are always a viable alternative to domestic supplies, and though the Coal Board can and does charge different prices to different users in different parts of the country, it must, like any other business, avoid offending its (in the short term) captive customers, who otherwise will make the investment necessary to switch to alternative sources of power.

The coal industry is threatened in the short term by imports of cheap coal. In the long term it could face a renewed threat from oil. After a decade of high oil prices, the reduction in world energy demand and development of non-OPEC sources of supply pose a serious threat to the stability of the OPEC cartel. The oil price fell sharply in 1986, and if it is stabilized in the range $15–$20 per barrel, implications for the coal industry are extremely serious. We will come back to this issue below.

UK industrialists frequently argue that they are handicapped, in competition with trading partners, by electricity costs that are higher than those prevailing abroad. The high cost of coal, the major primary energy source for electricity generation, is blamed. These claims are only partly justified, for it should be borne in mind that electricity costs are held down by the relatively high proportion of coal (which is cheaper than oil) used in this country to generate electricity compared with abroad. However, the UK coal industry is undoubtedly protected by the tax on fuel oil. And there is also evidence of informal official restraints on the CEGB when it has tried to burn more imported coal. This kind of protection preserves jobs in the coal industry at the expense of jobs lost in manufacturing.

6. Should we Subsidize Coal?

The Coal Board has a long history of making losses. To any businessman who believes in the role played by markets in allocating resources between competing claims, this is an important signal which suggests that the coal industry ought to contract. The resources absorbed by the marginal loss-making pits can be put to better use making or doing things that people will buy *without* being subsidized.

To the economist, things are rarely so simple. For example, a nationalized industry with a falling cost curve, and hence marginal cost lower than average cost, may make a loss because its prices are set on the basis of marginal costs. This, however, is not the case with coal. It is probably helpful to regard the long-run marginal cost of supplying additional coal as the average cost of the marginal pit. A supply curve can thus be constructed by lining up the pits in 'merit order', from least cost to highest cost. In this model the existence of losses certainly means that output has been expanded along the rising supply curve beyond the point where (long-run marginal) costs exceed prices.

Nevertheless, the closure of high-cost pits, which is being signalled by the market, imposes certain costs on the community at large. It is the failure of the market to take these costs into account that underpins the case for subsidizing the less efficient pits. Those who lose their jobs when pits are closed have to be supported at public expense and become more prone to make demands on the public health service, while the reduction in their incomes threatens the viability of businesses in the local community, with further knock-on effects on national output and employment.

A rather different sort of market failure concerns security of supply—as a nation we should be prepared to pay rather more than the market price for coal in return for the certainty that the supply will not be cut off. This argument has been used to justify the subsidy of agriculture in many developed countries around the world (together with an externality argument based on the supposed amenity value of agricultural land). The argument is complicated by the problem of variable supply conditions. Maintaining agricultural production capacity such that there is *always enough* food (even under the worst possible conditions) means that on *average* there will be *too much* food produced. The subsidy paid to the marginal producers in good years can be regarded as an insurance against starvation in bad years.

Coal-mining resembles agriculture in having variable costs of production from one year to another. Given this variability, it is not sensible to close a pit just because it makes a loss in one year. A loss-making pit this year may be profitable next. There is a strong case for keeping the most efficient tranche of loss-making pits in production in any given year if there is a reasonable chance that they will make a profit in future. But it is obvious that the coal industry has no amenity value, while subsidizing domestic production does little to increase security of supply as long as the industry is beset by industrial disputes.

Although the variability of costs and revenues is a good argument for keeping open pits at the margin of profitability, it is not these pits which were at the centre of the 1984 dispute. The Coal Board wished to close a number of pits that were making substantial losses, year in year out. The case for closing these pits in strict accounting terms was overwhelming. But a strong case was nevertheless put forward for keeping them open (for example by Glyn (1984)), based on the argument that the accounting cost of keeping the pits open did not represent the true economic cost.

What is an Avoidable Cost?

These arguments raise two rather different sets of issues. The first is a relatively minor technical issue: which of the costs of mining coal that appear in the NCB's annual accounts would in fact be avoided by closing particular pits? The Coal Board's definition of operating cost includes provisions for depreciation, for repair of surface damage, and past employee costs. If these costs are not *avoided* by closures, they should not be included in the comparison of marginal costs and revenues which is the basis of closure decisions.

Surface damage is the easiest case to deal with. Although it is true that the Coal Board cannot avoid paying out compensation for pit damage, it will avoid creating future damage liability if it ceases mining. It must therefore be right to include a provision for prospective subsidence claims in the cost of mining new coal. The cost of compensating for past subsidence is a reasonable estimate of the likely subsidence costs from future operations.

Depreciation is more difficult. The Coal Board has consistently to choose between buying new equipment or accepting higher repair and maintenance costs on old equipment. In the very short term, a repair may seem like a cost that is avoidable if a pit is closed, while depreciation is unavoidable. But over a longer time horizon, when the machine comes up for replacement, that too is avoidable. (Individuals face the same choice with domestic equipment: in the longer term the money set aside each month to replace the car when it finally wears out is just as much an avoidable cost as the money spent on running repairs. Both can be avoided by giving up motor transport.)

By contrast with these two cases, the money spent on past employee costs does not seem to be avoidable. Closing a pit cannot reduce the pension costs of past employees. These costs contribute some 8 per cent of total operating costs. It therefore seems appropriate to reduce operating costs by this amount in any calculation of the marginal effect of closures.

Social Costs

However, the question of the correct definition of avoidable costs pales into insignificance when compared with the wider issue of the *social* costs incurred by pit closure. These costs are not limited to the lost income and production of the miners themselves. Miners' wages are spent in the local community, generating additional income and employment. When these multiplier effects are taken into account, the increase in unemployment resulting from pit closures is likely to be more than the jobs lost in the pits themselves.

Against these wider costs must be set benefits, of which the chief (stressed by market economists) is the alternative output produced by resources freed from the uneconomic pits. However, these benefits cannot be taken for granted. Most of the marginal coal-mines are located in areas of industrial decline with high average rates of unemployment. Moreover, most mining communities are extremely cohesive and offer a quality of life to the inhabitants that is not easily obtained elsewhere. Miners who become unemployed have little chance of finding alternative employment in their own communities and may choose to remain unemployed in their home town rather than seek work outside.

Given these social realities, it must be recognized that the benefits of closing pits, in terms of alternative output, will be slow to appear. Moreover, as miners drift away from their communities they will incur additional social costs. The infrastructure in mining communities (housing, schools, roads, hospitals) will be under-utilized, with corresponding pockets of congestion in the areas to which the miners move. The cost of adapting the social infrastructure to the new pattern of employment should be set against the benefits obtained from the extra output.

If a given marginal pit is kept open, none of these social costs are incurred. Moreover, the pit will continue to produce a known quantity of coal with a definite market value. These are two large items to throw into the balance against the prospective benefits from closure—a stream of alternative output which may be very slow to materialize. These conclusions suggest that a pit would have to be *very* uneconomic before a full social cost–benefit analysis would show it to be worth closing.

The Case Against Subsidy

No economist, however great his faith in market processes, can dismiss these arguments lightly. However, there are powerful counter-arguments. Those who resist closure of uneconomic pits because of the disruption involved must admit that these disruption costs will be incurred eventually, if only through geological exhaustion. Putting off closure reduces the present value of these costs, but this has to be weighed against the (often sharply rising) costs of keeping an increasingly uneconomic pit open.

Secondly, the value of the marginal tonne of coal produced by an unprofitable pit is extremely hard to assess. At the limit it may have to be stockpiled or dumped on the export market, and is worth very much less than the market value of an 'average' tonne of coal. At present it certainly replaces an extra tonne of coal that could be produced at much lower cost from one of the efficient pits, which have been running at less than full capacity because of surplus production in the industry as a whole. This argument suggests that the loss of output from closing marginal pits would be much smaller than on conventional cost–benefit calculations.

Thirdly, it is extremely difficult to apply cost–benefit analysis to major economic, social, and technological changes, where the benefits are typically spread very thinly over whole societies and endure for many years, while the costs fall heavily on comparatively few people and for a relatively short period of time. A cost–benefit analysis of the introduction of the railways, taking into account the likely disruptions to existing communities and to the coach trade, could well have shown the enterprise to be unviable on social grounds. Or, to take a more contemporary example, many of the redundancies that occurred in the manufacturing recession of 1980–1 could have been avoided by public subsidies justified on cost–benefit grounds. There can be no question that a society that resists change because of its high social costs will in the short run be a more comfortable place to live. But in the long run it risks becoming a backwater.

Fourthly, given the extreme difficulty of agreeing on the appropriate criteria for cost–benefit studies (length of time horizon, appropriate discount rate) and the equally great difficulty of actually evaluating all the costs and benefits (about which no two economists, notoriously, would ever agree), the rough justice of the market-place has some attraction.

Finally, if a case for subsidizing mining employment can be made, then it can be generalized to all other threatened industries, in the private or the public sector. Mineworkers would have to take their turn and, given the high cost of keeping a marginal pit open, they would not be at the front of the queue.

This last argument acquires particular force if the cost–benefit framework of analysis is abandoned in favour of an overall limit on public spending (which may be justified on other grounds, e.g. the need for lower taxes to improve incentives or for lower public borrowing to control inflation and reduce interest rates). In these circumstances, the (opportunity) cost of subsidizing miners is the cash that is not available to spend elsewhere, for example in subsidizing jobs in the private sector or creating jobs in the National Health Service or in education. Within this framework, subsidizing inefficient collieries, viewed as a job creation scheme, is not at all cost-effective: the subsidy per man in the least efficient pit in 1981–2 was £14,000. For this amount of money, it would have been possible to meet the full salary cost of an extra two jobs at the average wage or to preserve more jobs in less capital-intensive industries by subsidizing marginal employees.

7. The Marginal Costs and Savings of Closing Uneconomic Pits

This brief consideration of the case for subsidy shows how the debate about pit closures quickly raises larger issues, which is why the 1984 dispute assumed enormous political importance. But too much of the debate was conducted in terms of broad aggregates—the total cost of subsidizing the Coal Board, the average price of coal, and the average cost of producing it. These averages are not what will, in the end, determine pit closures. It is marginal costs that matter.

At the margin the government faces a choice between subsidizing the Coal Board to keep miners at work in uneconomic pits or subsidizing the miners directly through the benefit system. Obviously any consideration of the marginal costs and savings that arise from pit closures must be based on the appropriate measure of cost. Part of the Coal Board's overall loss is attributable to sunk costs, notably interest charges, which have to be paid whether or not a particular pit is closed. Any proper assessment of the savings from closing a particular pit should ignore such costs. Thus if a pit is profitable taking into account only the *avoidable* costs incurred by keeping it in operation (wages and salaries, power, heat and light, necessary maintenance) it should be kept open, even if it appears *unprofitable* when made to bear its share of the unavoidable costs of the industry as a whole.

Fortunately there are in the MMC report (Appendices 3.3-3.5) data on colliery operating costs which come close to this definition of avoidable costs. Drawing on these data it is possible to construct a cost curve for the industry based on marginal avoidable costs. The idea can be explained most easily in terms of a concrete example. In 1981-2, according to the MMC report, the least efficient pit produced 62,000 tonnes of coal and made a loss of £104.8 per tonne. The total loss was thus nearly £6.5 million. The pit employed some 470 people, so the cost of keeping those men in work was nearly £14,000 per job (Table 17.4). Clearly at these rates of subsidy it is sensible to close the pit since the

Table 17.4. The scale of job subsidies in loss-making pits

	Output (000 tonnes)	Output per shift (tonnes)	Employment	Loss (£/tonne)	Total loss (£m)	Subsidy per job (£)
Least efficient pit	62	0.64	470	104.8	6.5	13,825
Typical loss-making pit	261	1.77	715	10.2	2.7	3,723

money saved is far greater than any possible combination of unemployment pay and lost taxes.

Consider on the other hand a marginal colliery, where total losses in 1981-2 were £2.7 million, with an implied subsidy of £3,723 per man to keep over 700 miners in employment. If the colliery were closed, the government would save some £3,723 per man in subsidies. But it would also lose the revenue from taxes and social security on the miners' income, and it would have to pay unemployment benefit and/or supplementary benefits to the miners as long as they stayed out of work. Any government concerned to control *total* public spending would have to think twice about closing such a pit even though the Coal Board might need to do so in order to hit its external financing limits.

What is clear from the example of the two pits shown in Table 17.4 is that the variations in subsidy from pit to pit are so large that any debate about the future of 'the industry' conducted in terms of national averages is liable to be highly misleading. Each pit has to be treated on its merits and any discussion of 'the industry' has to centre on its marginal cost curve.

This can be drawn from data on individual pits, as Figure 17.9 showed, but Figure 17.9 does not help in the present debate since it shows profits or subsidies in pounds per tonne. Since the crucial issues are employment and the required *subsidy per man*, the data used to construct Figure 17.9 have been transformed to produce a demand curve for labour in the coal industry. This is plotted in Figure 17.10 which shows, on the vertical axis, the cost per job of subsidizing miners in different pits in 1981-2, ranging from the most economic (where the cost was negative, i.e. the pit was profitable) to the least economic, where the cost was as much as £14,000 per job. The horizontal axis shows the number of jobs, so that we can read off from the figure the employment available at each level of subsidy. The zero line shows what total employment would be if all pits were required to make a profit.

The merit of Figure 17.10 is that it enables us to quantify the costs of employing workers in the coal industry *at the margin*. It reveals the interesting fact that in 1981-2 the number of miners employed in deep-mined collieries that were actually profitable was only 68,000. However, the figure also shows that the losses made on the next tranche of collieries were relatively small. Given the variation in costs from year to year, it is probably worth keeping many of these collieries open. The figure shows that, by using the surplus from the profitable collieries to subsidize the losses made by the most efficient of the unprofitable ones, the Coal Board could in 1981-2 have employed 164,000 miners, without requiring any outside subsidy. But note that even if the industry had been cut back to this break-even point of 164,000 jobs, the subsidy to workers in the marginal colliery would have been £4,745 per man in 1981-2. In other words, even if the mining industry as a whole is not receiving any subsidy, the cross-subsidy within the industry is very large—some very efficient pits are subsidizing some very inefficient ones. This point is presumably not lost on the miners of Nottinghamshire, where a high proportion of the profitable pits are located.

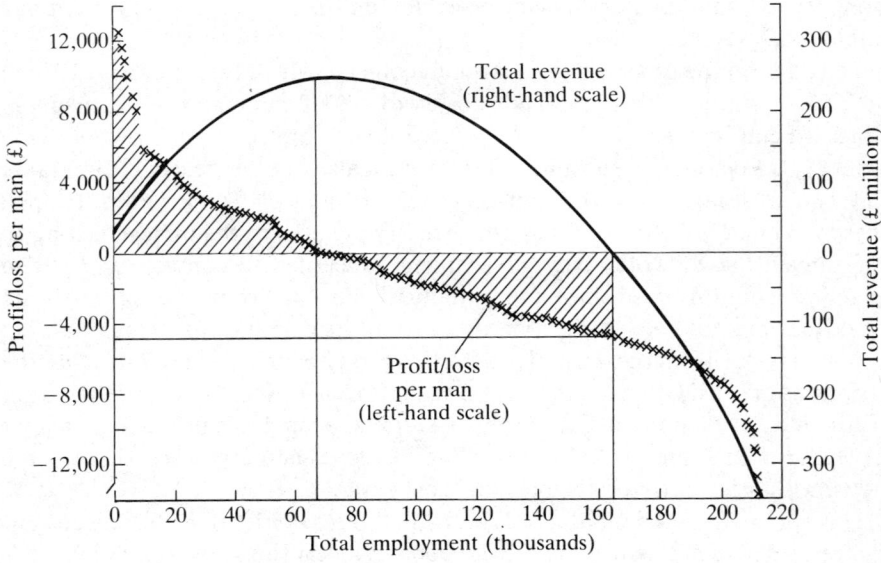

Figure 17.10 Mining employment and profit/loss per man, 1981-2

Figure 17.10 helps us to understand why the miners chose to make a stand in 1984. The industry had by then shrunk from nearly 220,000 to under 180,000 men since 1981-2, when the so-called 'tail' of uneconomic pits was so clearly identified in the Monopolies and Mergers Commission report. The 'tail' had thus been considerably thinned, if not entirely eliminated, and further cuts arguably threatened the body, not the tail. There is a point of inflexion on the cost curve somewhere around the 180,000 mark, beyond which the slope starts to change rapidly, and the case for closures becomes overwhelming. Those closures had in the main already occurred. The industry was left on a long and virtually straight cost curve with no obvious stopping-point. The National Union of Mineworkers (NUM) was therefore impelled to resist further closures. For if it accepted the logic of reduction below 180,000 to 160,000, what was to stop the Coal Board pushing it to 140,000, 120,000, or lower?

8. The Effect of Cost and Price Movements

Figure 17.10 illustrates clearly the choice that the Coal Board faces at the margin. But it suffers from a serious limitation. It is a snapshot of the industry taken in a particular year—and a year which is now some way in the past. In reality the position of the industry varies from year to year because it depends on

The Economics of Coal

the price of coal and the cost of producing it, both of which are subject to large fluctuations.

Table 17.5 illustrates the point. Between 1981–2 and 1983–4 the price of coal rose by $11\frac{1}{2}$ per cent but the avoidable cost of extraction increased by 17 per cent. Profits and losses, which are typically the small difference between two much larger numbers (costs and revenues), were dramatically affected: the loss of £84 million in 1981–2 had by 1983–4 grown to £410 million.

Table 17.5. Cost and price movements, 1981–2 to 1984–5, deep-mined coal

	1981–2	1982–3	1983–4	1984–5
Average selling price				
£ per tonne	35.6	37.9	39.7	41.7
Percentage change	8.5	6.5	4.7	5.0
Avoidable costs				
£ per tonne	39.5	41.0	46.3	48.6
Percentage change	13.2	3.9	12.9	5.0

Source: NCB Annual Reports.

Although differential cost and price movements have an important effect on the profitability of the industry, the effect on particular pits is more dramatic still. To illustrate the point, we show, in Figure 17.11, how the industry might have looked in 1984–5 if there had been no strike. The crucial assumption is that costs and prices in *each pit* moved in line with the average for the industry. The change between 1981–2 and 1983–4 is given in the NCB Annual Report. We assume no underlying change in the relationship between costs and prices in 1984–5 and have put in a notional 5 per cent increase to bring the figures close to prices of the day.

A comparison of Figure 17.10 with Figure 17.11 shows the effect of these adverse movements in the cost of producing coal compared with its selling price. In 1981–2 the industry as a whole could have broken even at an employment level of 164,000—though of these, only 68,000 were employed in profitable pits. By 1984–5 the increase in costs relative to the price of coal had changed the arithmetic. Only 58,000 jobs were available in profitable pits and the surplus earned in these was sufficient to cross-subsidize another 80,000 jobs. Without government subsidy, therefore, the industry would face a shrinkage from 164,000 to 138,000 men. This change between 1981–2 and our estimates for 1984–5 illustrates vividly the impact of variations in the coal terms of trade on employment prospects in the industry.

Figure 17.11 also shows that, even when the industry as a whole is receiving no subsidy, the cross-subsidy given to the marginal pit is substantial—over £5,000 per man. Figure 17.12 shows the effect on the profitability of the industry as a whole of raising employment from the profit-maximizing level of 58,000.

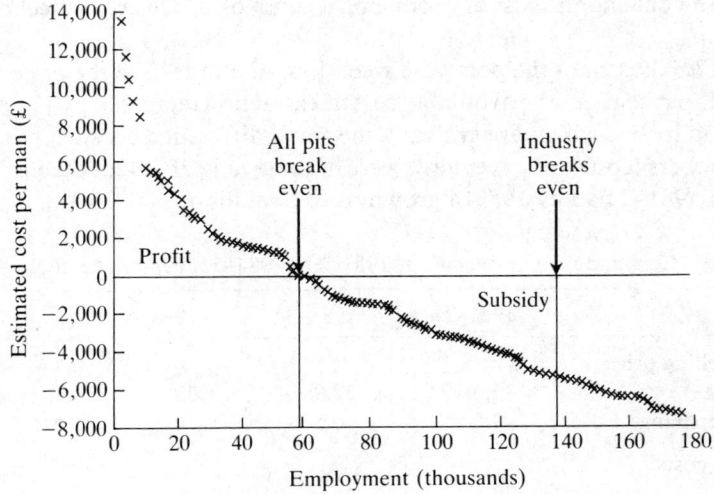

Figure 17.11 Mining employment and subsidies: estimated cost, 1984–5

Maximizing employment rather than profit would cost the industry some £400 million and cut the funds available for investment in new efficient pits. An employment-maximizing policy undoubtedly boosts jobs in the short term, but at the expense of the investment that produces secure long-term jobs in efficient and profitable pits. It is often asserted that the industry's problems are due to inadequate investment; but investment has been inadequate partly because profits that should have been available to plough back into the industry have been used instead to subsidize inefficient pits.

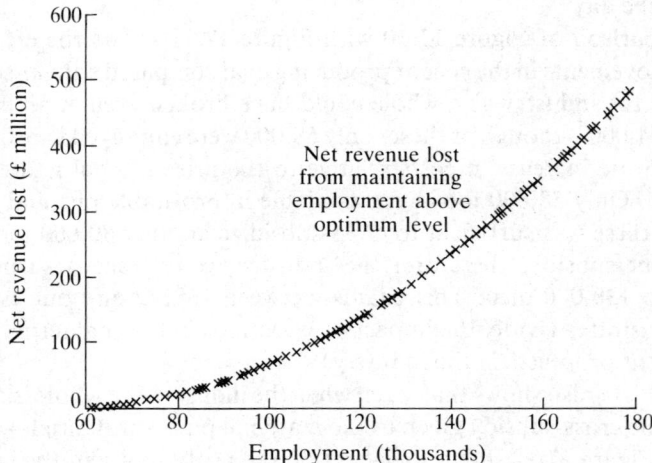

Figure 17.12 Mining employment and net revenue

The historical comparison between 1981–2 and 1984–5 is one example of the way in which cost and price movements can affect employment prospects. But the point is a very general one. Although the position of individual pits may change markedly at any moment for geological reasons, the overall shape of the industry's cost curve, as shown in Figures 17.10 and 17.11, has proved fairly stable over time. Given the data on individual pits which underpin the cost curve, we can simulate the effect on the industry (and especially on the marginal pits) of a wide variety of possible costs and prices.

The point is illustrated in Figure 17.13, reproduced from the Monopolies and Mergers Commission's examination of the industry in 1981–2 (1983a). In that year the volume of profitable production was just under 50 million tonnes, the point at which the cost curve crosses the price line. Clearly if the price is raised to £40 per tonne, the volume of profitable output rises to over 70 million tonnes. And if costs were cut by 10 per cent, volume would rise further to nearly 90 million tonnes. In other words, the volume of profitable output is highly sensitive to the coal terms of trade.

This sensitivity has been explored in a series of simulations in which coal prices and miners' wages have been changed from their (estimated 'normal') 1984 level. The results of these simulations are shown in Table 17.6.

The table shows the effect of varying coal prices and wages by 10 per cent. It illustrates the difficulties faced by those who have to plan the future of the industry. Variations in the price of coal can change total employment at

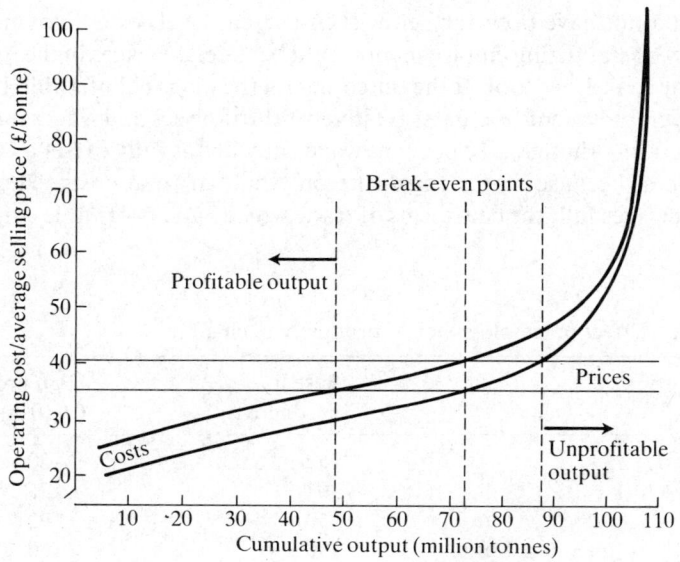

Source: MMC from NCB information.

Figure 17.13 Costs, prices, and profitable output

Table 17.6. Effects on employment of changes in the coal terms of trade

Coal price (£/tonne)		Miners' wages (£/week)		
		£156	£173	£190
38	A	52	29	18
	B	107	76	53
42	A	73	59	52
	B	172	138[a]	108
46	A	89	70	60
	B	>180	168	135

Notes: A—Employment (000s) in profitable pits.
B—Employment (000s) if industry as a whole breaks even.
[a] Central case—estimate of underlying position of the industry in 1984–5.

industry break-even points, by huge amounts in either direction. Under these circumstances any 'Plan for Coal' is likely, sooner or later, to become outdated. It is simply not possible to lay down rigid long-term plans for the volume of production or employment where there is so much uncertainty about relative prices. The table also illustrates the extent to which the level of profitable employment in the industry depends on wages. At a weekly wage of £156—which is what the miners might have been earning in 1984–5 if their average earnings had not grown so much faster than average earnings in the years following the first major increase in world oil prices—the industry break-even point would not have been far below then-current levels of employment.

One way of stabilizing employment would be to relate wages in the industry to the price of coal. If we look at the outcomes on the diagonal of Table 17.6, total profitable employment fluctuates relatively little if prices and wages move in the same direction—though a 10 per cent wage cut will not fully offset a 10 per cent fall in prices, because wages constitute only half of total costs. Thus as coal prices and wages fall, the coal terms of trade worsen and profitable employment falls.

Table 17.7. Effect on employment of productivity changes

			Base line productivity	High productivity (+10%)
Low coal price } −10%	A	52	80
Low wages		B	107	175
Base case		A	59	92
		B	138	>180
High coal price } +10%	A	60	90
High wages		B	135	>180

Note: A, B—See note to Table 17.6.

The Economics of Coal

In all the simulations shown in Table 17.6, productivity is the same. However, there have been substantial variations in productivity in the past. All things being equal, a 1 per cent rise in productivity reduces unit costs by 1 per cent (whereas a 1 per cent cut in wages reduces unit costs by only $\frac{1}{2}$ per cent). So in Table 17.7 we show the effect of re-running the simulations shown on the diagonal of Table 17.6, feeding in a 10 per cent productivity improvement.

The results are fairly dramatic in terms of employment, and emphasize the point that in an industry that is not governed by volume targets but is allowed to produce as much as it profitably can, productivity improvements *increase* employment. Such productivity increases may come about either through better working practices or through new investment. However, the case for new investment (which may also increase the level of profitable employment in the industry) is subject to the usual criterion that there is an adequate return on the capital employed.

The lesson of these simulations is that the future of the coal-mining industry is to some extent in its own hands. The size of the industry and the associated level of employment are highly sensitive to the level of wages and to the productivity performance. It is widely, though wrongly, believed that productivity increases in general threaten jobs—because the same amount of work is done by fewer people. These simulations of the coal industry illustrate clearly that the opposite is the case: the more productive the workers, the lower the level of costs, and the greater the opportunity for expansion.

However, it would be wrong to suggest that the future of coal is *entirely* in its own hands. The future of the industry depends crucially on the price and availability of alternative fuels—notably oil and nuclear energy. The sharp fall in the price of oil in 1986 has thrown the industry into a new crisis, which is not all of its own making.

9. The Consequences of Lower Oil Prices

Immediately after the year-long coal strike, the prospects for coal appeared remarkably bright. Normal production was resumed more quickly than expected. The fall in the exchange rate, especially against the dollar, had greatly lessened the threat from cheap imported coal. The number of miners employed in deep-mines was quickly reduced from 180,000 to 140,000, and the closure of uneconomic pits enabled production in the new superpits to be stepped up. The resulting sharp rise in productivity meant that the underlying position of the industry was close to break-even, or even modest profit.

The fall in the price of oil has plunged the industry into a new crisis. To preserve coal's market share, its price must follow oil prices downwards, which will drastically reduce the amount of coal that can be profitably produced in this country, with obvious implications for employment. We conclude the article by presenting some further simulations which illustrate the predicament of the industry in a world of lower oil prices.

10. Relative Prices and Market Share

The UK sterling oil price was fairly steady at around £20 per barrel in the period 1982–5 and coal prices have also been fairly constant over that period. In 1986 the price of oil on world markets fell from $26 per barrel to around $15. If this proves a permanent change, the price of the major fuels which compete with oil must also—eventually—fall.

The fall is not immediate because buyers cannot switch quickly from one source of energy to another. Costs of conversion are large, so although gas is much cheaper per therm than either coal or oil, many of those with (for example) coal- or oil-fired central heating continue to buy coal and oil rather than change their heating system. Moreover, the time taken for conversion between one source of energy and another means that the choice is not governed just by today's price. The key factor is the expected future price.

In the energy field, the most important set of price-determined choices are those between different types of power-station to generate electricity. A power-station can take up to fifteen years to plan and build, so the choice between a new coal- or oil-fired station must depend on the relative prices of coal and oil in the year 2000 and for many years (the life of the station) thereafter. The fact that today's oil price is well below the coal price is not a compelling reason to switch from coal to oil, and the fact that the coal price has fallen far less than oil shows that demand for coal is holding up well and users are not (yet) switching.

These considerations suggest that the market share of different fuels will change fairly slowly in response to price movements. Figure 17.2 above shows that such changes nevertheless occur, and Figure 17.1 suggests strongly that they are price-related. As long as coal was priced at parity with oil on a pence-per-therm basis, it steadily lost market share because coal is a distinctly inferior fuel from the user's point of view—it is costly to transport, costly to store, and creates costly waste. These disadvantages were sufficient to induce users to switch from coal to oil until the first oil price shock in 1973. That event sharply increased the relative attraction of coal by giving it a 40 to 50 per cent advantage. It also raised important questions about the long-term security of oil supply, giving coal a potential non-price advantage, subsequently eroded by the poor strike record of the industry.

Since the mid-1970s, the Coal Board has priced its coal so as to maintain a price advantage of around 35 to 40 per cent, and has with that policy succeeded in holding its market share. However, the dramatic fall in the price of oil threatens to push the coal–oil price ratio well above the levels prevailing in the 1960s. If the present ratio is maintained, the Coal Board must expect to start losing market share again, at the rate of about 3 percentage points per year. On that basis the industry would have shrunk to nothing before the year 2000.

These facts did not escape the Central Electricity Generating Board, which in 1986 started to negotiate a reduction in the price of coal. The CEGB's position was strengthened by its experience in using oil-fired generating stations during the coal strike. These stations, though uneconomical with oil at $25 per barrel, were commercially viable at $15 per barrel. The CEGB could thus, unusually, threaten an *immediate* and *massive* switch from coal to oil unless prices come down.

11. The Consequences of a Lower Coal Price

If the price of coal does fall, what are the consequences for the industry? To assess this we have to look at the cost curve, plotted in Figure 17.14. The vertical axis shows the estimated cost of production in 1986-7 for each pit. The horizontal axis shows the total production that is possible at or below that cost. The figure, which like Figure 17.9 has been constructed from data for 1981-2 supplied by the Coal Board to the Monopolies and Mergers Commission, has been updated to 1986-7 by applying Coal Board data on aggregate wages and productivity to each pit and taking account of closures. It may therefore be inaccurate in detail since there are bound to be pits that diverge from the industry norm, but as an overall picture of the UK supply curve for coal it is unlikely to be seriously misleading.

A crucial element in the calculation is the behaviour of productivity. In recent years output has fluctuated in the range 2.3–2.4 tonnes per man-shift, but since the end of the strike, accelerated closure of uneconomic pits has enabled the Coal Board to step up the output from the more efficient pits. The result is a sharp increase in productivity, and if progress is maintained the average for

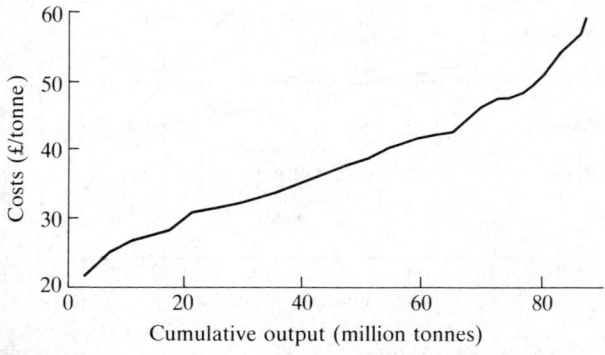

Figure 17.14 The supply curve for UK coal

1986-7 could be in excess of 3.0 tonnes per man-shift. We have used that figure in the calculations shown in the charts. (We also show below the consequences of making alternative assumptions.)

An important feature of Figure 17.14 is that the supply curve for coal is, over the critical range, very gently sloping—in other words, there are a large number of pits with a similar cost structure close to the margin of profitability. This means that comparatively small changes in the price of coal make a large difference to the number of pits that are profitable. This has important implications for the debate on closures. It implies that a comparatively small fall in the price of coal can threaten a relatively large number of jobs in the pits that are close to the margin. But it also means that a comparatively small amount of subsidy or borrowing could secure a relatively large number of jobs.

Given the available information on costs, prices, and employment in each pit, it is possible to calculate the implied cost per job of keeping loss-making pits open, and hence the cumulative subsidy to the industry associated with any level of employment. The results of these calculations are shown in Figure 17.15, which has been computed on the basis of an average coal price of £35 and the wage and productivity assumptions described above. Figure 17.15 is an estimate of the subsidy-employment trade-off for the industry in 1986-7, comparable to Figure 17.10 which showed the same trade-offs for 1981-2.

In 1986-7 the figure suggests that at the assumed wage, price, and productivity levels prevailing, the industry could have to shrink from 140,000 men to only 110,000 in order to break even. If it were required that every pit be profitable, then there would only be 53,000 jobs in the industry.

These estimates are surrounded by enormous uncertainty—uncertainty about

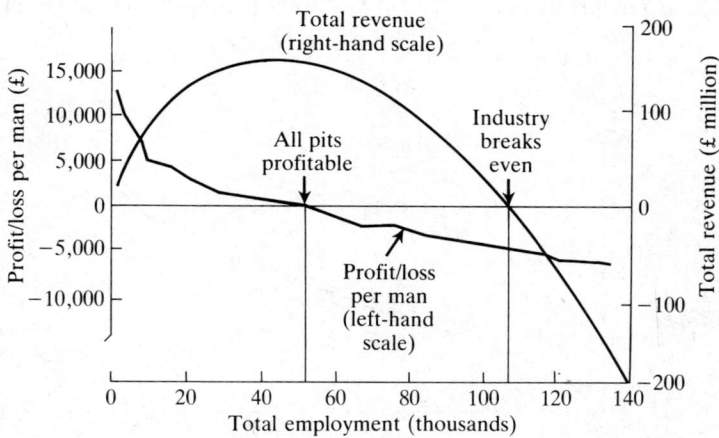

Figure 17.15 Mining employment and profit/loss per man, 1986-7

Table 17.8. The coal industry: prices, wages, and unemployment

	Coal price (£/tonne)	Productivity (tonnes/man-shift)	Wages (£/week)	Coal output (m tonnes)			Employment (000s)		
				All pits profitable	Industry break-even	Actual	All pits profitable	Industry break-even	Actual
1981–2	35.6	2.43	156.4	45	92	109	66	165	218
1984–5[b]	41.7	2.43	173.0	40	80	96	59	138	174
1985–6[c]	45.0	2.43	191.0	55	99		86	185	139
	45.0	3.00	191.0	87	—		151	—	
1986–7[d]	40.0	3.00	206.0	55	103		86	197	
	35.0	3.00	206.0	37	68		53	110	
	30.0	3.00	206.0	9	25		11	34	

[a] Estimates of profitable employment are generally overstated, because calculations ignore geological deterioration.
[b] Estimates of underlying levels. Actuals are strike-affected.
[c] Average conceals sharp changes through the year, with coal prices falling and productivity rising.
[d] Coal prices in likely range. Productivity assumption conservative.

official intentions (all pits break-even or industry break-even), uncertainty about productivity performance, and, above all, uncertainty about the coal price. To illustrate these uncertainties, we have carried out two simulations to show how the 1985-6 industry might have looked under two alternative productivity assumptions, and a further three simulations to illustrate prospects for 1986-7 under three different price assumptions. These simulations are shown in Table 17.8.

In 1985-6 productivity was still strike-affected, rising sharply during the year from below pre-strike levels to the new (much higher) post-strike norm. At the same time, expectations about future coal prices were being progressively revised downwards. The figures nevertheless bring out very clearly the extent to which the industry has benefited from an improved productivity performance. With coal prices at £45 per tonne and productivity at 3 tonnes per man-shift, our calculations suggest that the future of the industry was more secure than for many years.

The other side of the productivity improvements has been further reductions in employment, which had shrunk to 139,000 by the end of the 1985-6 financial year. If the coal price remained at the exceptional levels seen in 1985 and productivity gains were maintained, employment in profitable pits might even have expanded. If by contrast productivity fell back to pre-strike levels, there would be jobs for only 86,000 men in profitable pits—though the high coal price might make it possible for the industry as a whole to reach profitability at existing manning levels.

In fact it is highly unlikely that coal prices will remain much longer at their present levels. The 50 per cent fall in the price of oil in 1986 has as yet had

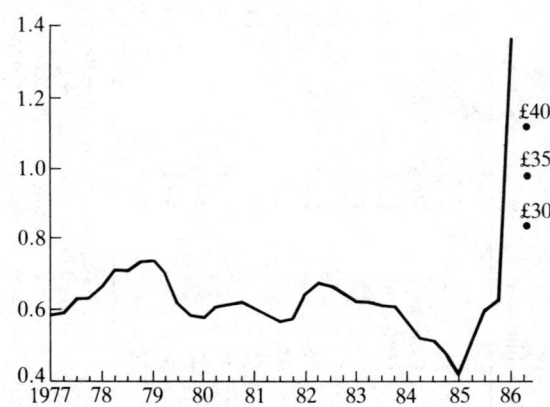

Note: 1986Q2 estimates assume oil price at present levels and coal prices as shown (£ per tonne).

Figure 17.16 Price ratio: coal/oil (pence per therm)

surprisingly little effect on the price of coal, even in the spot market, and its relative price compared with oil is now at a historic and clearly unsustainable peak, as Figure 17.16 shows. But as long as low oil prices persist, the question is not whether the coal price will fall, but how soon and by how much.

We have considered the effects on the industry of a (conservative) fall in the coal price to £30–£40 per tonne, which still leaves the relative price of coal at a historic high. Table 17.8 shows that even this effectively removes any hope of bringing the industry to the point where the current work-force all have secure jobs in profitable pits. With the price at the top of the illustrative range, the number employed in profitable pits shrinks to 86,000. As the coal price falls below £38, the industry as a whole moves into deficit, raising the prospect of a new wave of pit closures in order to break even. At £35 the required closures match those that provoked the 1984–5 strike. If the price falls to £30 then profitable production is really only possible in a few superpits and the industry shrinks to a small rump of some 30,000 miners, producing only a quarter of today's potential output.

12. The Policy Response

These gloomy predictions depend on the wage and productivity assumptions spelt out in Table 17.8. If there were no increase in wages (unlikely) or a much greater increase in productivity (quite possible), then the levels of profitable employment and output would be greater. But the scale of the probable fall in the coal price is unlikely to be offset by wage restraint or productivity improvements. So if oil continues to trade at around $10–$15 per barrel, the coal industry and its political masters face the following choices.

Strategy 1: sell coal at current high price

This would enable the industry to stay, in the short term, at around its present size without a government subsidy. There would, however, be a concealed subsidy from energy users, mainly paid in the form of high electricity bills. And as long as the disparity between coal and oil prices persists, energy users would convert to oil. If oil prices remain low, this would slowly but surely strangle the industry.

Strategy 2: sell coal at competitive prices

With no incentive to convert to oil, domestic demand for coal would be maintained. But the industry would make a loss if it remained at its present size. This would imply a choice between a further round of pit closures—which might be bitterly regretted if the oil price subsequently recovered—and borrowing to keep open capacity that is currently uneconomic. This could be regarded as an investment, which would reap a return if and when oil prices rise again. Like all investments, it would be kept under constant review and could be discontinued if ever the cost became excessive in relation to the prospective return.

In the present climate it seems likely that strategy 1 will be chosen. The government wants to eliminate the subsidy to the coal industry. The Coal Board would prefer the independence that it would then enjoy. It is not yet certain that oil prices will be low for ever, so that users will be slow to convert to oil even if coal prices remain high. These are good reasons for taking what is in any case the line of least resistance and hoping that the problem of low oil prices will disappear before too much damage is done.

The disadvantage of strategy 1 is that in the short term it places an arbitrary burden on coal users, handicapping energy-intensive industries which ought to flourish and create new jobs in a cheaper-energy world. In the longer term, it condemns the industry to inexorable decline if oil prices do not rise again.

There is a good case for keeping open pits that are not profitable at *present* energy prices as an insurance against much higher energy prices in *future*. This will mean a loss for the Coal Board, but as long as the industry is obliged to charge a competitive price for its product (e.g. by allowing the CEGB to use cheap imported coal or oil as it wishes) this does not distort the energy market. The external financing limits of the Coal Board ought to be extended to allow it to finance such a loss, which serves simply, at an explicit and transparent cost, to keep output and employment of the UK coal industry at a higher level than it would otherwise be. If the assumption that this will produce savings at some future date when energy prices rise again becomes patently unrealistic, the borrowing will become hard to finance, and can be discontinued.

For these reasons we believe strategy 2 would be in the long-term interests both of the industry and of the country at large, but its adoption would require jettisoning much ideological baggage, both by the Coal Board and by the government. The outcome is likely, as so often, to be determined more by political expediency than by economic rationality. The losers will be those who produce and consume coal.

Part VI

Oil

18 The Economic Implications of North Sea Oil Revenues*

PETER FORSYTH AND JOHN KAY

1. Introduction

The growth of North Sea oil revenues is the most important fiscal development in the British economy in the 1980s. Yet it is one that does not impinge directly on individuals—we do not see the activity that yields these returns nor are we direct recipients of them. Perhaps it is for this reason that there is little understanding or analysis of the impact of North Sea oil on the British economy.

The purpose of this paper is to set out a framework for such understanding and analysis. It is not concerned with the development of the oil sector itself, nor with directly oil-related activities—there are already several substantial discussions of these topics (e.g. Page (1977), Robinson and Morgan (1978), and Gaskin (1978)). Our concern is to elucidate the effects of the growth of the oil sector on the non-oil economy.

We shall show that the benefits of North Sea oil are very large, and generally underestimated. It is realistic to suppose that by the middle of this decade they will have raised the national income of the UK by at least 10 per cent. But this windfall arises in a highly unbalanced form. A consequence is that it demands extensive structural change in the UK economy. The adverse effects on manufacturing industry have been recognized, and deplored as 'the Dutch disease' or 'deindustrialization'. But it is not the case that pressure on manufacturing is an adverse side-effect of an undesirable, and perhaps temporary, over-reaction by the foreign exchange market to Britain's newly acquired oil self-sufficiency. Rather, the contraction of manufacturing output and an increase in domestic absorption of imported manufactures are—whether desirable or not—the only means by which the British economy can benefit from the North Sea. The rise in the exchange rate is simply the market's mechanism for bringing this about, and if it were suppressed, some alternative means of achieving the same result would emerge and would be required. However, we shall argue that the rise in the exchange rate is a beneficial mechanism for

---

* This paper was first published in *Fiscal Studies*, July 1980.

At the time of writing this article, Peter Forsyth was Senior Lecturer in Economics at the University of New South Wales, Australia. At the time of writing this article, John Kay was Research Director of the Institute for Fiscal Studies; he is now Director of the Centre for Business Strategy and Professor of Industrial Policy at the London Business School.

The authors are grateful to Alastair Ulph for comments and to David Morrison of Wood Mackenzie & Co. for advice and information on prospective oil revenues.

achieving this outcome, and one which confers significant indirect gains on the economy over and above the direct value of oil revenues.

It follows from these observations that the benefits from oil are uneven in their incidence—that although the net effect is substantially favourable, many sectors of industry and many individuals will be net losers. While efficiency demands that structural change should not be discouraged, equity demands that its consequences be made more tolerable for those directly involved. It is not only equity, but prudence, that demands this; for without it—and indeed even with it—there will be political pressure, from those whose existing interests are adversely affected or who misunderstand the nature of the economic future that oil offers, to dissipate this future by inappropriate policy responses.

2. The Size of Oil Revenues

How large are the benefits of oil, and what form do they take? Peak production levels, which will be attained in the mid-1980s, will be slightly in excess of 1 billion barrels per year. In 1979, UK consumption (inland deliveries plus refinery use) was 0.68 billion barrels and production 0.58 billion barrels; self-sufficiency is likely to be reached late in 1980. Estimated reserves are in the range 17–33 billion barrels, as shown in Table 18.1. Although we shall continue to talk about North Sea oil, it is likely that future major discoveries will be in other areas of the British Continental Shelf.

Table 18.1. UK oil reserves (billion barrels)

	Minimum	Maximum
Production to end-1978	0.8	0.8
Current discoveries—proven	10.3	18.6
Future discoveries, from present licences	2.6	5.9
Other reserves	4.1	7.4
	17.8	32.7

Source: Department of Energy, 1979.

What are the economic implications of a shift from a position where we import the whole of our oil requirements to one where we produce around 1 billion barrels per year? There are several measures of the benefits, each of which has relevance for particular purposes. We discuss their relationship and derivation further in the Appendix. The first measure, and the largest, is the gross value of oil output. At a price of \$35 per barrel, peak production is worth between \$35 and \$40 billion per year. Converting this to sterling at a rate of \$2.25 implies a value of £16 billion. The benefit to the balance of payments is somewhat less than this, because of imports required for North Sea operations and the remittance of profits and interest overseas. These may amount to £3 billion, implying a balance of payments gain of £13 billion. The direct value of oil to the British economy (value added from oil) is less than this because we need to subtract domestically incurred operating costs (including a return on capital that British companies have employed in exploration and development which might otherwise have been used elsewhere). These may be £2 billion, leaving a balance of £11 billion. Virtually all of this will accrue to the government as tax revenue; the balance is the profit earned by British companies over and above what they could have expected to earn from alternative investments. This is unlikely to be more than £1 billion, consistent with an annual tax yield of £10 billion. All these figures are at 1980 prices, and can therefore be compared with 1980 gross national income of around £180 billion and total tax revenue of £70 billion.

It is easy to calculate the effect on gross revenue from oil of changes in the real price of oil or the real exchange rate. We can then estimate the consequences of such changes for tax yields, since almost all the change in the value of output will be reflected in a change in government revenues. A complete model of the North Sea tax system is very complex, because of the interactions between the various taxes. Such a model has been developed by Wood Mackenzie & Co., and we discuss its conclusions in the Appendix. But it is both simpler and more realistic to estimate long-run revenue by describing government behaviour rather than the tax system. This is because the government has revenue objectives: effectively it is the revenue secured which determines the tax system rather than the tax system which determines the revenue secured.

The current target is 80 per cent of net revenue at the 1980 exchange rate and oil price. The estimates in Table 18.2 assume that the government achieves this target and imposes a marginal tax rate of 90 per cent on revenue changes resulting from variations of exchange rate or oil price. Since the marginal tax rate is now about 87 per cent, this assumes that (as has happened rather too frequently in the past) changes in the oil price lead to changes in tax rates. The range of figures shown in the table is quite wide; but this exaggerates the degree of uncertainty because it is unlikely that a high oil price will be associated with a low value for sterling, or a low oil price with a high exchange rate. A revenue projection in the range £9–£12 billion is therefore reasonably robust. This is only slightly less than the yield from value added tax, and is equal to half the revenue from income tax.

The figures we have presented are essentially illustrative in nature, but they do indicate the order of magnitude of revenues from oil and their considerable size—whether we compare oil output with national income or tax yield with public expenditure. The degree of uncertainty about these projections, especially of oil prices, is such that any precise numbers quoted should be distrusted. Moreover, our calculations relate to peak production, and for many purposes it is more appropriate to ask what is the size of a corresponding flow of revenue that is permanently sustainable? We return to this issue in Section 5. But

Table 18.2. Long-term estimates of North Sea oil revenue (£ billion, annual rates, 1980 prices)

	Real oil price			
	$25	$35	$40	$50
Real exchange rate: $1.75	9.3	14.4	17.6	22.2
$2.00	7.7	12.2	14.4	18.9
$2.25	6.4	10.4	12.4	16.4
$2.50	5.4	9.0	10.8	14.4

Source: Own calculations.

for our analysis in Section 3, we shall assume that oil production adds £10 billion to net output in the UK. This is not intended to be our best estimate of the gains from oil. It is simply the figure that we have chosen to illustrate the economic effects of a development of this kind, which is our principal objective. We believe we have said enough to demonstrate that it is of a realistic order of magnitude.

3. The Effect of Oil on Economic Structure

We begin by examining the basic structure of the pre-oil economy. Table 18.3 shows how production is distributed over the various sectors. The data refer to 1976—the last year before domestic oil production had significant effects on the British economy—but we have brought it to 1980 prices by using a general price index on all items. There have been important changes in relative prices since 1976, partly as a result of the influence of oil, but we defer discussion of these until Section 4.

The second and third columns of Table 18.3 also show exports and imports by

Table 18.3. The structure of the UK economy, 1976 (£ billion, 1980 prices)

	Production (as net output)	Exports (as gross output)	Imports (as gross output)
Primary production	9.3	0.8	11.5
Manufacturing	50.6	39.8	34.7
Construction and housing	23.3	0	0
Distribution and services	91.2	16.5	12.8
Public administration	14.0	0	0
Residual	−6.5	1.3	1.7
	181.9	58.5	60.7

producing sector. These indicate that manufacturing has a much greater role in relation to the traded sector than it does to the domestic economy. Manufactures account for well over half of both imports and exports, but little more than a quarter of domestic output. The comparison between column 1 and columns 2 and 3 is in some degree misleading, however. Column 1 records net output by sector, while exports and imports are measures of gross output. It is not surprising that most exports take the physical form of manufactured goods, but the tangible export usually includes the contributions of several different sectors. For example, there is little international trade in electricity, but refined aluminium is in a sense congealed electricity, and Norway and Canada, with considerable supplies of cheap electricity, are the major exporters of aluminium. More generally, although many services cannot be exported directly, they are exported indirectly by contributing value added to the goods that are exported; and Britain imports raw materials not only in their crude form, but also as part of the manufactured goods that we buy from abroad.

So we need to look behind exports and imports at the various production sectors that contribute to them. This we can do using input–output analysis, and Table 18.4 shows the composition of imports and exports in terms of value added.

Table 18.4. The structure of the UK economy, 1976 (£ billion, 1980 prices, by value added; residuals allocated pro rata)

	Production	Exports	Imports	Consumption
Primary production	9.0	− 1.2	+ 8.0	15.8
Manufacturing	48.9	− 24.9	+ 22.0	46.0
Construction and housing	22.5	− 0.2	+ 0.3	22.6
Distribution and services	88.1	− 18.8	+ 16.9	86.2
Public administration	13.5	—	—	13.5
	181.9	− 45.1	+ 47.3	184.1

Source: Our calculations, see Forsyth and Kay (1980b).

The picture presented is somewhat different. Both imports and exports are considerably smaller than in Table 18.3. This is because a high proportion of UK imports—some £13 billion—are ultimately re-exported after further processing in Britain. The approach we have adopted counts only the value added in the UK as an export, while the conventional trade accounts include the whole value of the export. Since the goods in question are not finally consumed in the UK, they are also excluded from imports.

The second difference is that the contribution of distribution and services to exports is significantly increased, reflecting the inclusion of indirect as well as direct imports. Distribution and services account for 28 per cent of exports by commodity but 42 per cent of exports by value added. It remains true, however, that manufacturing features much more in trade than do other sectors. Almost half of manufacturing value added is exported, as against 20 per cent for distribution and services, and less for other sectors.

This approach enables us to derive the fourth column of Table 18.4, which we have headed 'consumption'. This shows the composition, by producing sector, of the goods consumed in the UK. It is what the structure of the UK economy would be if we were self-sufficient (assuming, implausibly, that we could take over from abroad the technology required to produce domestically the goods that we currently import). Consumption is slightly higher than production, reflecting the balance of payments deficit that occurred in 1976. In that year, and in general, Britain consumed more primary products and fewer manufactures and services than it produced. We exported manufactures and services to buy primary products. It should be noted that although we have called this column 'consumption', it would be more accurate to call it domestic resource utilization. It includes not only private consumption but private and public investment as well.

Now we shall assume that the development of oil output creates £10 billion of additional value added in the primary sector. Assuming that other primary production remains constant, this will raise primary output to £19 billion. However, there is no reason to expect that demand for primary output will rise by this amount. An additional £10 billion of income raises total output from the UK economy by 5.5 per cent. If we increase domestic resource utilization in line with this increase in income, the resulting increases in demand will be spread over all sectors. The simplest assumption is that demand for value added in each sector rises by 5.5 per cent. In that case, consumption of primary output would be £16.7 billion and a deficit of £6.8 billion would have become a surplus of £2.3 billion.

If domestic resource use is to rise in line with national income, it follows that we must maintain a broad overall balance on external trade. This implies that the move from deficit to surplus on primary account must be reflected in a move from surplus to deficit in other accounts. If there were no change in the net payments position, then the non-primary balance of payments would deteriorate by £9.1 billion. As a preliminary assumption, suppose this change is reflected in an equal percentage increase in all non-primary imports and reduction in all non-primary exports.

Table 18.5 works back from these assumptions to the new levels of production that are implied. We start from consumption figures in which all items are increased by 5.5 per cent. We then impose the 11 per cent change on all imports and exports that is needed to avoid any change in the overall balance. And while in Table 18.4 we worked forward from the known production data by adding the net effects of trade, in Table 18.5 we work back from estimated consumption data by subtracting the net effect of trade to find implied production by sector.

Table 18.5. The post-oil economy (£ billion, 1980 prices)

	Production	Exports	Imports	Consumption
Primary production	19.0		(−2.3)	16.7
Manufacturing	46.1	−22.2	+24.6	48.5
Construction and housing	23.7	− 0.2	+ 0.3	23.8
Distribution and services	88.9	−16.9	+18.9	90.9
Public administration	14.2	—	—	14.2
	191.9		(+2.2)	194.1

Source: As Table 18.4.

Table 18.6 compares this production by sector with the original pre-oil figures. In the two non-traded sectors—construction and public administration—output expands in line with the overall growth of national income. In the primary sector, output grows much faster—it doubles. This is compensated for, however, by less rapid growth in the remaining traded sectors—services and manufacturing. Much the largest effect is on manufacturing, which not only declines in relative terms, but suffers an absolute contraction of 5.7 per cent.

Table 18.6. Production changes by sector (£ billion, 1980 prices)

	Pre-oil	Post-oil	% change
Primary production	9.0	19.0	+111.1
Manufacturing	48.9	46.1	− 5.7
Construction and housing	22.5	23.7	+ 5.5
Distribution and services	88.1	88.9	+ 0.9
Public administration	13.5	14.2	+ 5.5
	181.9	191.9	+ 5.5

The essentials of what is happening are very simple. North Sea oil adds considerably to the growth of the UK economy. However, this growth takes a highly unbalanced form: all of it occurs in a single sector whose size is, even then, small in relation to the overall economy. To use the additional resources that it makes available to us, it is necessary to convert them to a form in which they can be exploited domestically. But there is simply no way in which oil can be converted into houses, or restaurant meals, or retail and distributive services, either directly or through trade; and it is largely on items of these kinds that we shall want to spend our increased income. All we can do is to exchange oil for trade goods—predominantly manufactures—and redeploy the resources released from these sectors into the other, non-tradable, sectors of the economy.

There is no mechanism for deriving benefit from North Sea oil that does not, sooner or later, require this structural change.

The model we have presented is heavily over-simplified, and we shall discuss many qualifications to it in subsequent sections. But none of them will fundamentally alter the simple arithmetic logic of the argument. The discovery of North Sea oil involves a large expansion of the traded sector of the economy. Since, in the long run, the balance of payments must balance, that implies a contraction of the remainder of the traded sector, in which manufacturing is disproportionately represented.

It may seem paradoxical to those who see manufacturing as the heartland of the economy that economic growth should involve a contraction in its role. But what is the source of the common belief in the central role of manufacturing? Doubtless there are some people who think that large cars represent real wealth in a way that the plays of Shakespeare do not, but their views do not deserve much attention. The serious point is that because tradables are predominantly manufactures, the import of raw materials requires the export of manufactures in return, and other sectors of the economy cannot substitute in this role. But the corollary to this is that when, as is now the case, Britain emerges as a significant primary producer, the role of manufacturing is to decline. Traditionally, Britain has exported manufactures and services in return for primary inputs. In the future, this will be substantially less true and may even be reversed.

4. The Impact on the Exchange Rate

The analysis of the previous section was conducted in terms of physical quantities and we did not discuss the prices of the outputs of the various sectors. The mechanism that brings about the structural changes is, in a market economy, a rise in the real exchange rate. Externally, this reduces the competitiveness of British exports and increases the attractiveness of imports, thus restoring the balance of payments to equilibrium. Internally, it reduces prices, profits, and wages in the tradable sector—principally manufacturing—relative to those in less tradable sectors, and so brings about pressure for the reallocation of resources in the domestic economy.

These changes in relative prices have important implications in themselves. They also feed back into our earlier analysis. One simple way of seeing the problem they pose is this. Peak production of UK oil will, at current world prices, be worth perhaps $35 billion. At £1 = $2.25, this is worth £16 billion. At £1 = $1.85, this same output would be worth £19 billion; and if there were no oil, the exchange rate might well be $1.85 rather than $2.25.

At first sight, this suggests that as oil pushes up the exchange rate, so its value to the British economy falls. But in reality the reverse is the case. A shift in the exchange rate from $1.85 to $2.25 is equivalent to a revaluation of 22 per cent.

This reduces import prices in sterling terms. Primary product prices (other than for food) will fall by the full amount of the revaluation. The prices of imported manufactures will also fall, although by rather less, if Britain is a large importer relative to the size of the world market, or if foreign exporters can raise their profit margins on sales to the UK. Suppose import prices are 15 per cent lower than they would otherwise have been, and sterling export prices remain constant. On an import bill of £60 billion (Table 18.3), this implies a saving of £9 billion, equivalent to 5 per cent of national income. The gain from the reduction in the cost of imports is of the same order of magnitude as the direct gain from oil itself, and this much more than offsets the fall in the direct value of the oil. How do we actually receive these benefits? Because imports are around one-third of domestic consumption, a 15 per cent fall in import prices reduces domestic prices by around 5 per cent. The gain takes the form of lower prices at a given level of money incomes. The benefit is most obvious to those who travel abroad and find the pound buys more; but similar gains are being made by everyone who buys goods that are imported or contain imported components.

It is therefore important to estimate what the effect of oil revenues on the exchange rate is. We go on to consider how this affects relative prices in the domestic economy, and to estimate the size of the gains from exchange rate appreciation. We illustrate how the rising exchange rate shifts revenue from oil away from government tax receipts and directly into the pockets of the people; and that the amount of this indirect gain is substantially more than the amount of direct loss. We shall describe the difference between the equilibrium exchange rate with North Sea oil and without it as the 'oil premium'. If the current exchange rate is $2.25, but would have been $1.85 without oil, that implies that the current oil premium is 22 per cent. There are two ways of estimating the oil premium. One is to analyse the recent behaviour of the exchange rate, in order to see how its trend has been affected by oil. As shown in the Appendix (Table 18.A2), the real exchange rate rose by around 20 per cent between 1976 and January 1980. There are many other factors at work, and without oil there would have been some movement in one direction or the other, but this is one indication of the order of magnitude of the oil premium.

An alternative approach to assessing the oil premium is to ask what rise in the exchange rate is needed in order to worsen the non-oil balance of payments by the value of oil output. For this purpose, oil output includes domestically incurred operating costs as well as net value added, since these costs also represent a shift of domestic resources into the tradable sector. Set the value of oil output at $25 billion. Exports and imports of manufactures in the pre-oil economy (measured in terms of gross output, since it is on this basis that export and import elasticities are normally estimated) are between £35 billion and £40 billion at 1980 prices (Table 18.3), equivalent to $70 billion at a tentative pre-oil exchange rate of $1.85. Some deterioration of the trade balance follows automatically from the growth in national income and associated increase in

Table 18.7. UK terms of trade in manufactures

	Export prices (£) (1970 = 100)	Import prices (£) (1970 = 100)	Terms of trade[a] (1976 = 100)
1970	100	100	97
1971	110	107	100
1972	118	112	102
1973	128	127	98
1974	153	147	101
1975	194	184	102
1976	238	232	100
1977	287	265	105
1978 Q1	309	278	108
Q2	315	289	106
Q3	325	296	107
Q4	334	304	107
1979 Q1	340	306	108
Q2	340	296	112
Q3	346	293	115
Q4	352	304	113
1980 Jan.	358	306	114

[a]Terms of trade = ratio of export prices to import prices.
Source: *Monthly Digest of Statistics* (1970–6: machinery and transport equipment series; 1976–80: finished manufactures series).

demand for imports. If net imports rise by 10 per cent, the import bill rises by $9 billion. This leaves some $16 billion of adjustment, of which say $13 billion might fall on manufacturing, equal to just under 20 per cent of import or export volume. Central estimates of demand elasticities suggest that this might require a change in relative prices of 12 to 15 per cent.

This figure cannot immediately be translated into an exchange rate change, however. Although the real exchange rate has risen by more than 20 per cent since 1976, the export price of manufactures relative to the import price of manufactures has risen by somewhat less, as Table 18.7 shows. This is because domestic exporters have been forced to absorb part of the impact of revaluation, while foreign exporters to the UK have been able to increase their margins somewhat. The change in the exchange rate needed to produce a 15 per cent improvement in the terms of trade in manufactures is probably in the range 20–25 per cent. However, this 15 per cent reflects the change in the price of gross output: since the import content of exports is substantial, the terms of trade for value added have improved by an amount closer to the extent of the revaluation.

These calculations suggest that our preliminary assessment of the oil premium at 22 per cent, based on the difference between a non-oil exchange rate of $1.85 and a post-oil rate of $2.25, is a realistic order of magnitude. Tentatively adopting this figure, we now examine the implications of the oil premium for the analysis of Section 3. In converting 1976 data to 1980 prices, we assumed that all prices increased at the same rate; but the oil premium ensures variability in prices. The prices of some primary products (such as oil) are set internationally, so that when the real exchange rate rises, the prices of these items fall in terms of domestic currency by the amount of the revaluation, i.e. to 82 per cent of their previous level. Most agricultural goods are an exception, because of the way in which Britain has operated the Common Agricultural Policy, which has led partly to a diversion of oil income to domestic farmers and partly to its transfer to the EEC. The sterling price of imported goods and services will fall by most of the amount of the revaluation. The sterling price of exported goods and services will fall somewhat, because British exporters are unable to pass on the whole of the revaluation to their customers. Import-competing goods and services will also fall in relative price. We have illustrated these possibilities in Table 18.8. Primary products are assumed to fall in price by 15 per cent, and the same assumption is made for imports. Manufacturers absorb some of the revaluation by reducing profit margins: we have assumed 4 per cent for exporters and 2 per cent for domestic producers (who suffer intensified competition from imports). Smaller price reductions occur in domestically produced services. These precise numbers are not intended at all seriously; they are simply chosen to enable us to illustrate qualitative effects.

The final row and column of Table 18.8 represent implied average prices, given the structure of production described in Table 18.5. A feature of this is that the price of imports falls much more than the price of exports. This has the consequence that although trade as in Table 18.5 appears to cause a balance of payments deficit, if it took place at the prices implied by Table 18.8, the outcome would in fact be a balance of payments surplus. The difference results from a change in the ratio of export prices to import prices (the terms of trade). This means that the change in prices consequent on North Sea oil itself produces a gain in real income for the UK; activities that would previously have yielded a loss are now profitable. An alternative way of making the same point is to see

Table 18.8. Relative output, import, and export prices

	Production	Exports	Imports	Implied price of consumption
Primary production	0.85	0.85	0.85	0.850
Manufacturing	0.98	0.96	0.85	0.923
Construction and housing	1.00	—	—	1.000
Distribution and services	0.99	0.98	0.85	0.963
Public administration	1.00	—	—	1.000
Implied average price	0.976	0.969	0.850	0.951

Table 18.9. The post-oil economy with changed relative prices

(a) AT CHANGED PRICES

	Production	Exports	Imports	Consumption
Primary production	16.2	(−1.6)		14.6
Manufacturing	43.8	−20.0	+22.2	46.0
Construction and housing	24.3	− 0.2	+ 0.3	24.4
Distribution and services	88.4	−15.6	+17.1	89.9
Public administration	14.6	—	—	14.6
	187.3	(+2.2)		189.5

(b) AT CONSTANT (PRE-OIL) PRICES

	Production	Exports	Imports	Consumption
Primary production	19.0	(−1.9)		17.1
Manufacturing	44.7	−20.8	+26.1	50.0
Construction and housing	24.3	− 0.2	+ 0.3	24.4
Distribution and services	89.3	−15.9	+20.1	93.5
Public administration	14.6	—	—	14.6
	191.9	(+7.7)		199.6

that the price of goods that Britain consumes has fallen more than the price of goods that Britain produces. The value of British output has risen, measured in terms of its purchasing power for the goods Britain wants to buy. This change is itself equal to an increase in real income of about 4 per cent.

Such a rise in income allows a further rise in total domestic resource utilization. A possible final outcome is shown in Table 18.9. In Table 18.9, consumption is allowed to expand until the balance of payments deficit is at its initial level of £2.2 billion, as shown by the current price figures in (a). Table 18.9(b) shows that this deficit would have been £7.7 billion without the improvement in the terms of trade, but the higher exchange rate means that this is a purely hypothetical calculation. In Table 18.9(b), the outcome has been converted back to original pre-oil relative prices using the prices in Table 18.8, so that Table 18.9(b) can be compared with the pre-oil economy of Table 18.4. The impact on production is shown in Table 18.10. Alternatively, Table 18.9(b) figures can be interpreted as physical volumes of goods.

There are two principal points to note. First, overall production still rises by only $5\frac{1}{2}$ per cent, although national income increases by more. This is because the *value* of UK production has increased with the rise in the exchange rate. This effect is comparable in size to the immediate gain from the oil production itself. Second, the structural changes that are implied for the UK economy are now

Table 18.10. Output changes by sector

	Pre-oil	Post-oil	% change
Primary production	9.0	19.0	+111.1
Manufacturing	48.9	44.7	− 8.9
Construction and housing	22.5	24.3	+ 8.0
Distribution and services	88.1	89.3	+ 1.4
Public administration	13.5	14.6	+ 8.1
	181.9	191.9	+ 5.5

even larger. When we examined the direct impact of oil (Table 18.6), the difference in experience of non-oil sectors ranged from −5.7 per cent for manufacturing to +5.5 per cent for housing and construction. Now the range is from −8.9 per cent to +8.1 per cent. The reason is that again the increase in real income has occurred entirely in the traded sector. This implies a further expansion of the non-traded sector, compensated for by a further contribution in tradables—the magnitude of the required structural change is increased.

Table 18.7 showed the very considerable change in the terms of trade for manufactures that has occurred since oil production began. (The overall terms of trade do not display the same trend, but these are dominated by commodity prices which fluctuate for reasons unconnected with the UK domestic economy.) The terms of trade in manufactures have improved by around 14 per cent since 1976. This figure underestimates the true effect, because export prices are those of gross output, which, as we showed in Section 3, have a substantial import content. The corresponding rise in the relative price of exported manufacturing value added is around 20 per cent. Of course, oil was not responsible for precisely 14 per cent of the change, though it has been the major positive influence on the terms of trade during the period.

Many people will find it surprising that we should ascribe large benefits to what others would think a worrying decline in the competitiveness of British exports. But the proper attitude depends on the reason for the decline in competitiveness. A manufacturer is properly concerned if his prices rise because his costs rise relative to those of his competitors, and happy if they rise because market demand for his product forces them up. The flaw in this analogy is that in the case of British exports, it is the demand for one product—oil—which raises the price of another product group—manufactures—but from the viewpoint of the British economy as a whole, the effect is the same.

We have shown that the indirect gains from the improvement in the terms of trade are potentially large, and it is appropriate now to consider how large they are. As the exchange rate rises, the direct gains fall. This is because the sterling value of British oil is lower, and so are value added in the North Sea and government tax revenue from oil. But as the exchange rate rises, the indirect gains rise; and we can show that it is virtually certain that the increase in indirect gain more than offsets the fall in direct gain. It follows that the larger the effect of oil in pushing up the exchange rate, the greater the overall gain to the UK economy. The answer to the question 'should one assess the value of oil using a pre-oil or a post-oil exchange rate?' is therefore that in general the higher of these two figures—the pre-oil exchange rate calculation—will underestimate the total value of oil.

These points are illustrated in Table 18.11. In this table, we have assumed a non-oil exchange rate of $1.85, implying an oil premium of 22 per cent in a current exchange rate of $2.25. The value added directly attributable to oil is assumed to be $18.5 billion. The table shows the gains derived as government revenue and rents to oil companies, and as gains on the terms of trade, for various values of the oil premium. It implies that if the oil premium in the current exchange rate is indeed 22 per cent, the gain to consumers from the terms of trade improvement will actually be as large as the revenue directly attributable to the North Sea: the total gain is equal to 10 per cent of 1980 national income.

Table 18.11. Gains from North Sea oil

Oil premium	Implied current exchange rate	Gain to: (£ bn)		
		Companies and government	Consumers	Total
0	$1.85	10.0	—	10.0
10%	$2.04	9.0	3.4	12.4
20%	$2.22	8.3	6.8	15.1
30%	$2.41	7.6	10.1	17.1
40%	$2.59	7.1	13.5	20.6

Base assumptions: value added from oil $18.5 bn;
non-oil economy exports and imports £45 bn;
non-oil exchange rate $1.85;
¾ of exchange rate gain reflected in *value added* terms of trade.

Table 18.12 approaches the same question from a slightly different angle. It asks: starting from a current exchange rate of $2.25, what is the gain for alternative assumptions about the level of oil premium that this exchange rate incorporates? Under this assumption, the direct value of oil value added of $18.5 billion is £8.2 billion (since $2.25 is the actual exchange rate), but the indirect value depends on the assumed level of implicit oil premium.

We have shown that a major part of the gain from North Sea oil is incorporated in the terms of trade, and already has been. As the exchange rate rises, the benefits of oil are transferred from government revenue directly to domestic consumers of imported goods; and the higher the exchange rate, the

Table 18.12. Gains from North Sea oil

Oil premium	Implied non-oil exchange rate	Indirect gain (£bn)	Total gain (£bn)
0	$2.25	0	8.2
10%	$2.05	3.4	11.6
25%	$1.80	8.4	16.6
40%	$1.61	13.5	21.7
50%	$1.50	16.9	25.1

Base assumptions: value added from oil $18.5 bn;
non-oil economy exports and imports £45 bn;
current exchange rate (including oil premium) $2.25;
¾ of exchange rate gain reflected in *value added* terms of trade.

larger the total benefit and the proportion of it that is enjoyed immediately by consumers. On reasonably conservative assumptions, the total gain is nearly twice the direct gain and 10 per cent of current national income.

5. Questions and Answers

What Happens when the Oil Runs Out?

Britain's oil reserves will not last for ever. It would probably be a mistake to blow them all on raising current consumption during the limited period when oil production is flowing on a large scale. This argument gains particular force when the need for major structural change in response to North Sea oil is recognized. It would be very costly to make these structural adjustments only to be forced to unwind them again and restore the initial structure when oil production starts to fall.

The appropriate response is to invest sufficient of Britain's oil revenues to transform their windfall nature into a permanent increment to national income. We should consider how large such a permanent increment might be. Suppose we can earn a real rate of return on investment of 3 per cent. Suppose we can anticipate a production flow of 1 billion barrels a year for 17 years, followed by 17 years of production at $\frac{1}{2}$ billion barrels per year, after which UK oil production ceases. This implies total reserves recovered of about 25 billion barrels, which is a central estimate of available reserves (Table 18.1). At a real oil price of $35 per barrel, the present value of this production is, at a real interest rate of 3 per cent, some $600 billion. Investing this at 3 per cent yields $18 billion per year—just over half the value of peak production. If we consumed half of oil revenues and invested the rest, we could maintain that boost to our consumption level indefinitely.

This calculation is quite sensitive to the real interest rate assumed. At 1 per cent real interest, we should invest three-quarters of the revenue. At 5 per cent, we can afford to consume nearly all of it. It is also sensitive to the real price of oil. If oil reserves are currently being depleted at an efficient rate, oil prices would rise at the real rate of interest—by 3 per cent per annum, on our central assumption. If they did, the present value of the depletion plan described would be $900 billion and we could consume three-quarters of initial peak revenues. If oil is being depleted too rapidly, then the real oil price will rise faster and the value of UK reserves will be greater; conversely if depletion is too slow.

Why do we still have a Balance of Payments Deficit?

It is important to be aware that the foreign exchange market anticipates future developments. If Britain were likely to produce large quantities of oil in 1985, which would push the exchange rate up in that year, then people would buy sterling now in anticipation of exchange gains; so that even if Britain were not now producing a single drop of oil, the exchange rate would incorporate an oil premium. It follows that this would imply a balance of payments deficit at *current* levels of oil production. It is sometimes found paradoxical that Britain at present combines near self-sufficiency in oil, a current balance of payments deficit, and a high exchange rate; but for so long as oil output is increasing, that is precisely what one can expect to observe. The experience of Norway illustrates this clearly; in 1976 and 1977, when growth in oil production was expected but had not fully occurred, its balance of payments deficit was more than 10 per cent of national income. This was equivalent to a deficit of over £20 billion in the UK and was probably the largest deficit relative to gross national product ever incurred by a developed country. Yet the Norwegian krone strengthened before and during this period, as the foreign exchange market anticipated the prospective elimination of the deficit through oil (as has, in fact, subsequently happened).

Why can we not Run a Balance of Payments Surplus?

We can, and we shall suggest below that when peak oil production levels are reached we should. A balance of payments surplus on current account implies that Britain is acquiring assets abroad. There are two methods of doing this. One is by official intervention in the foreign exchange market, in which the Bank of England moderates the rise in the exchange rate by selling sterling and obtaining foreign currencies in return. By convention, such investment is limited to the short-term money market. The alternative mechanism is by straightforward acquisition of overseas assets, either through portfolio investment or by direct investment with British companies constructing factories or other property abroad.

We can avoid any requirement for structural change in the domestic economy by reinvesting oil revenues overseas, and by reinvesting the interest on this reinvestment, and the interest on the interest, and so on *ad infinitum*. The price of doing so is that we never raise the rate of domestic resource utilization above its pre-oil level. If we are ever to derive any benefit from oil revenues, it is necessary to reimport these assets, or the interest on them; and this we can only do by reducing domestic output of traded goods accordingly. Running a balance of payments deficit or surplus is a means of advancing or retarding the date at which structural change becomes necessary, but it is not a means of avoiding it; the only way of avoiding it is to throw North Sea oil revenue away.

What Kind of Investment?

We have suggested that it is appropriate for Britain to invest a substantial fraction of its oil revenues—perhaps £5–£10 billion per annum. What forms should such investment take? There are investment opportunities both in the domestic economy and overseas.

Investment in the domestic economy poses a number of problems. It is obvious that we cannot use oil directly for investment. It is therefore necessary to convert oil into investment goods. The simplest way of doing this is to export oil and import resources for investment goods. If the material for investment is to be domestically generated, it would be necessary to contract the domestic tradable sector further and divert resources to the investment sector. This change would be imposed on top of the contraction already occurring and would be essentially temporary in nature. This might be easy if the sectors whose export markets were disappearing were those equipped to produce items for investment; but it is much more likely that manufactured exports will be hit most in semi-finished and homogeneous products such as steel, while more sophisticated products will continue to make sales at higher prices. It is therefore desirable to expand domestic investment via imports. This is certainly possible for plant and machinery; but investment makes heavy demands on construction, and construction is the least tradable sector of all.

It is widely suggested that North Sea oil revenues should be used to support investment in manufacturing industry. We have seen that an inevitable result of these revenues is that growth prospects for UK manufacturing are worsened and those for other sectors are improved. It seems a perverse response to this change in relative fortunes to use the proceeds for investment in manufacturing. The effect of oil is to increase, over the long term, output of tradable goods relative to non-tradable goods, while it has little effect on demand for the two sectors. The ideal form of investment is one that uses tradables to produce non-tradable goods; imported machinery for use in service industries, for example. Manufacturing investment, which tends to use non-tradables to produce tradables, achieves precisely the reverse.

The same point can be made more forcefully in terms of prices and rates of return. North Sea oil implies a higher exchange rate and a lower rate of return in UK manufacturing than would otherwise be the case. Investment in manufacturing will therefore be unequivocally less profitable with oil than without it. Support for increased investment must therefore rely on the view that otherwise such investment would offer very high rates of return which are not obtained because of constraints on the availability of resources for investment. We know of no serious evidence to support such a proposition, and data on realized rates of return suggest the reverse—that investment is low because returns are low. Certainly it is hard to believe that any substantial fraction of the £5–£10 billion per year we have suggested should be available for investment could be added to the existing level of £7 billion per annum without driving returns down to nugatory levels. As past experience with British Steel and British Leyland has shown, there is no guarantee that investment in unprofitable industry will produce any return at all.

Investment abroad would seem to have considerable attractions. Firstly, it reduces the requirement for temporary structural adjustments and readjustments in the UK economy. Secondly, financial investment responds more easily and more flexibly than physical investment, and this is important when there is considerable uncertainty about the future of the real oil price and the equilibrium level of the oil premium in the real exchange rate. Thirdly, there is no serious doubt that the world economy can absorb these funds and yield a reasonable real rate of return on them; the capacity of the British economy to use an additional £10 billion of investment at acceptable real yields is much less clear.

Depletion Policy

An alternative method of investing North Sea surpluses, and of spreading North Sea benefits into the future, is to retard depletion of North Sea reserves. This can be done by slowing depletion of existing fields, or by discouraging further exploration or development either through tax policy or through licensing. British policy has so far been close to pursuing the maximum attainable production levels (the restrictions on flaring gas in the Brent field are the most important departure from this), while Norway has proceeded rather cautiously. The rate of return on conservation as an investment depends on the anticipated rate of appreciation of the real oil price. We have used a 3 per cent real rate of return as an illustration of what Britain might hope to earn on foreign investment. If oil prices are expected to rise more rapidly than this, then slowing depletion will be a profitable investment strategy. If they will rise more slowly, then the right investment decision is to deplete as quickly as possible and invest the revenues elsewhere.

It should therefore be recognized that two quite separate investment decisions are required. One is to ask what proportion of potential oil revenues should be exploited now, and what proportion deferred to the future. The second question is what form the investment implied by the answer to the first should take—domestic investment, capital export, or conservation. It is the second of these issues, rather than the first, which should govern depletion policy.

What about Protecting Manufacturing Industry?

We have painted a gloomy picture of the prospects for British manufacturing industry, especially for import-competing sectors. This will lead to pressure for protection against the 'cheap' imports which result from the 'unrealistic' exchange rate implied by the oil premium. We began our analysis by looking at volumes of imports and exports, and saw that oil necessarily implied the reduction and perhaps elimination of Britain's trade surplus in manufactures. The rise in the exchange rate is not the fundamental cause of this change; it is simply the market's mechanism for bringing it about.

It follows that one cannot stop this change by tariff or quota protection. What protection can do is to transfer the burden of adjustment from one sector to another. One reason why manufacturing will bear the brunt of the difficulties facing producers of tradable goods is that government policies effectively insulate agriculture and coal-mining from these problems. Import controls ease matters for the protected sector but, by pushing the real exchange rate up still further, increase the pressure on the unprotected sector. This is why industries will lobby for protection and it is also why such lobbying should be resisted. If the whole import-competing sector is protected by universal import controls, the burden of adjustment is transferred to the exporting sector. The rise in the exchange rate can only be prevented by capital export, but capital export would protect manufacturing industry anyway, whether one imposed import controls or not. Alternatively, one might seek to restrain the rise in the exchange rate by 'some complex combination of official sales of sterling, relation of fiscal and monetary policy and inspired rumour' (Cambridge Economic Policy Group (1980, p. 31)).

Why we should seek by these means to reduce artificially the value of UK assets and output relative to those of other countries is hard to understand. The proposal is about as sensible, and as rewarding, as a major shareholder spreading unfounded rumours of the imminent bankruptcy of his company. The primary effect of generalized import controls is to favour import-competing sectors at the expense of exporting sectors. This policy cannot succeed in its objectives and can only reduce industrial efficiency.

North Sea Oil and Employment

What are the effects of North Sea oil on employment? In recent years, the British economy has operated at increasingly high levels of unemployment, and the gap between actual and potential output levels has probably increased steadily. It is sometimes suggested that this reflects a balance of payments constraint which North Sea oil will relieve. With floating exchange rates, such a constraint could only reflect a policy of depressing domestic demand in order to push up the external value of sterling. Although the gains from improvement in the terms of trade are large, they are not so large as to make sense of such a policy; nor do we think it is an accurate description of what British governments thought they were doing. Demand has been depressed for other reasons of macroeconomic policy outside the scope of our present discussion.

Our principal concern in this paper has been with microeconomic effects and we return to these. Manufacturing is a relatively labour-intensive sector. Value added per worker in 1980 was around £7,000 in manufacturing and about £10,000 in the rest of the economy. If these ratios remain unchanged, each one per cent shift of total output from manufacturing to other sectors would imply a reduction in manufacturing employment of 250,000 and an increase in jobs elsewhere of 180,000, with a net loss of 70,000 jobs.

These structural adjustments will be difficult, and the reduction in wages in manufacturing relative to other sectors that they imply will be resisted. This makes it likely that there will be increases in structural unemployment. We might offset against these factors the small employment-creating effects of North Sea activity itself; against this, however, is the probability that labour-intensive industry will suffer particularly severely from the rise in the exchange rate. The overall effect of North Sea oil on domestic employment is almost certainly to reduce it.

Problems of Structural Change

Our simple model assumed that demand for the output of all sectors expanded in the same proportion when incomes did. This led to the largest requirement for increased output from the construction and public administration sectors.

However, these demands are largely determined politically rather than in the market, and expansion of them seems unlikely in present circumstances. Thus the main requirement for structural change is a shift from manufacturing towards services.

The incorporation of an oil premium in the exchange rate implies that British prices will seem high in relation to those of other countries. This will be particularly serious for industries where Britain has no cost advantage and where there is little difference between British output and that of other countries; steel is a good example of an industry whose future is permanently blighted. The more successful manufacturing industries will be those where goods are little traded—food manufacturing, for example—or where products are strongly differentiated—such as defence equipment. Britain's exports will move 'up-market', as has happened to other industrial countries, such as Japan, Germany, and Switzerland, which have experienced substantial changes in their real exchange rates. Some exporting service industries will also suffer—tourism is one; but financial services may be protected by inelastic demand for their product, and shipping and aviation may not suffer too severely because their value added in the UK is relatively small.

These adjustments will be difficult. The greater pace of structural change will imply an increasing level of unemployment, and as a result, the full benefit to potential output of North Sea oil will not be realized. Moreover, although it is certain that some contraction in the size of manufacturing output is required, uncertainty about the oil price implies uncertainty about the magnitude of the reduction; and the current oil premium simply represents the foreign exchange market's guesstimate of the real exchange rate appreciation required to achieve it, which may ultimately turn out to be too high or too low.

There is little cost to the foreign exchange market in an oil premium that oscillates around its ultimate equilibrium level, but these oscillations can have substantial adjustment costs for the domestic economy. There is therefore some case for damping initial movements toward equilibrium. The principal mechanism for doing this is capital export. It would seem wise to err, in the early years of oil production, on the side of excessive balance of payments surpluses and capital export, since it will be easier to move further in the direction of domestic structural change in future than to retreat from initial over-reaction.

Changes in the Real Oil Price

What is the effect on Britain of a change in the real price of oil? In 1981, Britain will probably be, for the first time, a net exporter of oil; and it would appear that a further rise in oil prices would be of positive benefit. However, it is probably unlikely that Britain will be a net exporter of oil indefinitely. We conjectured that British oil reserves might, at the current real oil price, have a value equivalent to $18 billion per annum, which is a little less than the current cost of buying Britain's oil consumption at the current price. This level of consumption has not increased since the first oil crisis of 1974. It is reasonable for Britain to be more or less indifferent to the real price of oil.

But we should look here also at the effect on the terms of trade. Rising oil prices do, in fact, seem to be associated with an increase in the real exchange rate for sterling. This is reasonable since Britain's principal trading partners are net importers of oil, and so a rise in the price of oil tends to depress the values of their currencies relative to those of countries that are self-sufficient in or exporters of oil. Even if a rise in the real price of oil is of no direct benefit to Britain, the improvement in the terms of trade to which it leads will be of indirect benefit.

It is appropriate at this point to note an important qualification to our earlier analysis. In 1976, the oil price was around £7 per barrel, or £12 at 1980 prices; in 1980 it was £15 per barrel but this would, assuming a pre-oil exchange rate of $1.85, be £19. On these assumptions, Britain would have needed to export an additional £4–£5 billion to pay for the increased cost of primary imports. Thus part of the contraction of the tradable sector that we have described is not an actual contraction but an escape from an expansion that in a non-oil economy would have had to occur. But if we think through the implications of this, the assumption of a 22 per cent oil premium in the current exchange rate may seem relatively modest.

Where has the Money Gone?

We have argued that the benefits of North Sea oil are larger than is generally realized. Moreover, a major part of growth in real incomes that they yield has already occurred; oil production is well over half its ultimate level and the terms of trade have already improved considerably. Yet many people would be surprised to be told they were experiencing a bonanza.

The answer to this is in several parts. First, a significant element of the benefit is not something that makes us positively better off, but something that prevents us from being as much worse off as we would otherwise have been. Real incomes in Britain have been protected from the oil price rise in 1979–80, and without indigenous oil they could not have been. Second, real disposable incomes did increase sharply over 1978–80: earnings rose by 35.4 per cent between January 1978 and January 1980 while the gross earnings deflator, which shows the rise in earnings required to maintain their purchasing power, rose by only 26.0 per cent.

The change is not properly reflected in aggregate statistics of economic growth. For technical reasons, national income accounting conventions do not cope well with major changes in relative prices, and very little of the benefits of North Sea oil will ever be reflected in the national accounts. Thirdly, although the growth of production and the rise in oil prices mean that major gains from the North Sea are now starting to be reflected in oil company profits, the structure of the tax regime means that it is not until 1983–5 that these receipts are fully reflected in government revenues.

Experience Elsewhere

There are very few precedents for the explosive growth of the primary sector in an already industrialized economy. The experience of the Netherlands is widely quoted. The analogy with the growth of mineral industries in Australia is less commonly recognized, but it has perhaps more to teach us, because of the widespread discussion of the need for structural change in the Australian economy that has resulted. This debate was inaugurated by Gregory (1976), and the arguments put forward by him have been an important influence on the thesis of this paper.

There are two aspects to the development of oil production in Britain which taken together mean that our experience is unique. There are no other commodities where the gap between costs of production and selling prices is as large as it is for oil. Hence the gain to income, and tax revenue, from the growth of output of oil is wholly exceptional. Gold is, at current price levels, the only comparable case. South Africa, as the principal gold producer, and Norway, benefiting from oil, are therefore likely to undergo similar changes. However, neither of these countries is, as Britain is, a major exporter of manufactured goods in relation to the world economy as a whole; and it is therefore principally in Britain that the further twist to the spiral implied by the appreciation of the terms of trade is fully effective.

6. Policy Conclusions

The method we have adopted in this paper is to propose a stylized model of the British economy and consider how that model is affected by oil. It is not intended for, or capable of, forecasting—there are many changes occurring in the British economy, and the growth of oil revenues is only one of them, albeit an important one. But this approach does, we believe, enable us to describe the effects and the policy options that result from this development. What are these policy implications? Perhaps the most urgent requirement is for a fuller public understanding of the main issues. The government is not contributing to such understanding by failing to publish realistic estimates of North Sea oil revenues or to state its plans for dealing with them. The present position is disconcerting to many people. The current level of the sterling exchange rate seems absurd in terms of purchasing power, or to the manufacturers of traded goods. Yet it is probably not a temporary aberration; it is the market's mechanism for imposing a cut-back in Britain's output of manufactured goods. It is possible to argue about whether this is desirable, but to do so is no more sensible than to argue about whether it is desirable that the sun should set in the evening; no doubt it would be more convenient if it did not, but that is neither here nor there. There is no other way in which we can benefit from North Sea oil revenues. The only serious alternative is to throw them away.

This should perhaps be considered as a serious alternative, and it seems to be what—consciously or unconsciously—the advocates of import controls or large programmes of manufacturing investment have in mind. It is possible that the costs of structural change are so high, and the capacity of the UK economy to adapt to new patterns of production is so low, that it is better to sustain the existing structure of output so long as national income does not actually decline. If we reduce the efficiency of manufacturing industry sufficiently, we can go on using the resources—labour and capital—presently employed there to produce a lower level of output. We are confident that a programme framed and justified in these terms would be unhesitatingly rejected. We are much less confident that a series of individual policy options that in aggregate amount to such a programme would be rejected.

If we do discard this alternative, then the most important policy implications are negative ones. We should accept the need for structural change in the British economy and, while we might try to ameliorate its social consequences, we should not seek to retard or reverse it. This means resisting inevitable pressure to support or protect declining sectors of manufacturing industry.

We have argued that a substantial proportion of North Sea oil gains should be used for investment, and that the bulk of this investment should be abroad. Domestic manufacturing is the least likely area to provide profitable scope for such investment. What kinds of policies are likely to stimulate capital outflows of the size required? It is clearly not desirable to encourage capital movement *to* the UK by maintaining relatively higher interest rates than other countries, or by actively stimulating inward investment. Outward investment—both direct and portfolio investment—has already been encouraged by the ending of exchange controls, and the high exchange rate will continue to make it more attractive. A policy that uses a major part of North Sea oil revenues to reduce the public sector borrowing requirement, and so puts the cash flows received into the hands of financial institutions, is likely both to encourage the use of a large proportion of these cash flows for portfolio investment and to facilitate their transfer overseas.

We are sceptical of the government's ability to outguess the foreign exchange market as to the appropriate long-term level of the oil premium, and would prefer to see foreign assets acquired by the private sector rather than by large-scale official intervention in currency markets. Nevertheless, there are costs to the economy if the market initially overestimates the oil premium, imposing excessive pressure on manufacturing which is ultimately relieved, which are greater than the costs imposed on the market itself. Speculators lose money equally from overestimation and underestimation; the economy stands to lose more from overestimation. There is therefore a case for some damping of the initial movement towards the new equilibrium by using official sales of sterling to reduce the rate of appreciation. But this pressure should in due course be phased out—which involves recognizing that it is an operation on which the Bank of England will lose money. It is, in fact, probably already too late for such a policy to make much sense. A considerable oil premium has already emerged, and this must be regarded as the market's assessment of its ultimate level.

We recognize that the approach and the arguments presented in this paper are ones that many people will find difficult to accept, although the logic of them is simple and hard to dispute effectively. We see two principal reasons for this. First, it has become all too common to believe that the end of economic activity is production rather than consumption. This seems to be born of spurious moralizing and the power of interest groups of producers. It is no doubt difficult to believe that one can become better off as a result of events that imply no merit of one's own and a reduction rather than an increase in hard and unpleasant labour; but the mineral royalties which once enriched undeserving landowners now enrich an undeserving population. It is worth stressing that there is no a priori moral or economic argument for preferring one industrial activity or structure to another; the appropriate pattern of production is the one best adapted to the needs of the society of which it is part and the trading opportunities open to it.

Second, much economic debate in the UK is conducted in terms that were set in the 1950s and 1960s. It is an over-simplification to contrast 'demand-side' and 'supply-side' approaches to economic policy, but it is certainly true that our analysis is motivated by a view that the principal constraints on economic growth in the UK are to be found on the side of supply rather than demand. Fifteen years ago, it was plausible to argue that Britain suffered from a fixed and overvalued exchange rate and that the level of domestic demand was depressed in order to sustain equilibrium at this level. But although the current exchange rate may appear overvalued in terms of its international purchasing power, there is no reason to regard it as other than an equilibrium rate. Loss of competitiveness is a serious matter when it results from a rise in domestic costs at a fixed exchange rate; but it has wholly different implications when it results from a rise in the exchange rate at a given level of domestic costs. Deindustrialization, as Sir Alec Cairncross has defined it, 'is a matter for concern if it jeopardises our eventual power to pay for the imports we need' (Blackaby (1979)) but when it is the product of an increased capacity to pay for these imports it is just the reverse.

Appendix: How Large are the Benefits?

It is necessary to derive two central measures of the impact of oil. One is the increase in net output that results from it—the additional value added in the UK economy. The second is the net effect of oil on the balance of payments; this requires us to see how the increase in net output is divided between the traded and non-traded sectors. Because (non-traded) domestic resources are sucked into the (traded) oil sector, the direct effect is that the balance of payments effect is larger than the increase in net output, and the non-traded sector contracts.

If we start with gross revenue from oil, it is appropriate to subtract remittances overseas, whether for the purchase of related imports or as the earnings of overseas companies. The balance consists of domestic operating costs, the profits of UK companies, and tax paid to the British government. All of this reflects an improvement in the balance of payments (an expansion in the output of the tradable sector) except to the extent that the resources involved would otherwise have been used for exports. Some of the domestic costs are spent on items that might otherwise have been exported, but the bulk of them represents net balance of payments gain. A major part of the capital invested by UK companies in the North Sea would otherwise have been used for oil production elsewhere in the world, and would therefore have formed part of the UK export sector in any event; only the return on the remainder represents net gain to the balance of payments.

In computing the net addition to national income, we begin by subtracting domestically incurred operating costs, which simply reflect the costs of transferring resources to the oil sector from the rest of the economy. Parts of oil company profits fall into this category also: the *net* gain to the economy is only the amount by which North Sea profits exceed the returns that could have been earned by investing the capital and other resources devoted to the North Sea elsewhere. The measure of benefit is the difference between the return that they are earning and the return that they would have earned by deploying the funds in alternative uses—the 'economic rent' that they are earning. It is clear that the returns in the North Sea are much higher than would have been obtained from investment in the British domestic economy—the safeguard provisions of petroleum revenue tax (PRT) remit the tax if the rate of return should fall below 30 per cent. It is less clear that these returns are higher than oil companies could have earned from investment in other parts of the world. It is impossible to make any scientific calculation of the rent earned by British companies, but it is unlikely that it would be more than £1 billion per annum. There is also some rent earned by those working in the North Sea or in onshore operations who are earning more than they would expect in alternative occupations, but this amount is quite small in aggregate. These items apart, the direct increment to national income that results from North Sea oil is equal to the tax receipts of the British government.

Thus the direct increase in national income that results from North Sea oil is a little larger than government revenue from North Sea taxes—larger by the amount of 'economic rent' earned there. The gain to the balance of payments is a little larger than that—larger by the value of domestic resources drawn into North Sea operations from non-exporting sectors. Precise computation of either of these figures depends on comparison with an unknown, and unknowable, hypothetical outcome.

Table 18.A1 gives estimates of the growth of oil revenues over the period to 1985 and their allocation. The calculations underlying such estimates are complex. This is largely attributable to the complexities of the tax system, which includes three different charges—royalties, which are based on output, PRT, which is calculated on a field-by-field basis, and corporation tax, which is levied on a company-by-company basis. These estimates of revenue, like all figures in this paper, are at 1980 prices, so that they can validly be compared with 1980 national income and 1980 tax revenues. But it is necessary to spell out what this means. We have assumed that the real price of oil is constant. By this we mean that the oil price rises at the same rate as prices in western economies generally. We have also assumed that the real exchange rate is constant. By this we mean that the effective international value of sterling depreciates by the amount by which British inflation exceeds the average of other countries. It would be convenient if there were a single international currency in which such prices could be quoted, but although artificial constructions exist, there is none that will mean much to many people. We therefore quote both prices in dollars, so

Table 18.A1. Allocation of oil revenues, 1980-5 (£bn)

		1980	1981	1982	1983	1984	1985
Operating costs		3.1	3.2	2.9	2.6	2.1	1.9
Earnings:	UK companies	2.3	1.6	1.5	1.0	1.1	1.6
	Foreign companies	2.8	3.4	2.5	2.7	3.2	2.5
	Total profits	5.1	5.0	4.0	3.7	4.3	4.1
Taxes:	Royalties	0.9	1.2	1.4	1.5	1.6	1.6
	Petroleum revenue tax	1.7	3.5	4.8	6.4	6.5	6.7
	Corporation tax	0.4	1.0	1.6	2.1	2.8	3.0
	Total tax	3.0	5.7	7.8	10.0	10.9	11.3
Total revenue		11.2	13.9	14.8	16.3	17.2	17.3

Note: All figures are at 1980 prices and assume a constant real price of oil, a constant real exchange rate, and 6 per cent domestic inflation.
Source: Wood Mackenzie & Co.

that we start from a real oil price of $35 per barrel and a real exchange rate of $2.25.

Table 18.A2 shows the behaviour of the real oil price since oil became available from the North Sea, by comparing the price of oil (in dollars). The real price was roughly constant over 1975-7, fell in 1978, and rose considerably in 1979-80. The table also shows the movement of the real exchange rate over the same period, by comparing the movement of UK prices (in sterling) with the cost of buying the consumer goods of OECD member countries (in sterling).

The tax yield also depends on the rate of domestic inflation. This may seem

Table 18.A2. The real exchange rate and the real price of oil, 1975-80 (1975 = 100)

	Real exchange rate			Real oil price		
	OECD consumer prices, in sterling	UK consumer prices	UK prices ÷ OECD prices	North Sea oil price ($)	OECD prices ($)	Oil price ÷ OECD prices
1975	100	100	100	100	100	100
1976	129	117	91	110	104	106
1977	147	135	92	121	114	106
1978	156	146	94	119	136	88
1979	164	166	101	203	154	132
1980	166	182	110	256	167	153

Sources: Prices—OECD *Economic Outlook*. Exchange rates—IMF *Financial Statistics*. Oil price—UK Offshore Operators Association.

paradoxical when we are making calculations in constant prices, but the reason is that inflation reduces the real value of allowances that are given on past capital expenditure. Thus rapid inflation increases government revenue at the expense of company profits. The 6 per cent rate implicit in Table 18.A1 is clearly too low, but this effect is not very large.

Yield is also sensitive to production levels. In the past, the growth of production has tended to fall below expectations, and if this is true in future then the growth of revenue both in aggregate and to the government will be slower than Table 18.A1 implies. But unless peak production is substantially cut back in order to delay depletion, this makes no difference to ultimate revenues—indeed it is likely to increase them slightly.

The revenue projections presented in the text are designed as long-run 'steady-state' estimates. They reflect basic assumptions of production of 1 billion barrels per year and annual operating costs of £2.5 billion (at 1980 prices). We assume that the tax system claims 80 per cent of net revenue at a price of $35 per barrel (this is the present target) and imposes a marginal tax rate of 90 per cent. The marginal tax rate in the existing system is 87 per cent (Table 18.A3); this implies that tax rates will be raised or lowered in line with oil price changes.

Table 18.A3. The marginal rate of tax on North Sea oil revenue

	Rate	Amount
Additional revenue	12½%	100.0
Royalty		12.5
		87.5
Petroleum revenue tax	70%	61.3
		26.2
Corporation tax	52%	13.6
		12.6
Implied marginal tax rate	87.4%	

Notes: Where the safeguard provision for marginal fields is operating, the marginal rate of PRT is 80% and the implied marginal tax rate higher.

Additional revenue may allow more rapid utilization of allowances against PRT and/or corporation tax; this would reduce the effective marginal tax rate.

Many companies with North Sea operations would otherwise be unable to offset ACT against any liability to mainstream corporation tax. Where this is true, it reduces the effective rate of corporation tax and also reduces the net gain to the UK exchequer from North Sea activities.

19 The Macroeconomic Impact of North Sea Oil*

CHRISTOPHER ALLSOPP AND JOHN RHYS

1. Introduction

The discovery and exploitation of North Sea oil and gas is one of the most important developments affecting the United Kingdom economy in the 1970s and 1980s. Despite this, there remains much disagreement over what the impact has been and about the implications of changes in production or price for macroeconomic policy—either retrospectively or for the future.

There are good reasons why the impact of oil should be hard to analyse. The discovery and build-up of production coincided with major movements in its price, notably the upward hikes of 1973-4 and 1979-80 and the equally dramatic fall in 1986. The UK faced the first oil crisis as an oil importer, the second as an approximately self-sufficient producer, and the recent fall as a substantial net exporter. Moreover, changes in the oil market do not just have direct effects on the UK, but, by affecting other countries differentially, have substantial indirect effects as well.

The impacts to be analysed are, therefore, highly complex. For analytical purposes they need to be broken down into simpler components. It is useful also, given the empirical uncertainties (especially about future production and price), to concentrate on the 'stylized facts'; otherwise a maze of estimation and accounting difficulties tends to obscure the essential points which, despite many good studies, are still commonly misunderstood.

This paper concentrates on two main issues, which interact. The first is whether North Sea oil added to the wealth of the UK in such a way as to make structural change, involving a decline in manufacturing, inevitable or indeed desirable. This section can be seen as a critique of the conclusions of Forsyth and Kay (1980a) reproduced in this volume (Chapter 18). (All references to page numbers, sections, and tables apply to the version in this volume.) The argument (which is not new: see for example Bank of England (1980, 1982)) is that their main theses are false, based on an important conceptual error. Thus North Sea oil was not a bonanza and the joint effects of the production build-up

---

* Christopher Allsopp is a Fellow in Economics at New College, Oxford. John Rhys is a Senior Consultant at National Economic Research Associates Inc.

The authors would like to thank Michael Clements, Michael Devereux, and Michael Saunders for very considerable help in assembling and understanding statistical information on the North Sea. Helpful comments from the editors of this volume are also gratefully acknowledged. Errors of fact and of interpretation are the responsibility of the authors alone.

and oil price changes do not imply a need for 'deindustrialization'.

The second part of the paper looks more directly at the macroeconomic effects of oil price and production changes. The focus is on the policy options available when oil market conditions change. Within the set of policy options, particular interest centres on offsetting or neutralizing strategies, if they can be identified. Thus actual policies followed can be compared with a hypothetical offsetting policy. This is a more natural and less arbitrary basis of comparison than the usual procedure of evaluating oil impacts on the assumption of 'no change in economic policy'.

Of course, fully offsetting policies—e.g. to the effects of an oil price rise—are not really possible. Nevertheless, since the United Kingdom was in the almost unique position of not being made much better off or worse off by oil market events in the 1970s and 1980s (due to near self-sufficiency on a longer-term basis), the main impacts on the non-oil private sector were financial and/or due to timing. Thus they could have been approximately offset using the instruments of monetary and fiscal policy. Impacts that could not easily have been offset (such as world economy effects) are themselves brought into relief. The analysis suggests that the policies followed, rather than the effects of oil *per se*, are the main explanation of the rapid decline of manufacturing in the early 1980s.

The structure of the paper is as follows. The next section briefly outlines the stylized facts of North Sea oil. Section 3 examines the structural change argument as put forward by Forsyth and Kay. Sections 4 and 5 outline the macroeconomic impact of price and quantity changes, concentrating on the question of the extent to which these impacts might have been offset or neutralized. The concluding section tries to assess in broad terms the extent to which policy towards oil in the 1980s accounts for the decline in British manufacturing.

2. The Quantitative Impact of North Sea Oil

The 'stylized facts' of North Sea oil production and price changes are outlined here and in the tables. In the early 1970s, the United Kingdom had no oil production to speak of, although gas was already coming ashore in significant quantities. The first major discovery—Forties—was made in 1970 and others soon followed. The bulk of existing proven reserves were established by 1976. Oil production started to build up from the mid-1970s to reach approximate self-sufficiency in 1980 (see Table 19.1). The output of oil from the North Sea continued to rise, so that at its peak in the mid-1980s, the United Kingdom was a substantial net exporter—of the order of 50 million tonnes out of a total production of about 130 million tonnes per annum. Estimates of the rate of decline of production vary, but it is expected to fall to about 90–100 million tonnes by 1990, and substantially further by the mid-1990s. In the longer term,

Table 19.1. UK crude oil production, consumption, and net trade (million tonnes per annum)

	Production[a]	Consumption[b]	Production[c] surplus	Net trade[d]
1970	—	104	−104	−100
1971	—	104	−104	−106
1972	—	111	−111	−101
1973	—	113	−113	−110
1974	—	105	−105	−110
1975	1	92	−91	−86
1976	12	91	−79	−84
1977	38	92	−54	−53
1978	53	94	−41	−42
1979	78	95	−16	−19
1980	81	81	1	−5
1981	89	75	14	17
1982	103	76	27	32
1983	115	73	42	46
1984	126	90	36	51
1985	128	77	51	53
1986	129	77	52	49
1987	120	77	43	
1988	113	77	36	
1989	106	77	29	
1990	97	77	20	
1995	56	77	−21	

[a] Illustrative forecasts based on figures from Wood Mackenzie, including production from 'probable' fields.
[b] Consumption projected forward on the basis of no change from 1985–6 levels for illustrative purposes.
[c] Derived as production minus consumption. These figures differ from recorded net trade due to variations in stock building and differences in statistical coverage.
[d] Crude oil only.
Sources: Production and consumption of oil up to and including 1986 are from the BP *Statistical Review of World Energy*. Production includes natural gas liquids and condensates.
Net trade figures are from Department of Trade and Industry, *Monthly Digest of Overseas Trade Statistics*.

forecasts range from practically nothing by the year 2020 to continuing self-sufficiency. Clearly, production in the longer term is, in any case, likely to be influenced by the price of oil on world markets, by the price of substitute sources of energy, and by government policy.

Consumption of oil grew rapidly in the 1950s and 1960s. In the 1970s and the first half of the 1980s, however, it fell from a peak of 113 million tonnes in 1973 to 72.5 million tonnes in 1983, before rising a little in 1984 and 1985. The marked change in trend reflected slower growth and changes in industrial structure in the 1980s, and, of course, the response to the high prices of the 1970s and early 1980s. Britain's position as a substantial net exporter during the 1980s largely reflects this decline in consumption. It is still not known with confidence how sensitive oil consumption is likely to be to the price falls of 1986—if they persist—but many of the changes (such as increased attention to conservation) are unlikely to be quickly reversed.

The most dramatic movements have, of course, occurred in the price of oil (Table 19.2). Before the first oil crisis, the price of a barrel of oil was about $3. It roughly quadrupled in 1973, and doubled during the second crisis of 1978-9. The mechanical impacts of the two price hikes on the world economy were, however, similar, at about $2-2\frac{1}{2}$ per cent of industrial countries' gross domestic product (GDP). At its peak in 1982, the price of Saudi Arabian light crude was $34 per barrel; in 1986 it declined to an average of about $14.

Table 19.2. Oil price trends

	World oil price[a] ($/barrel)	Real world oil price[b] ($1980/barrel)	Sterling oil price[c] (£/tonne)	Real sterling oil price[d] (£1980/tonne)	Real sterling export unit value for oil[e] (£1980/ton)
1970	1.7	6.4	5.3	20.0	27.3
1971	2.1	7.4	6.4	21.9	23.2
1972	2.3	7.4	6.8	21.1	21.7
1973	3.2	8.4	9.7	27.9	24.1
1974	10.7	20.1	33.6	82.6	77.1
1975	11.2	19.3	37.1	71.6	74.1
1976	12.0	20.3	48.5	81.8	90.4
1977	13.1	20.4	55.0	82.6	90.1
1978	13.2	18.6	50.2	67.2	71.8
1979	18.8	22.5	65.0	77.2	83.1
1980	30.9	30.9	97.2	97.2	109.8
1981	34.7	35.0	125.3	113.6	122.4
1982	33.6	35.3	140.7	119.2	120.0
1983	29.2	32.4	141.3	113.4	118.6
1984	28.6	32.3	156.9	120.3	123.0
1985	27.9	32.0	157.3	113.9	117.3
1986	14.4	13.8	72.1	50.6	53.1
1987	18.0	15.1	80.0	53.5	55.8

[a] Average unit value for oil exports from oil-exporting countries.
[b] Oil price deflated by the IMF index of manufactured goods export prices.
[c] World average oil price converted to sterling. This differs from the price of North Sea output both because North Sea oil typically commands a premium and because transactions prices differ from world prices.
[d] Sterling oil price deflated by the GDP deflator.
[e] Export unit value series from Department of Trade and Industry, *Monthly Digest of Overseas Trade Statistics*.
Note: Figures for 1987 are estimates by the authors.

The nominal prices of oil are, however, of limited interest. The table also shows indicators of the real price of oil, corrected for inflation. The real dollar price indicator is constructed using the world price of manufactured goods as a deflator. It indicates that, between the two oil price hikes, the real price tended to decline, and also that real prices were falling before the major break in 1986. The price fall in 1986 was sufficient to reverse most of the real price increases of the 1970s. The price of oil in sterling depends also on the exchange rate of the pound against the dollar. The real sterling price shown uses the UK's GDP deflator to deflate the nominal sterling price. As can be seen, variations in the UK's exchange rate give rise to a rather different time profile. The general picture of high prices in the 1970s and early 1980s, with a fall in 1986 to a level below that in 1974, remains.

The value of oil production obviously varies both with the level of production and with the price of oil. The figures in Table 19.3 show its increasing importance and the sharp fall due to the decline in price in 1986. At its peak in the early 1980s, the value of oil production was of the order of 7.0 per cent of GDP, with gas accounting for a further 1.5–2 percentage points. Turning to the balance of payments, net trade in oil became positive in the 1980s and, in value terms, accounted for over £8 billion per annum in 1984–5. (The contribution of oil to the balance of payments is often, somewhat misleadingly, assessed in a different way—that is, by considering the contribution as compared with a situation with no oil. In a period of high prices, such calculations would indicate a much greater impact.)

Finally Table 19.3 indicates the government's dependence in the 1980s on tax revenue from oil and gas. An important feature, further discussed below, is that the rise in tax revenues lagged the build-up in the value of production, largely because the producers had unutilized tax allowances in the early 1980s which postponed the effect of price rises on tax revenues. At the peak, the North Sea accounted for some £12 billion of tax or over 9 per cent of government tax revenues. The tax flow was about 60 per cent of the total value of oil production. Tax revenue was, in any case, expected to fall as production started to decline: in the event, the decline was much sharper, due to the price fall in 1986. Projections of tax revenue from oil are extremely sensitive to the assumptions made about its world price and about the UK exchange rate (see Devereux (1986 and Chapter 20, this volume)).

Summarizing the complex impacts of oil in the UK, the major discoveries were made in the early 1970s, production built up from 1976, to peak in the mid-1980s, after which it was expected to decline. Prices first rose before the UK had significant production, but even then the production could be anticipated. The further major price rise occurred as self-sufficiency was achieved, and the recent fall has coincided with peak production. Net exports from oil were highly significant in the 1980s, during the hump of peak production. Their contribution in value terms has been much reduced by the price fall but would, in any case, have fallen as production tailed off. The profile of tax revenues is

Table 19.3. Impact of North Sea oil

	Value of oil production[a]		Value of net trade[a]		Tax revenues[b]		
	£ billion	As percentage of GDP	£ billion	As percentage of GDP		£ billion	As percentage of GDP	As percentage of total government tax revenue
1970	—	—	−0.6	−1.3	1970–1	—	—	—
1971	—	—	−0.7	−1.3	1971–2	—	—	—
1972	—	—	−0.8	−1.4	1972–3	—	—	—
1973	—	—	−1.1	−1.7	1973–4	—	—	—
1974	—	—	−3.5	−4.7	1974–5	—	—	—
1975	—	—	−3.4	−3.5	1975–6	—	—	—
1976	0.6	0.5	−3.8	−3.4	1976–7	0.1	0.1	0.2
1977	2.1	1.6	−3.0	−2.3	1977–8	0.2	0.2	0.5
1978	2.7	1.8	−2.1	−1.4	1978–9	0.6	0.4	1.0
1979	5.1	3.0	−1.0	−0.6	1979–80	2.3	1.3	3.2
1980	7.9	4.0	0.1	—	1980–1	3.7	1.8	4.4
1981	11.1	5.1	1.8	0.8	1981–2	6.5	2.9	6.5
1982	14.5	6.1	3.8	1.6	1982–3	7.8	3.2	7.2
1983	16.3	6.3	5.9	2.3	1983–4	8.8	3.4	7.5
1984	19.8	7.2	5.6	2.1	1984–5	12.0	4.3	9.5
1985	20.1	6.7	8.0	2.7	1985–6	11.4	3.7	8.3
1986	9.3	2.9	3.7	1.2	1986–7	3.6	1.1	2.4
1987	9.6	2.8	3.4	1.0	1987–8	2.1	0.6	1.3
1988	9.0	2.4	3.0	0.8	1988–9	1.8	0.5	1.1
1989	8.8	2.3	2.5	0.6	1989–90	1.4	0.4	0.8
1990	8.5	2.0	1.8	0.4	1990–1	3.4	0.8	1.8

[a] Authors' estimates based on Tables 19.1 and 19.2. Values estimated using sterling world oil price series. For 1988–90 the real sterling oil price was projected forward as constant at estimated 1987 level. Projections of nominal GDP from Oxford Economic Forecasting. Value of net trade approximated as value of surplus of production over consumption.
[b] Tax revenues from Devereux (this volume; see Table 20.1).

even more peaked, due to the lagged response and the recent fall in price. During the period of high prices and peak production, the value of oil production was about 6-7 per cent of GDP, net trade was about 2.5 per cent of GDP, and tax revenues about 4 per cent of GDP.

3. The Structural Change Argument

The thesis developed by Forsyth and Kay in their article depends upon the United Kingdom being substantially better off due to oil. Their argument, an adaptation of the analysis used by Gregory (1976) in discussing the effect of the Australian natural resources boom, is in essence very simple. Additional wealth accrues as oil, which must largely be exported to turn it into other goods for domestic absorption (consumption and investment). In the process, other exports, mainly manufactures, must decline to ensure external current account balance. The resources of labour and capital released from manufacturing are then used to expand other sectors of the economy (including non-traded sectors) in line with increased demand—the increase in demand (spread across the economy) being itself due to the accession of oil wealth. The mechanism for bringing these changes about is a rise in the real exchange rate which, far from being a problem, implies further benefits due to an improvement in the terms of trade.

This argument, when the article appeared in 1980, was enormously influential and extremely controversial. Naturally enough, it tended to be welcomed in government circles, since it seemed to show government policy in a favourable light. The (real) exchange rate was very high and manufacturing was suffering a rapid contraction. Apparently this was not only inevitable, but desirable. It obviously came under attack, especially from those on the Keynesian wing of the profession, who saw North Sea oil as releasing a 'balance of payments constraint', thus allowing a higher level of demand and utilization of resources.

It could be argued, however, that though the criticisms might have practical force, they were basically missing the underlying point. The Forsyth/Kay analysis was comparative static in spirit: full utilization of resources was assumed. It was simply not about shorter-term cyclical effects or about unemployment (though of course unemployment might arise due to frictions during the course of adjustment).

A much more fundamental set of criticisms came from the Bank of England (1980, 1982), which examined the premises of the argument. The Bank focused on the first proposition—that the United Kingdom had received a windfall gain because of oil—on which the whole argument depends. If the United Kingdom could be shown not to have gained in the way assumed, then the structural change argument as put forward by Forsyth and Kay was simply false as applied to the UK. This is indeed what the Bank showed. Because of North Sea oil, it suggested, the UK largely avoided the loss imposed on oil importers (such as

Germany or Japan) by the oil price rises of the 1970s. But the avoidance of a loss applying to others is quite a different thing from receiving a windfall gain or 'bonanza'. The latter would imply structural changes of the Forsyth/Kay type; the former would not.

The question of whether the Forsyth/Kay analysis is relevant to the UK or not is basically factual and it might seem that it should be easily settled. But the situation to be assessed is, as noted, complex, due to the major changes in production and in price that occurred in the 1970s and 1980s. It also matters whether a long-term or a short-term view is taken. Beyond that, there is a fundamental conceptual problem about what should be compared with what. Thus there is plenty of room for error and confusion. Generally it now appears accepted that the structural change argument as originally put forward is false (Bank of England (1980, 1982), Rhys (1980), Byatt, Hartley, Lomax, Powell, and Spencer (1982), Whiteman (1985), and Allsopp (1986)). Perhaps more importantly, the idea that North Sea oil was a bonanza for the UK is at last beginning to be questioned in public debate. Nevertheless it is still worth going through some of the intricacies—in part because they bear equally on the macroeconomic assessment of the next section.

The Forsyth/Kay Analysis

We are concerned here with a particular aspect of the analysis: that part that leads to Forsyth and Kay's main conclusion. There is much in the rest of the article that is extremely valuable, and to the extent that it does not depend on the structural change argument, or on the proposition that the UK was absolutely better off (as opposed to relatively better off compared with other countries), there is no problem with it—though some of the details could of course be questioned. It is also worth stating that we are not concerned with the details of estimates or forecasts of the size of North Sea production, costs, or revenue. Clearly these can and do differ, but that is not the point at issue. (Thus their stylization of the situation to be analysed is broadly similar to that of other researchers (e.g. Bank of England (1982)).)

The core of their argument is contained in their Sections 2 and 3 and Tables 18.3 to 18.6. The conceptual error occurs right at the beginning of their Section 2 and is applied in Section 3. They discuss the benefits of oil as compared with a situation where no oil is produced and it is imported instead. Since they assume a 1980s price of $35 per barrel, and a mid-1980s production level of 1 billion barrels per year (equivalent to 136 million tonnes per year), it is hardly surprising that the benefits (compared with not having any oil and importing for domestic comsumption) are large. In gross terms, they suggest a figure of £16 billion (1980 prices) or about 9 per cent of GDP. This figure is then reduced by allowing, perfectly correctly, for expatriate profits and domestic operating costs. The final figure, which they stress is illustrative, is a benefit of £10 billion, or 5.5 per cent of GDP. This is the figure used for their calculations in Section 3 of their paper.

There are several conceptually different reasons why this figure is not the appropriate one to use. The most important is that the comparison of North Sea oil at $35 with the hypothetical situation of a UK without oil is irrelevant to the UK's own past history. When the UK did import all its oil, in 1970 or in 1976 (the base for Forsyth and Kay's later calculations), the price of oil was much lower. Specifically, oil was imported in the early 1970s at about $2-$3 per barrel and in the mid-1970s at about $12. In real terms, the dollar oil price more than doubled in 1973-4 and approximately doubled again by 1981. Thus when the UK was importing oil, oil was much cheaper than $35. Thus the calculation, which assumes that oil for domestic consumption would be obtained at $35, is highly misleading. The problem is that changes in production and changes in prices occurred together, and get mixed up in the analysis.

This relates to another difficulty. The 1 billion barrels of oil per year (136 million tonnes) assumed for the mid-1980s is by no means all net exports. About 80 million tonnes, or nearly 60 per cent, was for domestic consumption. It is quite reasonable that net exports of oil should be valued at world prices (net of the costs of producing it and of repatriated profits) since that is an indication of the benefit to the economy in terms of the goods and services that could be obtained from abroad. But what about the value of home-consumed oil? Since consumption of oil did not change very much between the mid-1970s and the 1980s, how could an increase in the *price* of the oil be regarded as a benefit? In fact, whatever benefit or welfare the consumers of oil achieved should depend upon the quantity of oil, *not* on the price. It is evident that there is something wrong with the basis of the calculation of net benefit.

The easiest way of seeing what the problem is, is to start by considering the comparison between the situation in the early 1970s—when the UK imported all its oil and before the major price rise—and the situation in 1980-1—when the UK was almost exactly self-sufficient and when the price was approximately $35. (We return below to the comparison between the mid-1980s and 1976 used by Forsyth and Kay.) This is the comparison used by the Bank of England (1980). It is adopted here primarily for analytical convenience, since it eliminates the problem of dealing with net exports.

In the early 1970s, Britain obtained virtually all its oil on world markets at a price of less than $3 per barrrel, or about $7 if the value of the dollar is adjusted to its value in 1980 (see Table 19.2). The price of $7 (1980 dollars) was the opportunity cost of a barrel of oil, and represented the cost in terms of real resources used in the exporting industries to obtain the oil.

A decade later, Britain was more or less exactly self-sufficient in oil production (though it exported and imported for technical reasons due to the type of oil produced) and net trade in oil was zero. Oil for domestic needs was obtained from the North Sea. The resource costs of extracting that oil are hard to pin down exactly, but were estimated to be of the order of $10-$12 per barrel (again in 1980 dollars). Thus, in the early 1980s, the resource costs were higher than the costs of importing oil in the early 1970s. In sterling, the costs rose from about £23 per tonne for imports in the early 1970s to about £35 per tonne in the

1980s. (See Bank of England (1982, pp. 58–61) for a discussion.)

If the estimates are accepted, Britain was worse off in economic terms in the early 1980s, in spite of the build-up of North Sea oil and the change from being an importer to self-sufficiency. The UK was devoting a larger amount of real resources in the early 1980s to obtain a smaller quantity of oil than in the early 1970s. Over the decade as a whole, there was no North Sea 'bonanza'. On the contrary, Britain was worse off.

The reconciliation of this result with the type of calculation performed by Forsyth and Kay is straightforward. Their assessment of benefit would compare the value of oil production in 1980 with the notional costs of importing it at $35 per barrel. (This would give a figure of about two-thirds the one they used, which was based on a higher production assumption.) But what matters to the British economy is not the price at which oil might be bought or sold, but the *costs* of supplying domestic consumption needs. These, as we have seen, were far lower than $35, and changed in a quite different way over time. In the late 1970s, prices rose. Apparently the value of North Sea oil, domestically consumed, also rose. The rent element in the price rose and with it the likely tax take for the government. This increase in tax take (as well as any extra rent retained by the domestic producers) is computed as a benefit by Forsyth and Kay. So it is. But to set against that benefit there is a *loss to consumers* of about equal magnitude. (It is well known that, in the short term at least, oil consumption is rather insensitive to price.)

Essentially, therefore, the problem can be seen as arising over the treatment of the rent component in domestically produced and consumed oil—i.e. the difference between its final sales value and its production cost. Forsyth and Kay treat this as a benefit, which it would be as compared with importing the oil at $35. In any comparison over time, however, a distinction needs to be made between variations in rent that arise due to changes in costs as opposed to changes in price. Lower costs do imply a benefit to the British economy. Higher prices, on the other hand, do not. Though the government gains, the consumers pay for it. (The rent on net exports is a gain, since it is the 'foreigners' who pay.)

Thus, in general, a comparison of how much better or worse off the economy is between any two time periods needs to treat domestically consumed and produced oil differently from net exports or imports. For net imports, the relevant cost to the economy is the world price. For domestically produced and consumed oil, the relevant opportunity cost is the real resource or production cost. And for net exports, the gain is the world price minus the domestic costs.

The importance of making these distinctions can be appreciated if a rise in the world price of oil is considered. For a country that imports all its oil, the loss is measured by the change in price (though it will be somewhat lower than that implies, due to induced reductions in the use of energy and substitution against oil). For a self-sufficient economy, there is no gain or loss. The government may gain tax revenue, but the consumer loses. (Again a caveat is necessary: any reduction in consumption that is induced leads to net exports so that, at the margin, the value to the economy is the world price minus production costs.) Finally for a major oil exporter, the gain is measured by the change in the value of net exports (minus any change in production costs).

Alternative Comparisons

It is sometimes argued that the difference between the Bank of England results and those of Forsyth and Kay is largely due to the different basis of comparison: that while the economy may not have gained between 1970 and 1980 from the development of North Sea oil, it did between 1976 (the base for Forsyth and Kay's further calculations in their Section 3) and the notional mid-1980s situation of production at 1 billion barrels per year with a price of $35. This, however, is not really true, though the question of the actual situations to be compared is, of course, of great importance.

That the Forsyth/Kay results do not depend upon the actual years being compared can be seen if it is noted that the degree to which the British economy is calculated to be better off does not depend upon a comparison with 1976—or any other year. The benefit of oil, as we have seen, is the value of oil production at $35 minus production costs. This can of course be seen as the benefit as compared with a notional situation of importing oil at $35, in the sense that the calculation of the change would then be correct. Thus one way of denying the relevance of the structural change argument to the United Kingdom is to point to the irrelevance to the UK of the hypothetical situation of having no oil and importing it at $35. The high price of imported oil was highly relevant to other countries without oil—which would need to industrialize—but not to the UK. The structural change argument is formally correct, but applied to the wrong country.

Alternatively, the calculation of benefit should be changed. In particular, the rent on domestically produced and consumed oil, largely accruing to the government as tax revenue, should not be treated as a benefit due to oil. That would only be sensible if the alternative were to import the oil at $35. Again the implicit comparison is with an irrelevant counterfactual.

These arguments show that the structural change thesis is wrongly applied. To go any further requires specification of relevant bases of comparison. One possibly relevant basis of comparison is of the mid-1980s with 1976, when the UK actually did import oil. It is important to note that this is not what Forsyth and Kay do. They value oil imports at $35, not at the actual price of imports in 1976. (Though later in the article they do note that, for this reason, some of the structural change they describe 'is not an actual contraction but an escape from an expansion that in a non-oil economy would have had to occur' (p. 365).) Nevertheless, it could be argued that if they had performed the relevant calculations, then comparing the mid-1980s with 1976 there would have been a substantial gain to the economy, requiring structural change of the type they describe.

It is true that between 1976 and the notional mid-1980s situation, the UK can be argued to have gained from oil. In 1976, the opportunity cost of obtaining oil was the world price, which though much lower than $35 was substantially higher than that in the early 1970s. It was indeed somewhat higher than the production costs involved in obtaining oil for home consumption in the mid-1980s. So there is a (small) gain here. There is also a gain since, in the mid-1980s, production is assumed to be well above consumption needs and net exports are some 40 per cent of total production. For that 40 per cent, the calculation made by Forsyth and Kay is conceptually correct. Thus the economy appears better off in the mid-1980s than in 1976, though not to the extent implied by the figures in the original article.

But the relevance of this comparison is denied by Forsyth and Kay themselves within their article. They well recognize that it is not the situation in a given year that should matter, but the 'permanent' value of the increase in wealth. Later in the article (in Section 5), they argue that only about half their additional value of oil should be regarded as sustainable. Allowance for this would reduce the relevant production flow to somewhat less than normal domestic consumption needs. That is, on a long-term basis, the UK is probably less than self-sufficient in oil. The present value of net exports is likely to be less than zero.

What this means is that it is not appropriate to compare 1976 with a future position of peak production, involving substantial net exports. At best the situation in 1976 should be compared with approximate self-sufficiency. If that were done, however, the notional benefit from oil would be very substantially reduced (as compared with 1976). It would depend upon the degree to which production *costs* were less than the costs of importing oil in 1976. This is not a large number. Indeed in the future, the costs of obtaining oil from the North Sea are more or less bound to rise as low-cost sources are exhausted and production shifts to more marginal fields.

There is another objection to the comparison. The position in 1976 was itself temporary, and known to be temporary since North Sea production could be anticipated. On a longer view, there is no doubt at all that the costs of obtaining oil for domestic consumption have risen, and are likely to go on rising. Particularly if structural adaptation is being considered, the comparison with 1976 is misleading. Given the anticipated build-up of oil production, it would have been wrong for the UK to adapt its industry to the mid-1970s prices for oil, and in practice it is unlikely that it had so adapted.

Oil Wealth, the Balance of Trade, and Manufacturing

If the UK on a longer-term basis was not becoming better off because of the development of oil, it follows that the structural change argument, which depends upon a gain in oil wealth, does not apply. (It does apply, in reverse, to oil-importing countries which needed to increase their manufacturing base to 'pay for' higher-priced oil; that was well understood in the countries concerned.) That, in essence, was the argument of the Bank of England. It is useful, however, to go a little further into the structural change argument. Evidently there is more to it than just a rise in wealth—which would not normally require structural change.

The argument (in Section 3 of Forsyth and Kay) depends upon an accession of oil wealth in the form of foreign exchange—i.e. oil wealth accrues as net exports or a reduction in imports of oil. Thus in their tables, the whole impact of oil (£16 billion) is applied to the primary production sector, and affects its net exports. The rise in the exports of the primary sector means that the net exports of other traded goods have to decline if trade balance is to be maintained. More fundamentally, since the additional wealth is foreign exchange, the only way in which the economy can benefit is if it is repatriated in the form of goods or services for domestic absorption. As they put it, 'All we can do is to exchange oil for trade goods . . . and redeploy the resources released from these sectors into the other, non-tradable, sectors of the economy. *There is no mechanism for deriving benefit from North Sea oil that does not, sooner or later, require this structural change.*' (Their italics, p. 351-2.) Further on in the article they add 'Running a balance of payments deficit or surplus is a means of advancing or retarding the date at which structural change becomes necessary, but it is not a means of avoiding it; the only way of avoiding it is to throw North Sea oil revenue away.'

These statements are, however, clearly false if oil does not accrue as foreign exchange. In fact, most oil production from the North Sea, even in the period of peak production, is not exported. It directly meets domestic consumption needs. For that oil, there is no need to export it and no need for structural change. What is more, in the longer term it is, as we have seen, unlikely that the UK is an oil exporter at all: oil surpluses during the period of peak production provide resources for future oil imports. Present oil is converted into future oil. For that, international markets are required, e.g. so that foreign assets can be built up and then run down (an alternative would be to limit production to needs year by year), but there is still no need for structural change, and it is certainly not true that the only way to benefit from oil is to export it and consume something else. On the contrary, the way to benefit from oil in the longer term is to consume it all as oil.

Finally, it is necessary to reconcile this result with the way in which the structural change argument actually works. For Forsyth and Kay, £10 billion of net exports are taken as accruing to the primary goods producing sector, and so displace manufactured and other traded goods exports in order to maintain balance of payments equilibrium. The displacement is nearly complete, in that consumption of primary goods is taken to rise only in proportion to the rise in wealth of the economy (5.5 per cent). In fact, only 40 per cent of their assumed oil production should actually be treated as net exports, and it has been argued above that, on a longer-term basis, net exports should be taken as zero or less (which is apparently also the view of the authors). Thus if something like self-sufficiency were assumed, the impact would be lower.

The rest of the impact is due to domestically consumed and produced oil, valued at $35 minus production costs. This, as we have seen, is the benefit as compared with importing at $35. The benefit as compared with importing at any other price would be valued by the import price minus production costs, and would be lower than implied by their calculation if actual import costs were less than $35, which in practice was the case. Compared with 1976, there would, as noted, still be some gain, but a small one if net exports on a sustainable basis were taken as zero. On a longer-term basis—say making a comparison with the situation in 1970—there would be no gain at all; rather, there would be a substantial loss.

There would, however, be structural change of a kind as the economy moved from importing to self-sufficiency, even if there were no change in the costs of obtaining oil. Resources that previously had been used to produce goods (such as manufactures) for export to obtain oil would (notionally in terms of the calculations) be switched to the production of oil.[1] Imports of oil would fall, and so would exports of manufactures. There would be no impact on the balance of payments.

A rise in the price of oil when the economy was self-sufficient would have no impact on the balance of payments, no impact on wealth, and carry no direct implications for structural change. It would simply be like a rise in domestic taxation on oil. (And of course there would be induced effects as consumers of oil economized or switched to other energy sources.) The Forsyth/Kay analysis, by contrast, would treat the rise in price as implying increased wealth, increased foreign exchange earnings, and a reduction in manufacturing production. None of these implications is correct—as seems to be recognized by Forsyth and Kay themselves, since they state (in Section 5) that 'It is reasonable for Britain to be more or less indifferent to the real price of oil.'

Some Implications

The structural change argument put forward by Forsyth and Kay is wrong. It might nevertheless have been pointing in the right direction if the UK were a substantial net exporter of oil on a long-term basis. This, however, is denied by Forsyth and Kay themselves.

It follows that their analysis can have nothing to contribute to an explanation of the rapid decline in the share of manufacturing in the 1980s. (Moreover, it was never intended as an explanation of unemployment.) The explanation for manufacturing decline thus needs to be sought elsewhere. It may of course have

---

[1] Evidently the statistical category of manufacturing could decline and that of primary production could rise as resources switched (though, in practice, many of the costs of producing oil would appear within manufacturing). This element of structural change is not the one that Forsyth and Kay are concerned with. It is excluded since their impact of oil on the primary sector excludes production costs.

a great deal to do with oil and with the way in which it was managed—as is suggested in the next sections.

The basic reason the structural change argument does not apply to the United Kingdom is that Britain did not in absolute terms become better off due to North Sea oil. The development of the North Sea did, however, allow the UK to avoid the losses—and the need for structural change—imposed on oil-importing countries. The implication that the discovery and development of North Sea oil did not constitute a 'bonanza' for the UK is extremely important in assessing the policy options available as oil market conditions change.

Similar arguments suggest that recent price falls for oil, if they are maintained, improve the position of oil-importing countries, whilst leaving the UK more or less unaffected in a long-term sense. Indeed, to the extent that the UK in the mid-1980s was (in forward-looking terms) less than self-sufficient, a lower oil price is of direct *benefit*, though the benefit is less than that applying to oil importers.

None of the above implies that important economic variables such as output, inflation, or the exchange rate would not be affected by oil market events. If they are, however, this may pose a problem for policy. In particular, if the exchange rate is affected by oil price changes or by net exports in the short to medium term, this should be seen as a problem, and not as a beneficial concomitant of desirable structural adaptation—as suggested by Forsyth and Kay.

4. Oil Market Events and the UK Economy

Early analyses of the impact of North Sea oil (such as that of Forsyth and Kay) tended to evaluate the total impact of oil production as compared with a notional situation of having no oil, or, as in the Bank of England studies, in relation to some past situation judged relevant. This is an awkward procedure since, as we have seen, price effects, production volume effects, and international impacts tend to get mixed up.

For analytical clarity it is better to separate out the different impacts—at least initially. Thus there have been three major price impacts since the early 1970s which impinged not just on the UK but on other countries as well (see Table 19.2). Moreover, as shown in Table 19.1, the profile of production is very uneven, posing quite different questions about the macroeconomic effects through time and about how they might be offset or mitigated.

Fortunately, in analysing the impact of oil events on the United Kingdom, there is one major simplification available which does not do too much violence to the 'stylized facts'—namely the assumption that the UK can be treated as self-sufficient. This was, as seen, approximately true on a short-term basis at the beginning of the 1980s, when the second oil price rise was having its impact. It is also a reasonable assumption for the longer term—though if anything it probably overstates the true position. A situation of long-term self-sufficiency (looking forward from say 1980) is particularly convenient in analysing the implications of the 'hump' of production and net exports in the 1980s.

It needs to be stressed that impacts do, in practice, interact. Thus production and consumption profiles depend on prices, including those expected for the future. Consumption will depend upon the development of the economy which, *inter alia*, will be influenced by the policy response to oil events. The degree of self-sufficiency is a slippery concept (for a discussion see Bank of England (1982)) and sensitive to assumptions made about discount rates, about reserves, and about costs. It would also depend upon prices. (Thus, at a high enough price for oil, presumably any country would make do with substitutes.) Thus, although free use is made of simplifications and stylizations, they should not necessarily be regarded as 'realistic'. Their purpose is to avoid having to talk about all aspects at once.

The Exchange Rate

Even though the structural change argument as applied to the United Kingdom is wrong, this does not mean that the exchange rate would be unaffected by North Sea oil—in the short run or in the longer term—or by oil market events such as price changes. What it does mean, however, is that if the exchange rate does, for example, rise, inducing deindustrialization, this could well be regarded as undesirable, presenting a policy problem for the authorities and indicating the need for offsetting policy action.

Forsyth and Kay, as part of their general analysis of structural change, suggest that the real exchange rate needs to rise to induce the structural change and restore external payments balance. Given the former, the exchange rate rise is desirable, and produces further benefits to the economy (and more structural change) due to improvements in the terms of trade. They suggest an 'oil premium' of about 22 per cent due to the coming on stream of North Sea oil. Of course, if the structural change is not needed, such a rise in the exchange rate would have quite a different implication: it would tend to induce undesirable deindustrialization and an undesirable deterioration in the external payments position, which would later have to be reversed.

There have been a number of attempts to estimate the exchange rate effects of the North Sea, using, for example, some of the UK macroeconometric models (such as those of HM Treasury, the National Institute, or the London Business School). Often these are concerned with a somewhat different (and better specified) impact—namely the effect of a rise in the price of oil. They are not, however, very informative, since they all track actual developments in the exchange rate poorly. A Treasury study suggested 15 per cent as an upper bound to the effect of the price rise after 1978 (Byatt *et al.* (1982)). A more recent survey of a number of different estimates suggested a range for the elasticity of the exchange rate with respect to the oil price of 0.15 to 0.45 with a central estimate of 0.3. This would imply that a doubling of the oil price (roughly the case in the late 1970s) would raise the exchange rate by some 30 per cent. (See Powell and Horton (1985, annexe by Vernon).) Such estimates should, however, be treated with considerable scepticism. Effectively they are reduced-form estimates which conflate a large number of different possible effects and take no account of alternative possible policy responses, at home or abroad. Most treat the UK more or less in isolation, given the rest of the world, though some simulations include world recession effects.

Concentration on the effect on the United Kingdom is unsound. Oil price rises affect the position of other countries as well, and indeed, if the UK is regarded as approximately self-sufficient, it is effectively isolated from oil price shocks: a rise in the price of oil does not affect the balance of payments or the 'wealth' of a self-sufficient economy.

At the very minimum, a three-country model is required even to begin to analyse the effects. The impact of a rise in the international oil price can be seen as analogous to a tax, levied by oil producers on oil-consuming countries, with the United Kingdom unaffected if self-sufficient. The question is, what happens to the United Kingdom's exchange rate if country A taxes country B with associated resource and income transfers?

Without knowing much about countries A and B, the simple answer is nothing. The more complicated answer is that the effects would depend on the relative propensities to import, directly and indirectly, from the United Kingdom, and on the relative portfolio preferences for sterling-denominated assets if the transfer is financial. Even the sign of the effect is ambiguous until empirical assumptions are made. In the short term, moreover, the effect would depend on the policy responses in gaining and losing countries, as well as in the United Kingdom. (Thus, if world recession were induced, the UK's balance of payments position would worsen and the exchange rate would tend to fall unless recession occurred in the UK as well.)

Thus analysis of the effect of an oil price rise or fall on the exchange rate is inherently complex and conditional. The response should not be expected to be the same whenever oil prices change: it would depend upon circumstances and policy reactions at home and abroad.

Nevertheless, there are reasons for expecting an effect in practice. We consider the situation in the late 1970s/early 1980s first. One possibility is that transfers of real resources to OPEC (Organization of Petroleum Exporting Countries) producing countries increase the world demand for UK exports. This would be the case if the United Kingdom gains disproportionally from OPEC demand. This is a long-run effect and is likely to be small. If it occurs, it is entirely beneficial and the implied terms-of-trade improvement could produce a mini-Forsyth/Kay effect (Byatt *et al.* (1982)).

More importantly, in considering the effect of an oil price rise, there is, in the short term, before the OPEC surpluses lead to increased spending, a substantial asset transfer to oil producers. There may be reasons for supposing that the demand for sterling-denominated assets would rise (Bank of England (1982)). (It is important to note that if self-sufficient, the oil price rise does not affect the UK balance of payments on current account, so that the supply of sterling-denominated assets does not change.) It is possible that the United Kingdom's exchange rate was affected for this kind of reason after both the first and the second oil price shocks (though in the former case the UK did run a deficit). Such effects, if they exist, should mostly be temporary.

As noted, reactions overseas, such as recession in consuming countries, should (if the United Kingdom were unaffected) pass some of the deficit problem to the UK due to lower export demand, tending to weaken the exchange rate. In practice, however, the United Kingdom's recession in the early 1980s was both earlier and deeper than in most other OECD countries, which would tend, if anything, to raise the exchange rate. Monetary policy in particular was tightened early, which may have raised the exchange rate and caused 'overshooting' (Buiter and Miller (1981)). This illustrates the general difficulty that it is more or less impossible in practice to separate out the effects of oil market events from other influences on the exchange rate.

Returning to the asset or portfolio side of the picture, the UK's position in the early 1980s as a major oil producer does mean that the risks perceived by the market to be attached to sterling assets might be different from those applying to other currencies—especially in the face of oil price shocks. This in itself is a reason for diversion of international portfolios in favour of sterling-denominated assets, and a reason for sterling's strength (Bank of England (1982)).

The factors so far discussed suggest relatively minor effects from international oil price changes on a self-sufficient economy. In the UK, however, changes in production were equally marked in the period under consideration, as the UK moved from being an oil importer, to short-term self-sufficiency in 1980, and to substantial net exports in the mid-1980s. The question arises as to whether, at a given price, these production changes would account for the major changes in the real exchange rate that occurred. It is easiest to assume that the UK may be taken as approximately self-sufficient on a long-term basis—that is, that the value of net exports, suitably discounted, is about zero. (Note that the realism of this assumption depends on the date used as a base for the forward-looking calculation, as well as on discount rates and the inherently uncertain profile of future production.)

One possibility is that financial and exchange markets are fully rational and forward-looking, and that operators are unconstrained by liquidity, imperfect capital markets, and the like. On such a view (which of course was influential in policymaking in the early 1980s), variations in production and net exports should have no effect so long as they are anticipated. Since most oil reserves were established by the mid-1970s and the production build-up was more or less as forecast, the uneven profile of oil production should not have affected the exchange rate. Temporary imports before the oil build-up and net exports during the 'hump' of production in the 1980s should have been discounted by the market. Only the longer-term situation should matter.

On such a view, oil market events that could not be anticipated (that is, 'news') should nevertheless affect the markets. The price changes in the late 1970s and in the mid-1980s could both fall in that category. If, however, in the late 1970s the UK was, in a long-term sense, approximately self-sufficient, the price change at that date should not have much effect either, since the underlying (long-term) payments position and the wealth of the economy would not be affected. On this view, therefore, the analysis of the effects on a self-sufficient economy should apply in the long term as well. Furthermore, to the extent that by 1986 the UK was less than self-sufficient (and it was certainly nearer to the period when net imports of oil would resume), the fall in the oil price, anticipated to continue into the future, should have improved the UK's position and if anything strengthened the exchange rate.

It is, however, highly unlikely that the above assumptions apply in practice. (This is not to deny powerful effects from expectations.) Exchange markets do not seem to behave in accord with this kind of theory. To the extent that exchange markets are myopic and react to short-term payments flows and other influences, powerful upward effects on the exchange rate could be expected in the period of net exports—and indeed before it, so long as the further future is heavily discounted. And reductions in the current value of net exports due to price falls in the mid-1980s would tend to weaken rather than strengthen the exchange rate.

The conclusion must be that the exchange rate effects of oil price changes on the UK 'should' be relatively minor, especially in the longer term. But potential upward pressure arises from possible asset market and current account effects over the medium term. Looked at in relative terms, current account effects in particular may have been rather large in the early 1980s. In 1981, the United Kingdom's current account surplus was about $12 billion at a time when most other OECD countries were running substantial deficits. Indeed, relative to Germany, the United Kingdom was in surplus by about $20 billion in both 1980 and 1981. The UK's relative surplus reflected, of course, both the absence of oil effects and the domestic recession.

The policy problems posed by upward pressure on the exchange rate in the late 1970s and early 1980s, which perhaps should not have been there on a longer-term view of the UK's prospects, are further considered below. Certainly the very high real exchange rate was a major reason for recession in the UK and for the decline in the exposed sectors of the economy, such as manufacturing.

Domestic Financial Implications

Oil market events have also had large direct effects on the domestic economy, most of which can be seen as fiscal impacts. These fiscal impacts could have been largely offset, but were not.

It has been seen that the tax revenue effects of North Sea oil have been large (Table 19.3). Again, considering the United Kingdom as self-sufficient (i.e. as in 1980) the tax flow arises because the world price, which was approximately matched in the United Kingdom, was substantially above the resource cost, including the profit of oil producers. The difference accrues as 'rent' to the government.

When the world price rises, as in the late 1970s, the rent element also increases. The marginal rate of taxation on oil was, in the early 1980s, very high. Something like 80 to 90 per cent of the effects of a price rise would eventually accrue as increased tax. We have already noted that this does not make the economy 'better off' if it is just self-sufficient, since domestic consumers and businesses pay for the higher-priced oil. (The economy does, however, gain to the extent of the rent element in net exports if production is above consumption.) In fact the effect of an oil price rise for a self-sufficient economy is exactly like the effect of an increase in taxation levied on oil. Indeed, to all intents and purposes it *is* a rise in taxation and may be analysed accordingly. This can be seen most clearly if it is noted that a self-sufficient economy would in principle have the option of not letting the domestic price of oil rise when world prices increased. This could be seen as an increase in the taxation on oil, balanced by an equal reduction in taxation on oil, and would obviously fully neutralize the direct domestic effects of a world oil price increase.

In practice, leaving the oil price unchanged as world prices changed would seldom be an optimal policy. Given the amount of trade in oil, to alter the mix of different types for technical reasons, problems could be posed at the frontier. More seriously, any economizing on oil that was induced by the high price would have a value to the economy at the margin of the world price (minus production costs) so that, on resource allocation grounds, it would normally be better to match the world price domestically. (Though adjustment costs and expected future price changes should also be taken into account.)

An approximately neutralizing or offsetting policy is still available, however. Other indirect taxes, such as value added tax, could be lowered in a compensating manner, which would have the effect of retaining the real relative price increase for oil but offset any overall fiscal effect. Such a policy would neutralize any demand-deflationary effects on the non-oil private sector, as well as eliminating the price-raising effect (which in practice is highly inflationary if expectations and nominal wages react). Not to do this involves, in effect, a policy decision to tighten fiscal policy in a price-raising way as international oil prices rise.

In 1979 and 1980 this type of offsetting policy was not adopted. Thus the oil price impact in those years can be seen as expenditure-reducing and inflationary exactly as in other countries that imported their oil. The OPEC 'shock' was in effect internalized in a self-sufficient economy. This policy response goes a long way to explain the apparently paradoxical fact that the effects of the two major oil price rises were rather similar in the UK, in spite of the change from being an oil importer to self-sufficiency. The effects were similar because the impacts were similar. There is, of course, no necessary presumption that in the United Kingdom, any more than in the world economy, fiscal deflation was necessarily undesirable: it might be justified by the need to curb inflation, though the initial price-raising effects of the non-offsetting policy are unfortunate to say the least.

There was, however, a complication. The initial tax revenue effects were delayed, due to unutilized tax allowances in the North Sea. This meant that immediate offset would have, temporarily, raised the public sector borrowing requirement, which was subject to government targets (or more strictly, 'consistent projections') under the government's Medium Term Financial Strategy (MTFS). There are several possible views of policy at that time. One is that the government should nevertheless have acted to offset the oil price rise so as to smooth the impact on the non-oil private sector, i.e. that it should have modified the MTFS to take account of oil. (Thus oil impacts may have interacted in a most unfavourable way with the MTFS (Allsopp (1985)).) Another is that it did, and that the MTFS was formulated taking account of the effects of oil—i.e. that the additional deflation was desired, and without the oil price rise, other taxes would have been raised. A third, which is perhaps most consistent with government pronouncements at the time, is that the non-oil private sector could be expected to see what was happening and to discount the initial adverse effects on its financial position in the knowledge, or expectation, that the tax effects (or the 'fiscal adjustment') would come through in the future. This latter view seems very optimistic about private sector behaviour and about the perfection of capital markets. If it were not justified, then the policy followed could be seen as either intended or mistaken deflation. (It may be noted that an offsetting strategy would have been safer if deflation (and inflation) were not intended. Presumably if the private sector were rational enough to discount a disturbance to its own financial position, it would have been rational enough to discount an offsetting strategy.) Domestic recession would add to any exchange rate rises due to asset market effects and effects on the current account.

The fiscal effects of the oil price rise in 1979-80 are of some importance in view of the many attempts that have been made to account for the major recession of the early 1980s in the UK. (See, for example, Bank of England (1981) and Artis, Bladen-Hovell, Karakitsos, and Dwolatzky (1984).) The difficulty is that the main fiscal deflation apparently occurred in 1980, too late to contribute much to the recession of that year. Thus macroeconometric model simulations that look at fiscal policy in terms of changes in tax rates and expenditure programmes suggest that the timing and extent of the recession are hard to account for. If, however, the fiscal impact of oil on the non-oil private sector were included (which amounts to comparing actual policies with an offsetting strategy) then the additional fiscal impact of some 2 to $2\frac{1}{2}$ per cent of GDP which came through early in 1979-80 would suggest a substantially greater role for fiscal policy in causing the recession.

Similar analysis suggests that a passive response to the oil price fall of 1986 was a far from 'neutral' response. As far as the non-oil private sector was concerned, the oil price fall was like a cut in indirect taxation, which boosted incomes and lowered the price level. A neutral policy—not that it was necessarily desirable—would have involved raising other indirect taxes to keep the public sector borrowing requirement (PSBR) roughly unaffected.[2] (There was a lag effect here as well: much of the initial effect of lower oil prices fell on oil companies rather than on the government's tax receipts.)

The analysis needs substantial modification when the effects of the 'hump' in production are considered. In the period when production exceeds permanent consumption, a substantial extra rent accrues to the public sector. On the face of it, this would justify reduced fiscal deficits over that period, which could rise again when production declined and tax revenues fell. It is important to take into account, however, that the additional rent accrues to the government as a result of sales of excess oil production abroad—that is, the revenue accrues effectively as foreign exchange. The implications of this are outlined in the next section, where a more elaborate offsetting strategy to fiscal and current account effects is outlined.

5. Towards a Neutral or Offsetting Policy

This section is mainly concerned with ways in which the joint impact of price rises in the late 1970s and substantial net exports of oil in the 1980s could have been mitigated or offset. Much of the analysis is also relevant to the effects of the price fall in the mid-1980s.

It will be argued that a neutral or offsetting policy towards the impacts of North Sea oil would have involved offsetting price increases by lower taxes elsewhere—in the way outlined above—and beyond this, offsetting the effects of the 'hump' in production by a policy of public sector accumulation of foreign financial assets which could then be run down in a subsequent period of net imports. Such a policy would, it is argued, minimize but not eliminate the effects on the economy. Some impacts, especially those arising from the effects of price changes on other countries, obviously could not easily be offset. It needs to be stressed that the main reason for considering an offsetting strategy is to serve as a standard of reference; there is no implication that a neutralizing set of policies would have been desirable in the circumstances of the time.

As argued above, a neutral policy towards North Sea oil and price shocks for a self-sufficient economy would have involved altering indirect taxes elsewhere, so that the fiscal impact was zero and the effect on the price level was zero. As we

---

[2] The mechanical impact of the oil price fall in 1986 on domestic expenditure on oil consumption was of the order of £6 billion, or about 2 per cent of GDP. This is a rough indication of the fiscal impact and of the impact on the price level.

have seen, a special case would be to leave domestic oil prices unchanged, though we have suggested that this would not be optimal and that, generally, relative prices should be allowed to change with world prices. There could, however, be a case for smoothing domestic oil prices, especially if world oil prices fluctuate.

If, however, the economy were more than self-sufficient in the long run—say production were expected to exceed consumption for a long period—the public sector gains to the extent foreign consumers of oil are taxed, and this is a gain to the economy. This gain could, in principle, be passed on to consumers by lower taxes. Since the benefit would accrue as foreign exchange, current account balance would require some reduction in traded goods production, unless all the extra income were spent on imports. To induce the change, some rise in the exchange rate, with favourable terms-of-trade effects further improving welfare, would be likely to result. Thus the rent element *in net exports* would be expected to produce structural changes of the kind outlined by Forsyth and Kay. With continuing net exports, but with rising production costs in relation to the world price (perhaps a more realistic case), the rent element and the implied fiscal adjustment would attenuate over time; alternatively the permanent value of the fiscal adjustment could be assessed and taxes lowered to that extent only.

The practically interesting case, however, is where the time profile rises and falls, with a period of net exports followed by a period of oil deficits. For clarity, assume that the economy is self-sufficient in a long-run sense, so that the specific problem of the time profile can be analysed. (Extension to cases where the economy is more or less than self-sufficient is straightforward.)

In this case, when production is above domestic (permanent) consumption and additional tax revenues accrue, there is no reason to pass on the addition. The tax is paid by foreigners, and is a benefit to the economy, but is balanced by the future costs of oil imports. As noted, the tax would accrue as foreign exchange and would appear as a current account surplus, as compared with the base-line of production according to needs and all other things being equal. But these factors are also balanced by future oil deficits and a future need for foreign exchange.

An obvious neutral policy during the period of net oil exports would be for the public sector to accumulate foreign assets to the extent of the current account and tax surplus. What would be happening is that a reduction in the public sector's wealth—its claim on the North Sea—would be balanced by an increase in another asset, in this case foreign financial assets. Both North Sea oil and foreign assets, unlike domestic national debt, are 'outside assets' for the British economy. The neutral policy outlined would involve the running-down of one and the building-up of the other.

If the economy really were just self-sufficient in present value terms, consumption of oil by the private sector (its capitalized value) exhausts the wealth in the North Sea. Thus the wealth of the public sector in the North Sea is zero, as it should be. But when production runs ahead of consumption, the wealth in the North Sea is being run down; this can be balanced if the public sector accumulates other assets instead. These assets are then used to support the consumption of oil by the private sector when production in a future period is less than consumption. This policy of using the international capital market thus drives a wedge between the consumption profile and the production profile.

It should be noted that it is the 'store of value' function of financial assets that is particularly important in balancing supply and demand over time. One store of value (oil in the North Sea) is being replaced by another (e.g. US Treasury bonds). Obviously, the store of value should have as high a rate of return as possible.

Such a policy would, looking at the economy in isolation, largely offset not just the fiscal effects of the hump in oil production, but the exchange rate effects as well. The current account impact—which is zero in the long run—of oil surpluses would be balanced by capital account transactions, due to the build-up and run-down of foreign assets. We note at this point the possible importance of the public sector performing this balancing act: the public sector can simply take a decision to carry out such a policy; the private sector, whilst it could also perform this function in the absence of an offsetting policy by government, would need to be induced to do so via exchange rate changes.

Whilst the above seems the most obvious way of carrying out an approximately neutralizing or offsetting policy, there are some alternatives. One, which is even more transparent, is via depletion policy. Production of oil could be limited to permanent consumption, in which case, assuming self-sufficiency, all that is necessary is the tax adjustment. There would be no effect on the current balance when prices changed, and no effect from the uneven time profile since, by assumption, it would not be uneven. This would not normally be optimal, for all sorts of technical and economic reasons. Thus it is generally better to use the (international) capital markets to allow production to exceed or fall short of consumption.

Another alternative is that excess oil production should be balanced within the current account by additional *imports* of capital equipment. The build-up of these assets is then the counterpart of the run-down of the North Sea assets in the period of excess production, and their contribution to output can be used in the future to obtain oil when production falls. In the first instance, the choice between this policy and the alternative of accumulating foreign assets would seem to depend only on the (social) rate of return on additional capital installed in the United Kingdom, as compared with the return obtainable in international markets. In terms of balancing the flows, however, fixed capital may appear somewhat inflexible. Alternatively, it may not seem right that the time profile of (additional) investment should depend on the vagaries of the oil production profile. Indeed, if the economy really is self-sufficient in the longer term, it is hard to make a case for additional capital accumulation just because (say) the oil price rises. (This is not to say that there may not be a case for additional investment in the UK; if there is, however, the case is not altered by an oil price rise if the economy is just self-sufficient.) In general, even if additional domestic investment is undertaken, there is still a role for the international capital markets in the smoothing process.

The kind of policy prescription that this line of reasoning leads to is that, in the period of excess production, the public sector should accumulate foreign assets as a counterpart and, over time, translate some of these assets (via imports) into domestic capital. That is, indeed, just what is done by the major surplus countries. The only difference is that, since the United Kingdom is at best approximately self-sufficient, the resource of oil, instead of being converted into a stream of consumption goods (directly and indirectly via the accumulation of imported capital), is best seen as being converted into future oil or oil substitutes.

Public Borrowing and National Debt

The fiscal consequences of such a policy need to be examined further. In the first place, we have already seen (Section 4, above) that if there is a lag in the response of tax revenues to (say) a rise in oil prices, the effects on the private sector could be smoothed by a policy of anticipating future tax revenues, which would involve a rise in the public sector deficit (and PSBR) in the short term. The uneven profile of production has further effects on the normal indicators of fiscal stance.

During a period of excess production, the public sector gains, owing to its increased revenue, and the public sector deficit and the PSBR fall. But the benefit to the public finances is used to accumulate foreign assets (take these for simplicity to be reserves of currency, though they might be other foreign assets or a reduction in indebtedness to foreigners). The result is that though the PSBR falls, domestic holdings of national debt (gilt-edged stock and other public sector liabilities) remain unchanged. Similarly, in a future period of oil deficits, the public sector deficit rises, but is financed by a run-down in holdings of foreign assets. Again, domestic holdings of national debt, and indeed the private sector's wealth, are unchanged. The policy amounts to stabilizing the financial position of the non-oil private sector.

Alternatively, the policy may be looked at in terms of the sectoral flow of funds. The non-oil private sector's net acquisition of financial assets (NAFA) is stabilized. Policy towards a lag in the collection of taxes amounts to mirroring variations in the oil sector's position by variations in the public sector's NAFA. Smoothing the time profile of production by public sector accumulation or decumulation of foreign assets involves equal and opposite movements in the public sector deficit and the current account surplus (i.e. the foreign sector's deficit). Thus whether looked at in terms of flows or stocks, the neutralizing policy outlined amounts to offsetting the impact of oil on the non-oil private sector's financial position by making use of the domestic and international capital markets.

Looking in more detail at the international aspects, compare the policy of using the tax receipts on excess oil production to buy foreign assets, with the situation that would result if the same fiscal policy were adopted, but where the public sector did *not* accumulate foreign assets. (In practice, in the 1980s, the policy was approximately one of a so-called free float, with no build-up of reserves.) In this case it is useful to simplify and to imagine that the public sector, in effect, uses its receipts of foreign currency to buy assets from the domestic private sector. In particular, assume it uses its receipts to retire its own financial liabilities, such as gilt-edged stock. What would be the effects?

It is evident that since the balance of payments and the public sector balance are the same, the private sector's net acquisition of financial assets is also unchanged. The overall wealth position of public, private, and overseas sectors is unaffected by the alteration in policy. What does have to happen, however, is that the private sector has to run down its holdings of national debt (domestic assets) and raise its holdings of foreign assets. That is, the difference is that the private sector's portfolio of assets has to change: in the period of oil surpluses, foreign assets rise, domestic assets fall; with oil deficits, foreign assets have to be repatriated and replaced by domestic assets—that is, by (say) gilt-edged stock supplied through the government deficit. Clearly, in accounting terms, one would see the balance of payments surplus due to (net trade in) oil, balanced by a private sector capital outflow—and an equivalent reversal when net trade in oil turned negative.

The main reason for concern is that, in order to induce the private sector to switch its portfolio in periods of current account surplus, the exchange rate may need to rise (or alternatively, domestic interest rates may need to fall relative to yields obtainable abroad). Thus this policy is not neutral, and its consequences would tend to be a (possibly substantial) rise in the exchange rate. This point needs to be stressed since, in policy discussions, it is often argued that a policy of 'paying off national debt' when oil surpluses arise increases potential borrowing power for the future. This is quite true—this is the fiscal aspect—but neglects the international dimension. It is foreign assets that must be accumulated and then repatriated. To induce the private sector to do this, movements in the exchange rate may be needed.

Another possibility is that the public sector 'uses' its receipts of foreign exchange to buy back sterling-denominated assets from foreigners. In this case, the currency composition of the domestic private sector's portfolio would be unaffected. There would, however, still be effects on the exchange rate as compared with a policy of foreign asset accumulation, as foreigners would need to be persuaded to give up sterling-denominated assets in exchange for 'dollars'.

Generally, the government would not know (until the statistics came in) whether it was domestic or foreigners' portfolios that were switching. The important point, however, is that the total supply of sterling-denominated assets to be held in portfolios at home or abroad is diminished owing to the impact of oil surpluses on the current account, unless the public sector increased its holdings of foreign assets. The effect on the exchange rate depends on the relevant elasticities of substitution: in the short term particularly, when changes are being effected, the exchange rate change necessary to induce the substitution may be quite large (unless relative interest rates change), especially as oil effects themselves may increase the attractiveness of sterling-denominated assets to market operators.

The neutral policy can be seen as amounting to exchange market intervention. It is equivalent to the non-neutral policy (i.e. with no build-up of public sector holdings of foreign assets) plus intervention to the extent of the current account impact. What is being suggested is that intervention of a generalized kind (i.e. not just in terms of building up the reserves) should be used to smooth the impact of the oil profile. It is clear that the amount of intervention might have to be large during the 'hump' of oil production.

Monetary Policy and the Feasibility of Large-scale Intervention

The above has concentrated on the fiscal effects of oil market developments. The question arises whether the offsetting policies outlined would have adverse monetary effects, especially on £M3, the government's principal monetary target during most of the period under consideration.

A difficulty is that the interactions between monetary and fiscal policy are themselves controversial in the UK, and indeed £M3 (which pays interest and is subject to financial innovation) is widely regarded as a difficult or inappropriate target (see Allsopp and Mayes (1985), Laidler (1985), Allsopp (1985), and Goodhart (1986)). These problems are not the principal concern here. To an extent, they can be avoided by the device of considering an offsetting strategy, since there is then little reason to expect monetary effects of either a positive or negative kind. Beyond that, however, it is necessary to analyse the possible effects of exchange market intervention on monetary holdings, since these are frequently misunderstood.

In the case where there is a domestic fiscal offset to an oil price rise and the government accumulates foreign assets during the 'hump' of production to run down later, there is no mechanical impact on monetary aggregates such as £M3, or, for that matter, M3. This is immediate since the effects on the financial position of the private sector are offset. The government's holdings of foreign assets are no more part of £M3 than holdings of oil in the North Sea.

Looking at the situation in terms of the counterparts of £M3—the starting-point for many official analyses (see Fforde (1983))—the same may be true even if the government does not accumulate foreign assets, but relies on the private sector to do so instead. The reason is, as seen above, that the effect on private sector portfolios is likely to be a reduction in sterling-denominated assets balanced by a rise in holdings of foreign assets. Even if these foreign assets are holdings of foreign currency deposits in banks, these are not part of £M3 and do not affect the aggregate. Thus if, for example, the private sector lowers its holdings of gilt-edged stock (not part of £M3) to raise its holdings of foreign currency (also not part of £M3), there is no change in £M3.

This, however, is not the only possible result. If the government does not accumulate foreign assets, there is likely to be a *favourable* effect on £M3, i.e. £M3 will tend to fall. This can be seen as follows. If the private sector reduces its holdings of sterling bank deposits in order to increase its holdings of foreign currency or any other asset, £M3 declines. The effect would be seen as a negative external contribution to the growth of money. (Note that M3 would be unchanged even though £M3 declined.) Thus a policy of not accumulating foreign assets by the public sector during periods of current account surplus will, if anything, be helpful to the control of a monetary aggregate such as £M3.

It is important to get the perspective right. The offsetting policy of public sector accumulation of foreign assets has no direct effect on money. On the other hand, a policy of not doing this may drive up the exchange rate, inducing favourable switches in private sector behaviour, tending to lower £M3. Thus, comparing exchange market intervention with no intervention, intervention may be unhelpful. But driving the private sector into holding foreign currency is a very odd way of trying to control money. (If they are regarded as close substitutes, it would be natural to target M3, in which case the effect disappears.) It would be very hard to argue that the effect on money *per se* was important (though it might save political embarrassment); the effect on the exchange rate would, however, influence the economy and, in particular, would be favourable to the control of inflation though damaging to international competitiveness.

The effects of exchange market intervention on money may be looked at more generally. In principle, intervention on any scale to lower an exchange rate (compared with what it would have been otherwise) is possible, though intervention to support the currency is limited by the amount of reserves and borrowing power available to the authorities. To see the feasibility, we have only to observe the massive accumulation of foreign assets by the low-absorbing OPEC oil producers. It is, however, often objected that intervention interferes with money and that the only external policy compatible with domestic monetary targetry is a free float.

This argument is false. It has already been seen that a policy of offsetting oil surpluses need have no effect on money. The point is even clearer if an exogenous shift in the external demand for sterling assets is considered—such as an increase in OPEC demand for sterling. In principle, the government can supply the additional sterling assets and take foreign assets in exchange. £M3 is unaffected. Even if the increased demand is for sterling bank deposits, and these are created via the exchange equalization account, the end-point is that externally held sterling balances rise. These are not part of the domestic money supply: not part of £M3 or M3. Particularly if gilt-edged stock were sold to foreigners for foreign currency, neutralization of the effects on the money supply—or 'sterilization'—is not only possible but automatic.

The problem, or supposed problem, lies elsewhere. When the authorities intervene, they may not know at the time whether they are transacting with foreigners or domestic asset holders. In the latter case, they may be taking foreign currency from the domestic private sector, and replacing it with £M3 deposits. £M3 rises, though M3 remains unchanged.

In fact, it may be asserted, the problem of sterilizing foreign exchange intervention is as easy or difficult as controlling domestic money. Intervention *per se* makes little difference. There may, however, be important indirect effects. Thus, as argued, a high exchange rate may induce negative external effects on money, though hardly in a predictable way, and may have effects on activity and prices that do affect money. But whether or not a high exchange rate policy is desirable in the circumstances of the time is a question that goes far beyond the feasibility of sterilized intervention.

Finally, the neutralizing policy outlined, even though it would have involved large-scale intervention, is quite different from a policy of fixing exchange rates according to some target. The suggestion is that known disturbances should, as far as possible, be neutralized. Apart from that, the exchange rate could be left to float. Thus the market would not be given one-way bets, which are damaging and may affect the money supply as domestic residents switch in and out of foreign currency. The problems associated with fixed exchange rates, or targets, should not be taken as implying that intervention, even on a very large scale indeed, is never justified.

The main problem with intervention is thus not the implied monetary effects. Rather it is that the policy might not work. Market operators, it could be argued, would see through the policy, and, what is more, the relevant elasticities might be large. There are also the circumstances when markets might be expected to 'get it right' without intervention. But in these circumstances, intervention would not do any harm; it would just be ineffective. On the other hand, if markets are as imperfect as they appear to be, a policy of non-intervention in the face of oil shocks could lead to substantial upward pressure on the exchange rate, as, in fact, occurred.

6. Conclusions

What would be thought of a government which, on assuming office, raised indirect taxes by an amount equivalent to some $2-2\frac{1}{2}$ per cent of GDP, using the proceeds to cut public borrowing, and then, some seven years later, cut those taxes by a similar order of magnitude? Presumably, the effects expected from the first move would be a substantial rise in measured inflation (particularly if other prices and wages were to react) followed by recession, which might then lead to a deceleration in price rises over time. The effects expected from the second move would be a fall in measured inflation and a substantial boost to wages and real incomes. There would be a consumer boom, and the effects of lower prices might or might not feed through into lower nominal wages.

Without attempting an answer to the rhetorical question, this is exactly what a passive policy to oil price impacts in the UK since 1979 would imply. Not to offset the fiscal impact of oil price changes within the United Kingdom amounts to a price-raising fiscal impact when oil prices rise, and a price-lowering one in the opposite direction when they fall. The main reason this is not generally appreciated is that fiscal policy is usually assessed by looking at the actions of the public sector. Fiscal policy should be looked at in terms of the position of the private sector. Alternatively, a change in taxation that is the automatic result of an exogenous change in the international oil price is as much a policy change as an explicit and well-publicized alteration in domestic tax rates—such as the reduction in direct taxes and rise in value added tax that occurred in 1979 (which also added to the price level in 1979–80).

It has been suggested in this paper that a neutral policy towards the fiscal impact of oil market events would have involved the public sector offsetting the impacts on the private sector by reductions in other indirect taxes, such as VAT. The main point of analysing such a strategy was not to suggest that a neutral policy should have been adopted, but to make it clear that a neutral policy was not adopted. Compared with such a policy, actual policy tightened dramatically in 1979, and prices were substantially raised. The effects on the British economy seem entirely predictable, and there is no paradox any more about the apparent similarity of response after the first oil shock (when all oil was imported) and after the second (when the UK was self-sufficient). The impact was, to all intents and purposes, the same. Fiscal policy thus appears as a major reason for inflation and recession in the early 1980s, and fiscal policy is a major reason for the consumer boom in 1986 and 1987.

So far so good. But that is not the end of the story. A principal feature of the recession of the 1980s was the rise in the real exchange rate of sterling, which concentrated the impacts on the exposed sectors of manufacturing industry. Two questions arise. The first is whether the rise in the exchange rate was due to Britain's position as an oil producer. The second is whether the exchange rate impact could have been offset. Both are inherently difficult to answer, due to the large degree of uncertainty that persists over what it is that does determine exchange rates.

One account of the exchange rate rise can be dismissed. The argument that North Sea oil constituted a bonanza for the UK and that the exchange rate rise was nothing more than a desirable concomitant of the need for structural change away from manufacturing is false. Unfortunately, however, though the structural change argument does not stand up, this does not mean that the exchange rate was unaffected by oil market events. If it was, however, then this was undesirable from the point of view of the longer-term health of the British economy. Though the factors influencing the exchange rate are not well understood, there is no such uncertainty about the consequences. In broad terms, a high exchange rate makes the traded goods sectors uncompetitive internationally, and though good for inflation and real wages, is likely to lead to just the kind of pressure on the manufacturing base that was observed. Unemployment and a deterioration in the underlying trade balance are the more or less inevitable result. And in the long run, that underlying deterioration has to be put right if the UK is not a net oil exporter on a long-term basis.

It was suggested above that the effects on the UK's exchange rate are better seen as resulting from the *international* impact of oil price rises, rather than due to the effect on the UK itself (which, unlike most countries, is not much affected by oil price changes). Changes in the world distribution of income and wealth towards oil-producing countries could lead to upward pressure, especially in the short term. (In particular, *relative* current account effects and asset market influences could favour sterling in international markets.) If this occurs, the real exchange rate is driven to the wrong level from the point of view of longer-term equilibrium. (It may be noted that this is not something that could easily be put right by lower nominal wage demands and lower domestic inflation; on the contrary, such a response would, if anything, increase the attractiveness of sterling to asset holders.) It is not easy to be at all precise about how large the effect would be; but it seems reasonable to suppose that the effects 'should' be small in the longer term, once the 'rest of the world' had adjusted.

Potentially more serious effects on the exchange rate arise from the extraordinarily uneven and 'humped' profile of North Sea oil production. Again, if markets were perfect and rational and forward-looking, with a clear 'view' of the longer-term 'fundamentals', then they 'should' discount the uneven profile of current account effects, so that production changes (and price changes) should have little effect. The history of what has happened to the exchange rate since North Sea oil first started to contribute substantially to the current account in 1976 suggests that markets do not behave in this way. (Thus the build-up of oil production should have been anticipated; nevertheless the exchange rate started to move up as production built up, not before.)

This paper has suggested an obvious 'neutralizing' or 'offsetting' strategy, which would have involved the public sector in accumulating foreign assets when oil production was above the longer-run sustainable level, which could then have been used to buy oil in the future when production fell below consumption needs. (This is not the only way in which it could have been done—depletion policy could have been used to limit production to sustainable consumption, or imported foreign-produced equipment could have been installed in the UK. It is, however, conceptually the most straightforward.) The appropriate amount to invest overseas in this way during the 'hump' in production is effectively the government's tax receipts on net exports. Such a policy amounts to balancing the current account impact of oil (the varying impact) with public sector capital account transactions of an offsetting kind.

Such a policy, especially if combined with the fiscal offsets as described above, can be seen as isolating the private sector's financial position from the vagaries of the oil profile. There is an immediate implication. If the private sector's finances are unaffected by oil, there should be little effect on other financial indicators, such as the quantity of money. Compared with what actually happened, however, it would have involved a major build-up, during the period of oil surpluses, of foreign assets held by the public sector (cumulating to many billions of dollars) to be run down later. (In practice, the public sector did move in this direction by paying off previously incurred foreign borrowings; otherwise policy amounted to a relatively free float.)

A policy of not offsetting the current account impact of variations in net trade in oil would be likely to impose severe strains on the rate of exchange. Evidently, a capital outflow must balance a current account surplus in some way or other. If the public sector does not accumulate the foreign assets, the private sector must. This involves switching private sector asset portfolios out of sterling assets and into overseas assets, and/or persuading foreign holders of sterling to take 'dollars' instead. The exchange rate has to move to the point where these adjustments take place willingly; the danger is that the exchange rate might have to move quite a lot and overshoot to bring them about. Similarly, in the future when oil again has to be bought from abroad, with net exports turning negative, a large fall in the rate of exchange might be necessary to induce the 'repatriation' of foreign assets and to finance current deficits at that time. (Note that the abolition of exchange control in 1979, stimulating or facilitating private sector switches to holdings of foreign assets, mitigated upward pressure on the exchange rate. It has the disadvantage, however, of making it harder to control outflows or induce inflows when conditions change.)

Thus an offsetting policy of public sector accumulation of foreign assets has several advantages. First and foremost, it would work against likely upward pressure on the exchange rate during the period of net exports. Second, the public sector would have control over the stocks of foreign assets and could use the interest on them or run them down to finance deficits and smooth adjustments (e.g. when expectations of oil prices changed), stabilizing the exchange rate in line with longer-term objectives. Third, and this may seem somewhat paradoxical, the main objection to public sector intervention is that it would not work, in that markets might see what was happening and 'discount' the policy. This would be likely to happen if markets were indeed able to judge the fundamentals correctly and were subject to no market failures. But, in that extreme case, the public sector policy would not do any actual harm: the authorities would neither gain nor lose. In the more likely case, intervention on the scale suggested would lower the exchange rate compared with a policy of non-intervention.

Again, it needs to be stressed that the purpose of considering the form of an offsetting strategy is to bring out the non-neutrality of the policies followed. In practice, offsetting exchange rate impacts is unlikely to be at all simple. To start with, anticipations are important (even though they may drive the market to undesirable positions) and a policy of building up reserves and other public sector holdings of foreign assets would probably have had to start well before the major current account effects. It was, after all, the Labour government that 'uncapped' the exchange rate in 1977, and there was a substantial rise before 1979. Second, some impacts, as we have seen, emanate from abroad. Nonetheless, these too could have been mitigated, and the direction in which policy would have to move is clear enough. Thus world recession effects could be partially offset by expansionary policies at home, low interest rates, and a policy of supplying sterling assets and taking foreign assets in return. All this suggests that, in practice, policy would have had to go further than the simple balancing strategy outlined if major rises in the real exchange rate were to be avoided. In practice, on every count except the abolition of exchange controls and repayment of public sector foreign borrowing, it went in the opposite direction. Thus interest rates rose relative to other countries, and the recession in the UK was (in part due to not offsetting the impact of oil market events) deeper and earlier than elsewhere.

Thus the contention here is that oil market events and the policies followed combined to produce results that are entirely predictable. A price-raising fiscal tightening (with a further impact on the price level from the switch to indirect taxation) accounts for much of the initial inflationary/deflationary response. High interest rates and a policy of not offsetting the temporary current account impact of net exports of oil allowed the exchange rate to become grossly overvalued, far above its long-term equilibrium level. The combination of price inflation, real wage increases, unemployment, and 'deindustrialization' is just what would be expected. By contrast, an offsetting policy would have neutralized the inflationary/deflationary impact, and stemmed at least some of the rise in the exchange rate. Moreover, the implied build-up of foreign assets by the public sector would have acted as a cushion, both to exchange rate volatility and to future problems of adaptation to lower oil production.

Does this mean that North Sea oil is to blame for the decline in the manufacturing base and the deterioration in the underlying balance of payments position in the 1980s? Surely not. Though North Sea oil was not a 'bonanza' in the sense that it did not make the UK better off than it was before, it did mean that the UK was, in a longer-term sense, substantially isolated from the losses that applied to competitor countries. Thus the set of policy options was much wider in the UK. Effectively, the UK was given, through an accident of nature, a long time to adapt, denied to others.

But the impacts of North Sea oil as oil prices and production changed were large. The interaction of these essentially short-term impacts with the policies followed does go a long way to explain what happened.

One sense in which oil market events might be 'blamed', therefore, is that the policy problems posed were so acute or so unusual that policy was bound to be passive and to 'get it wrong'. This is to see what happened as essentially accidental and unintended. More plausibly, perhaps, the interaction between the kind of policy framework in operation at the time and oil market events can be seen as unfortunate. Thus an emphasis on internal financial targets when the exchange rate was being affected by North Sea oil may have meant that a policy that was meant to be gradualist turned out to be draconian (Laidler (1985)). Moreover, a concentration on the public sector borrowing requirement when the non-oil private sector was hit by oil before tax revenues accrued to the government—so that when the 'fiscal adjustment' came through it was effectively pre-empted to finance the resulting unemployment—may have led to an unintentional degree of fiscal squeeze. Certainly the view that markets would automatically get it right in the absence of policy action, even in the face of major shocks, has hardly survived the events of the early 1980s.

Yet another view is that the policy stance was intended, and justified due to the need to curb long-term inflationary pressures. In some versions, the shock administered to manufacturing industry and to the labour market is seen as the main purpose of the policies followed; in others, it was the result of a situation which turned out to be much worse than imagined. On these views, policy took account of North Sea oil. Without its impact, policy would have been different, to produce similar results.

Whether this last view is correct or a specious *ex pòst facto* justification of events, the policy framework of the early 1980s has now been more or less completely abandoned. It was this framework, interacting with oil market events, that accounts for deindustrialization, unemployment, and the deterioration of underlying trade performance. Thus it is possible to argue that the decline of manufacturing industry was an accidental result of North Sea oil. It is possible to argue that it was intentional, or even that it was desirable. It is not, however, possible to argue that it was inevitable.

20 The British Experience of Taxing Oil Extraction*

MICHAEL DEVEREUX

1. Introduction

The discovery of oil on the UK Continental Shelf in the early 1970s raised two important questions: whether, and by what method, the government should control the exploitation of the resource; and how the potential profits on its extraction should be shared between the government and the companies undertaking the extraction. One possibility was to follow the example of the gas and electricity industries by creating a publicly owned company to exploit the oil. A second was for the government to sell the rights to exploit particular areas of the North Sea. A third was to rely on taxation. Although all three of these possibilities have been tried at various times since, the government's principal role in the North Sea has been through taxation.

In 1975, just as the first oil was being extracted from the North Sea, the UK government introduced an Oil Taxation Act. This created a new tax, petroleum revenue tax (PRT), which was designed to capture a share of North Sea profits for the government. It was introduced alongside two existing charges, licence royalties and corporation tax. Over the next twelve years, that tax structure changed significantly no less than sixteen times. Two other taxes have been introduced and abandoned. Royalties have been abandoned for fields starting development after April 1982. PRT has become progressively more onerous, with lower reliefs and higher rates. At the same time it has become more complex. As successive Chancellors have adjusted the system in line with their short-term aims, the taxation of North Sea oil has become one of the most complicated fiscal regimes in the UK.

It might be supposed that successive alterations to the system have been creating a finely honed instrument, carefully attuned to the problems facing oil producers. One of the central arguments of this paper is that this is far from the case. In fact, as problems arose with the existing structure, more and more *ad hoc* adjustments were made to correct them. Not surprisingly, such *ad hoc* adjustments eventually created their own problems. That this process continues is vividly illustrated by the 1987 reforms—10 per cent of development expenditure on new fields can now be offset against tax liabilities on existing fields. This move was born out of fears of new developments being too few. It does nothing to correct the underlying fundamental weaknesses of the tax

* Michael Devereux is a Programme Director at the Institute for Fiscal Studies.

system which discourage new developments; it does, however, create new distortions between companies with and without existing tax liabilities.

This paper is concerned with the relatively narrow question of the appropriate taxation of North Sea oil production. It is concerned with the objectives that the government might have for collecting revenue and for the rate of development and production, and the ways in which the tax system might be used to achieve those objectives. It does not address wider issues relating to the appropriate policy for energy as a whole. For example, it leaves to one side issues concerning the appropriate taxation of oil consumption relative to the consumption of coal and gas. Also, it only briefly deals with other methods of government control of North Sea activities.

The next section examines briefly what economic theory suggests are the optimal arrangements for the taxation of the extraction of mineral resources. It considers arguments for neutrality and the widely supported resource rent tax. Section 3 then outlines the history of the development of the UK taxation of North Sea oil. Section 4 merges the themes of these two sections to ask how well the tax system has met the requirements and objectives that might be asked of it. The paper ends with a brief conclusion.

2. Optimal Taxation of Production

There is a substantial measure of agreement between economists concerned with energy and fiscal issues as to the appropriate taxation of a non-renewable resource. There are two central features of this consensus. First, the tax should be neutral—that is, the imposition of tax should not affect producers' decisions on the speed and scope of their exploration, development, and production. Many authors have argued in favour of a neutral system (for example, Part Committee (1981), Kemp and Rose (1982), Garnaut and Clunies Ross (1983), and Newbery (1985)). The aim of a neutral tax system does not necessarily imply that the market outcome which would occur in the absence of taxation is optimal. However, it does imply that the tax system should not be used to induce a change in the market outcome. Although it may be the case that oil companies do not act in a way that optimizes social welfare, using the tax system to induce a socially beneficial change in their behaviour is unlikely to be successful.

A second area of consensus is in favour of a resource rent tax. The basic features of such a tax are as follows. All expenditure is deductible in computing the tax base. In particular, there is no distinction between current and capital expenditure (equivalent to 100 per cent first year allowances for investment expenditure in the corporation tax system). In its pure form (sometimes referred to as the 'Brown tax': Garnaut and Clunies Ross (1983)), a resource rent tax is a symmetric tax on cash flows. Any negative flows (e.g. during developments) would yield taxable losses and, in turn, a negative tax payment. Positive flows would be taxed at the same rate. In practice, such a tax would make the government equivalent to a shareholder in the company, or partner in the field, the size of the shareholding being determined by the tax rate. When a field is developed, the government effectively contributes part of the cost. When the returns are earned, it collects the same proportion of the profit. This implies that the size of any project for the company is scaled down by the size of the tax rate. However, since all flows are scaled down, the internal rate of return earned by the project is unchanged. The drawback with this form of the tax is, of course, the high degree of risk for the government. This is because it contributes to the cost of development of all fields, irrespective of their profitability. It shares equally in loss-making, as well as profit-making, enterprise. (Low risk might be considered to be a subsidiary aim of a tax.)

The more popular form of the resource rent tax is one in which taxable losses can be carried forward, with some interest mark-up, to be offset against future profits. This eliminates the need for the government to contribute directly to development costs. The crucial feature of such a system is the rate of mark-up in carrying forward taxable losses. If the mark-up is equal to the internal rate of return (IRR), then the post-tax IRR will equal the pre-tax IRR. If it is equal to the firm's discount rate, then the tax will be neutral in that a positive present value of a project cannot be made negative after tax. Since the project remains profitable in the presence of tax, therefore, we treat the tax as neutral. (This would not necessarily be the case if there existed other profitable opportunities for investment, but scarce resources available to the company.) In fact, the present value will simply be reduced by a proportion equal to the tax rate. However, if the mark-up is lower than the firm's discount rate, it is possible that the tax may not be neutral in the sense defined. It is therefore crucial to set an appropriate level for this mark-up.

Below, we attempt to assess the history and structure of the UK taxation of North Sea oil in the light of these consensus views. We shall be mainly concerned to assess how neutral the system is, and has been. However, before considering how the taxation of North Sea oil developed in the UK and analysing its faults and weaknesses, it is necessary to raise some questions concerning the consensus. Two main questions arise. First, is neutrality the appropriate aim for the UK government in the mid-1980s (and should it have been in the 1970s)? Second, if neutrality is an appropriate aim, is the resource rent tax the best way of achieving it?[1]

Neutrality

A persuasive case against neutrality would require two arguments. First, it must be shown that there exist some market failures which the government should legitimately aim to correct. Second, it must be shown that the tax system is the most appropriate method of correction. This paper argues that, although it is

---

[1] These questions are more fully discussed in Bond, Devereux, and Saunders (1987).

possible that the first argument may hold, the UK experience of taxing oil production casts doubt on the second.

There is a large economic literature on the possibility of market failures in the energy industry (for a brief discussion see, for example, Newbery (1985)). Indeed, what is more at issue is not the existence, but rather the significance, of such failures. We do not attempt here an exhaustive discussion of such issues. Rather, we confine ourselves to a somewhat selective list of market failures which may lead to the need for corrective action against the behaviour of the oil industry by the government in the UK.

Perhaps the best case for government intervention into the exploitation of oil is the possibility that market imperfection may lead companies to under-invest. One reason for this is that risks to society are objectively lower than risks to individual investors. Individual investors are concerned with the success of particular projects in which they have an interest, whereas society is concerned with all such projects. Suppose, for example, that there are two potential oilfields that may be explored. It is known that only one contains oil. If a single company explores both, it can offset the cost of failure in one against the success of the other. However, if the fields are controlled by different companies, the possibility of failure in each company's project may deter any exploration. Since society as a whole has no interest in the individual control, it would prefer both to be explored. Thus, if the second case holds, there may be 'too little' investment.

Strictly, this argument would be insignificant if there existed insurance markets in which companies could insure the risk of failure. In practice, however, such markets do not exist, perhaps because of information asymmetries. A related argument is that, because capital markets are imperfect, companies are obliged to discount potential earnings on risky projects at a higher rate than society as a whole does. This would lead companies to undervalue potential projects, and hence to under-invest.

The scale of this problem for the UK is difficult to judge. The principal oil companies already diversify considerably, and are therefore able to spread their risk (indeed, by diversifying across countries, BP may be less concerned about a single dry hole in the North Sea than the UK government is). In addition, their credit-worthiness is extremely good. This suggests that the problem is small for the UK. On the other hand, however, companies do tend to discount future earnings at very high rates. If the government did wish to correct what it believed was under-investment, it has several different channels open to it. First, through its licensing policy, it could concentrate control of projects in the hands of the larger, more diversified, companies. It could increase public ownership. It could offer a subsidy in the form of a grant, or through the tax system. At first sight, the last option seems attractive. However, this paper aims to point out that the simple concept of paying a subsidy through the tax system is, in practice, likely to conflict with the other aims of tax legislation, and to result in a confused and complex tax system.

There are other reasons why the government may wish to influence the behaviour of those operating in the North Sea. An implication of the neutrality argument, spelt out more fully in Newbery (1985), is that the government should aim to maximize the economic rent earned in the North Sea. Subsidiary to this aim is the question of the distribution of this rent between the government and the oil companies. A commonly held view is that, given that the resource is, in effect, nationally owned and that it is possible to tax away a high share of the rent without distorting behaviour, the government should aim to maximize its share of the rent (see Allsopp (Chapter 19, this volume)). If this is accepted, then the production profile preferred by the government should depend, among other things, on its expectations about future prices of oil. If it expects that in the early 1990s the world oil price will quadruple, it should aim to delay production until then. On the other hand, if it believes that the price will remain at $15 until the end of the century, it should aim to produce as quickly as possible. In general, the rate of depletion should depend on the relative size of the government's expectation of price movements and the social discount rate.

However, companies do not share the same objective. Let us assume that they wish to maximize the present value of their share of the economic rent available world-wide. In the absence of differential taxation in different countries, they would develop the cheapest areas first. This is true irrespective of price expectations, as long as there are limited resources available to companies for development, although the speed of extraction world-wide would vary with company expectations of the world oil price (among other things). If the UK government believed prices would remain low, it might wish to compete with other governments to attract companies which would otherwise use their scarce resources to develop other areas. It might therefore be rational for the government to consider a system of taxation that allowed relatively high post-tax profitability, or even effectively to subsidize company returns, allowing post-tax returns to be higher than pre-tax returns. This latter course does not imply that the government would not gain any share of the rent from the North Sea, only that its own effective internal rate of return would be lower than that of the companies. If the government wished to delay production, it could do the opposite.

Of course, the situation is unlikely to be as simple as described. In particular, we might expect other governments to retaliate by introducing similar incentives. Taken to its logical conclusion, with perfect knowledge, governments would end up competing away the whole of the rent available from taxation world-wide—involving no change in company behaviour, but allowing companies to claim all of the rent. Such a course of action by a government is therefore rather dangerous. In any case, the argument here is not that the government should aim to adjust the market outcome, merely that there might be cases in which it would wish to do so. However, this is not a sufficient argument for non-neutrality of the tax system because of the complexities which inevitably arise in tax systems.

Neutral Taxes

The second question raised above was whether a resource rent tax is the best way of achieving neutrality. Before considering alternative tax systems, it is worth briefly examining two other methods by which the government could capture a share of the economic rent of the North Sea. One possibility would be to extract the resources of the North Sea directly, through a pubicly owned company. To some extent this was done in the late 1970s, through the British National Oil Corporation (BNOC) and, until recently, was the case for gas production. The rent earned by this company could be directly channelled to government. Depending on the degree of independence given to it, the government could intervene directly to control production levels—unlike operating through the tax system. However, if, as Rees (Chapter 5, this volume) argues, nationalized industries act as output, rather than profit, maximizers, then the level of economic rent earned under such a system may be lower. Such a policy would also involve higher risk for the government, since it would have to finance all exploration and development, with no guarantee of a profitable return.

Another alternative to a resource rent tax is some form of fee paid for the rights to exploit a particular area. There are a number of ways in which this could be operated; the crucial feature for neutrality of such a fee is that its size should be independent of the actions of the companies that pay it. If it is not, companies could alter its size by their actions; the extent to which there is an incentive to do so introduces a degree of non-neutrality. However, as mentioned above, neutrality is not the only aim of governments; they can also aim to capture the highest possible share of the rent subject to not reducing the total rent. This adds an important requirement to what is required of the fee.

It has been suggested that if licences to develop and produce from an oilfield were auctioned in a competitive market, then the government would be guaranteed virtually 100 per cent of the expected rent, without distorting companies' actions. This is because each company would be willing to pay just up to the present value of its expected rent. The most efficient company would have the highest expected rent, and would therefore gain the licence, paying its present value to the government. If this were true, such an auction would be an extremely simple way for the government to capture the rent. However, two disadvantages have been suggested. First, the uncertainty faced in developing an oilfield is immense—the size of the resource and ease of recovery are not known with any certainty, nor are movements in the oil price up to twenty years ahead. This uncertainty will reduce the amount that the bidders in the auction will be prepared to pay for the licence. Given that fields are eventually successfully developed, this would reduce the government's share of the eventual rent. It has been suggested that such an auction could therefore be combined with a resource rent tax. While this could, in principle, increase the share of the rent collected by the government, and reduce the problems caused by a very high rate of resource rent tax, two factors will further reduce the amount that companies would be willing to pay at auction. First, they would obviously only be willing to pay an amount equivalent to their share of the rent after the resource rent tax. Second, as Newbery (Chapter 2, this volume) and others have pointed out, uncertainty will be increased further if companies believe that the government may increase the rate of taxation in the future. Offsetting these points, it should be noted that a fee will be paid for all potential fields, whether eventually profitable or not. However, since it would be inappropriate to auction rights after exploration (otherwise there would be no incentive to explore), it is likely that the amount raised would be considerably below eventual economic rents.

A further problem that has been raised with such an auction is that the winner might not be the most efficient company, but the most mistaken company. It is possible that the most optimistic company values the field above its actual value, and is therefore willing to pay more than the total rent available from the field. However, it is interesting to note that this need not affect actual development. By the time the company realizes its mistake, the fee it paid at the auction would be a sunk cost, and ought not to influence its future decisions. The auction system therefore remains neutral.

There is, therefore, a case for having some form of fee to be paid for the right to extract a resource. However, it cannot replace further taxation if the government wants to maintain a large share of the rent. As long as there is a competition between companies, an auction is likely to raise a larger sum than a fee determined by the government. If the fee were set higher than the highest price offered at auction, no company would be willing to pay it.

In considering alternative tax systems, we can immediately put to one side a number of issues that arise in the literature concerning the appropriate taxation of non-renewable resources, on the grounds that they concern possible tax rules that are not, and are not intended to be, neutral. Examples are investment incentives or the granting of tax holidays as a means of encouraging oil companies to develop in a particular country (see, for example, Galenson (1984) for a discussion of such measures). Here we are concerned only with possible alternative neutral systems of taxation.

A possible argument against a simple resource rent tax rests on a potential conflict with the aim of maximizing the government's share of the rent. As shown below, some fields in the North Sea earn extremely high internal rates of return—Forties, for example, has a rate of over 50 per cent (we ignore, for the moment, problems of using the internal rate of return as a measure of profitability). Others have a much lower rate. However, it is commonly suggested by companies that they require a relatively low return of only 10 or 15 per cent if they are to develop a field. If this is true, then there is no need for the government to treat Forties in the same way as a marginal field which would only earn, say, 15 per cent in the absence of tax. The government could claim a higher share of the rent from Forties, and allow the internal rate of return earned by BP to fall substantially. As long as the rate did not fall below BP's cut-off rate, it would continue to produce from the field. Marginal fields, however, require at least a neutral tax, or possibly a tax that increases their return while allowing the government a smaller share of the rent.

This is, in effect, an argument for a progressive tax, whereby the government can capture a larger share of the rent from the more profitable fields. The aim of progressivity does not necessarily conflict with that of neutrality. Using our definition of neutrality that the tax does not cause a positive present value (or rent) of a project to become negative (see Garnaut and Clunies Ross (1983)), it is still possible for the government to capture a higher share of the rent for more profitable fields, without introducing non-neutralities. Schemes have been suggested to introduce progressivity into the structure of a resource rent tax (e.g. Part Committee (1981), Devereux and Morris (1983), Sumner (1978), and Garnaut and Clunies Ross (1983)). The basic principle is that a higher tax rate is imposed on fields with proven higher profitability.

We have argued, then, that in theory there are occasions on which a simple resource rent tax would be inappropriate—either because the market outcome is suboptimal, or because the government does not maximize its share of the rent. There is, however, a powerful argument in favour of relative simplicity of the tax system. That is, that the UK government has, over the last twelve years, failed to create even the relatively simple, neutral, and progressive system that it has aimed to. There can be little doubt that successive governments since 1975 have aimed to avoid discouraging development. The 1975 Oil Taxation Act was launched by Robert Sheldon, as 'a very workable scheme, which is understood by the oil industry and which gives the industry a fair share of the proceeds to make sure not only that it receives the fruits of equity, but that it continues and increases risk investment in the North Sea'. This approach was continued by Geoffrey Howe, who claimed in 1980 that 'I am confident that none of these changes will deprive companies of a fair return on the North Sea projects and exploration' and in 1983 that 'I believe that my proposals will provide the industry with the right fiscal incentives for the further successful development of the country's North Sea resources'.

It is also clear from the tax system that various reliefs are intended to introduce some progressivity into the system. However, the evidence that even such relatively simple aims have not been met is given below. It suggests that if we cannot successfully create a neutral system, we should be wary of the dangers of trying to create a beneficially non-neutral system.

3. The British Experience

At the time of writing, it is just over a decade since the first production of oil from the UK Continental Shelf. In that time the North Sea tax regime has been changed significantly at least sixteen times. There have been five separate taxes, each with its own set of rules, exemptions, and special conditions. It is extremely complicated and contains many anomalies. It has been argued that 'in defence, it might be claimed that the tax was being adjusted in the direction of greater efficiency and effectiveness, so that each adjustment reduced the fear of large future adjustments' (Newbery (Chapter 2, this volume)). However, such a defence would be entirely mistaken.

Petroleum revenue tax (PRT) was introduced in 1975 and designed as a resource rent tax. As applied, however, it had two serious deficiencies. It was imposed with two other taxes—licence royalties and corporation tax—the former, in particular, being a wholly unsuitable method of taxing mineral exploration activities. In addition, progressivity was introduced by arbitrary allowances which were intended, but in practice did not always turn out, to be of disproportionate assistance to relatively unprofitable fields. As the resulting weaknesses of the system emerged, they were tackled, not by returning closer to the resource rent tax, but by a never-ending series of *ad hoc* modifications to the tax base. It is rather as though, having attached a fifth wheel to a car, you then attempt to deal with the resulting problems by compensating adjustments to the steering mechanism. In this section I shall briefly trace through what happened to the tax system, and why.

The tax system as it began in 1975 included three taxes. Licence royalties were charged at $12\frac{1}{2}$ per cent of all revenue less some costs for transport and treatment of oil. There was and is no allowance for the bulk of costs of development or extraction. Licence royalties had been in existence for offshore activities since 1964. Corporation tax was charged on a similar basis to that for onshore companies. The 1975 Oil Taxation Act introduced a 'ring-fence', separating North Sea activities from those on shore. The main reason for the creation of the ring-fence seems to have been to restrict the opportunity to reduce profits made in the North Sea by losses made elsewhere.

The third tax, which was entirely new, was PRT. This was modelled on a resource rent tax, although there were some differences. All capital expenditure was deductible for tax, together with an 'uplift', of originally 75 per cent (i.e. 175 per cent of capital expenditure could be offset against tax). The uplift was partly in compensation for not allowing interest payments to be deductible. Losses could be carried forward, but only at historic cost—so that their real value was eroded before offset. Two other features of the original PRT system were designed to reduce the burden on marginal fields. First, there was an oil allowance, which allowed 1 million tonnes of oil per annum to be exempt from PRT, up to a cumulative limit of 10 million tonnes. Second, there was a 'safeguard' provision, which limited the PRT liability in any six-month chargeable period to 80 per cent of the amount by which gross profits exceeded 15 per cent of cumulative capital expenditure.

A number of comments can be made about this structure, which was left virtually unchanged until 1979—the longest period of stability in the history of the North Sea tax system. First, it is necessary to ask why licence royalties and corporation tax were used. Given that oil production was to be taxed separately from other company profits, why was PRT not introduced to stand on its own?

A large part of the reason was a desire to ensure that substantial sums would be paid in tax. Oil companies had avoided paying significant amounts of corporation tax for at least ten years before production in the North Sea, due to manipulation of double tax treaties (see Committee of Public Accounts (1972–3)). While the government seemed to be unsure about the revenue-raising potential of PRT, it was clear that it was impossible to avoid paying at least licence royalties. The retention of corporation tax had less of a rationale. Partly it was kept for the same reason of ensuring at least some revenue, and partly, it has been suggested, because otherwise companies would not be able to offset their advance corporation tax (ACT). Neither of these reasons is at all convincing—particularly the second, which implies that an unnecessary tax should be kept simply so that a further tax can be offset against it.

The retention of licence royalties is of particular importance for the neutrality of the system. Since they are a charge on revenue, with little allowance for costs, they are unrelated to the profitability and rent of a field. Its effect is to drive an arbitrary wedge between pre- and post-tax profitability, which for some fields could be a decisive factor in whether or not development should proceed. The main case against corporation tax is not that it is non-neutral, but that it adds unnecessarily to the complexity of the system, with the result that it is extremely difficult to judge the effects of any particular measures in PRT.

The two other features of the 1975 system that require comment are the oil allowance and 'safeguard' provision. These were clearly intended to introduce some progressivity into the system. This progressivity was required, at least partly, because less profitable fields were hit more severely by the retention of licence royalties. However, they did so in a haphazard way. There is only a low correlation between small fields (helped more by the oil allowance) and fields with low profitability. Further, the concept of the safeguard provision, which implies that there is a critical cut-off point for all fields determined by the relationship between gross profits and cumulative expenditure, is hard to defend.

The first major changes came in 1979 from the Conservative government, following 1978 proposals by the Labour government. They arose because of the perceived need to raise more revenue from the North Sea: it was felt at the time that not enough of the rent was accruing to the government. Consequently, the rate of PRT was raised from 45 per cent to 60 per cent, uplift was reduced to 35 per cent, and the oil allowance was halved. By 1980, the government was again in need of short-term revenue. In the 1980 Budget, the PRT rate was raised again, to 70 per cent, and the payment mechanism was accelerated. November 1980 saw the introduction of the fourth tax, supplementary petroleum duty (SPD).

SPD was charged on a field-by-field basis at a rate of 20 per cent of gross revenue less an oil allowance of 1 million tonnes per annum. Like licence royalties, it therefore penalized marginal fields relative to profitable ones, although this effect was mitigated by the oil allowance (which nevertheless was not aimed directly at marginal fields). Some fields paid SPD early in their lives, long before development expenditure was fully recouped. The fields hardest hit by SPD were therefore large, high-cost fields. The reason for introducing a tax on revenue constituted a central problem in the development of the tax system—the desire for short-term revenue. Since in the early years of production fields paid little PRT because of accumulated development expenditures brought forward, it was necessary to introduce SPD as an alternative way of raising money quickly.

The next change to the system reflected most of the themes of this section. The government considered that it could collect a higher share of the rent from the North Sea. An Inland Revenue paper (December 1980) suggested that the reliefs, introduced in 1975, were 'not doing what they set out to do', reducing tax more than intended. Instead of changing the system of reliefs, a further feature was added. This was the concept of 'payback'—the moment at which cumulative field income first exceeds cumulative allowable expenditure plus licence royalties and SPD. The new provisions, introduced in the Budget of 1981, were that uplift on capital expenditure was allowed only until payback, and that the safeguard provision should apply only for the period up to payback and half as long again. It is difficult to see how these changes could be described as 'in the direction of greater efficiency and effectiveness'. Certainly they led to higher government revenues, but their effect on the system's neutrality is difficult to assess even with complex computer models. Certainly the government seemed to have misjudged the response it expected from the oil companies, which began to announce delays and cancellations of development plans. Soon after the Budget, SPD became a temporary measure, to last only eighteen months.

By the 1982 Budget, however, the government expected its revenue to drop because of falling oil prices. As a result, the PRT rate was raised again, to 75 per cent, and payments were further accelerated into a monthly, rather than six-monthly, system. SPD was extended until the end of 1982, to be replaced then by advance PRT (APRT), which was almost identical to SPD. The only innovation in the switch was that any APRT that had not been offset against PRT liabilities by the end of the field's life was to be refunded. As with the rest of the system, APRT was to be carried forward only at historic cost.

This system, which was introduced following the 'careful study of proposals for the long term future of the fiscal regime' called for by Geoffrey Howe a year earlier, lasted three months. By then, company threats concerning new developments had gained the concession that APRT was to be charged for only five years, and would be refundable at the end of that period if not offset against PRT. A small change in the other direction saw APRT disallowed in computing payback for PRT.

The vacillation between the desire for short-term revenue on the one hand, and threats by companies concerning the effects of taxation on marginal fields on the other, finally switched in favour of the latter in 1983. By 1983 the government was finally persuaded that it was doing significant damage to the prospects for new developments in the North Sea—and this is borne out by the figures presented below. Consequently, it finally agreed to abolish licence royalties—but only for fields licensed after April 1982. At a stroke, this removed one of the fundamental weaknesses in the system, although by applying it only to new developments it did nothing to encourage full exploitation of existing fields when their costs increase towards the end of their life. In addition, the oil allowance was restored to its original level for new fields, exploration expenditures became offsettable against PRT, and APRT was to be phased out over four years.

From 1983 to 1986 there was a period of relative stability in the fiscal regime. 1986 saw the collapse of the world oil price, which once again threatened many developments in the North Sea. With a neutral tax system guaranteed not to reduce profitability below the cut-off level, there would be no need to change the tax system in the light of changing prices. In November 1986 the government accelerated refunds of APRT, but introduced a limit on the size of those refunds. It is difficult to see how this change could be seen to encourage further development of new or existing fields, other than to improve companies' cash flow positions. In the 1987 Budget, companies were given the option of offsetting 10 per cent of development expenditure on new fields against PRT liabilities on existing fields. This clearly benefits companies that currently pay PRT, and finally eliminates the strict ring-fence around each field for PRT purposes. This Budget also allowed general expenditure on research and development to be offset against PRT—a move aimed at helping to reduce development costs.

Successive governments have had two principal aims from the taxation of North Sea oil: to maximize government revenue, but to avoid discouraging new development. The importance of each has varied with different circumstances over the years. As perceptions of the relative importance varied, numerous additions and other changes were made to the tax system. It is difficult to believe that a tax legislator coming afresh to the area would propose the system with which we have ended up.

4. The Effects of the British System

Three particular questions arise from this description of the history of the British North Sea tax system. First, what has been the relative importance, in revenue terms, of each of the taxes? Second, how neutral and how progressive is the current system? Third, how has the neutrality and progressivity of the system changed over time? I intend to answer the first question partly by referring to published sources, but the others mainly by using a computer model of the tax system (described in Devereux and Saunders (1988)). The model uses data on production, capital expenditure, operating costs, and the ownership of each field. The rules of each tax system are applied to these data to estimate tax liabilities. Pre- and post-tax profitability, and hence an effective tax rate, can be calculated.

Table 20.1 shows the revenue collected from each tax from first production of oil until the mid-1990s. The most striking feature of this table is the rapid rise and decline in the importance of North Sea taxes for the government. A peak was reached in 1984-5, just nine years after first production, when these taxes accounted for 9.5 per cent of total government revenue. Since then, a combination of falling production and much lower prices has dramatically reduced revenues. It is unlikely that they will ever recover to their previous levels. Since the early years of production, PRT has raised the largest amount of revenue. Corporation tax, against which all the other taxes can be offset, took longer to reach a substantial figure. The delays in payment of corporation tax lead it to respond more slowly to changes in the oil price.

It was argued above that successive governments had aimed for a neutral tax system in the sense of wanting to avoid discouraging development. At the same time they wanted to maximize tax revenue. This led to attempts to create a progressive tax system, where marginal fields were given various reliefs in order

Table 20.1. UK tax revenues from the North Sea (£ million)

	Royalties	SPD	PRT[a]	Corporation tax	Total	% of total government tax revenue
1975-6	20	—	—	5	25	—
1976-7	71	—	—	10	81	0.2
1977-8	228	—	—	10	238	0.5
1978-9	289	—	183	93	565	1.0
1979-80	628	—	1,435	250	2,313	3.2
1980-1	992	—	2,410	341	3,743	4.4
1981-2	1,396	2,025	2,390	681	6,492	6.5
1982-3	1,632	2,395	3,274	521	7,822	7.2
1983-4	1,904	—	6,017	877	8,798	7.5
1984-5	2,426	—	7,177	2,427	12,030	9.5
1985-6	2,082	—	6,380	2,923	11,385	8.3
1986-7	476	—	1,680	1,400	3,556	2.4
1987-8	279	—	893	881	2,053	1.3
1988-9	248	—	1,144	454	1,846	1.1
1989-90	317	—	878	202	1,397	0.8
1990-1	730	—	2,504	194	3,428	1.8

[a]Including APRT.
Sources: Up to 1985-6: North Sea taxes—Inland Revenue Statistics. Total revenue from taxes and social security contributions—Economic Trends.
1986-7: Financial Statement and Budget Report (FSBR) 1987-8 and own estimates.
1987-8 onwards: North Sea taxes—own estimates. Total revenue from taxes and social security contributions—FSBR, 1987-8.

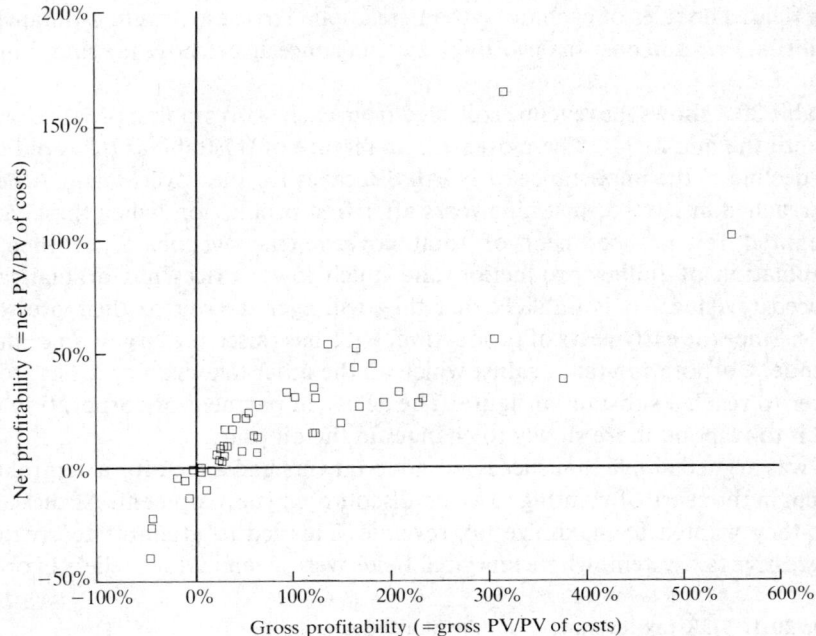

Note: PV = present value discounted at a real rate of 10% to start of life.

Figure 20.1 Gross and net profitability of oilfields

to reduce the gap between pre- and post-tax profitability. Figure 20.1 examines how well the tax system met these aims. It is based on the tax systems and prices actually faced by fields during their lives.

Figure 20.1 plots, for each of 51 oilfields, a measure of pre-tax profitability against post-tax profitability. The profitability measure is based on the expected net present value of the field over the whole of its life. To enable comparisons between fields, the net present value is scaled by the present value of development expenditure and operating costs. The net present values are calculated at a real discount rate of 10 per cent. The points on the graph indicate the different positions of the fields.

The dispersion of the points in Figure 20.1 indicates the arbitrariness of the system. No consistent pattern emerges whereby either all fields face the same tax rate (equivalent to all points being on a straight line through the origin), or the tax rate increases with profitability (equivalent to all points being on an upwards-sloping curve that flattens at higher rates of gross profitability). Rather, fields with the same profitability face very different tax rates. The degree of progressivity is therefore minimal. Neutrality is less easy to demonstrate. At the discount rate chosen, there are 4 fields with a positive gross profitability and a negative net profitability. If this rate and measure are reasonable, we can conclude that the system is not neutral. However, even ignoring these fields, the scale of dispersion of points suggests that it is unlikely that the system is so refined as to be neutral. It might also be noted that in no case is the net profitability higher than the (positive) gross profitability. Thus, the system never acts as a subsidy to development.

Part of the variation in the tax rates in this figure is due to the fact that fields that begin production at different times faced different tax systems. It is interesting also to examine the individual tax systems that have been in operation to see if the same features can be noted. This is done in Table 20.2, where each field is assumed to face the same constant real price over its life. In addition, we apply a number of the different tax systems to each field, on the assumption that the system applied for the whole of the field's life. We consider seven different systems, over the period 1975 to 1987. We present an average tax rate for the 28 fields considered, and also the averages of the 5 most profitable and 5 least profitable fields. The tax rate is defined using the same measures of profitability as for Figure 20.1. On these measures, the tax rate is simply the proportionate fall in profitability arising from taxation. Finally, a measure of the variability over fields is shown by the standard deviation.

Several points should be noted. First, in virtually all the systems shown up to 1987, when all the revenue-based taxes had been abandoned (for new fields), the 5 most profitable fields faced a lower tax rate than the least profitable. This implies that the progressive elements within PRT were more than outweighed by the regressive nature of the revenue-based taxes. This regressive feature was lessened as the burden of profit-based taxes grew over time. This pattern has now been reversed, since new fields face only PRT and corporation tax.

Another feature of the table, again up to 1987, is that the overall average tax wedge increased, as tax rates were increased and allowances reduced. The first move in the opposite direction was in 1983 with the abolition of royalties for new fields. At the same time, the variation in tax wedges also increased over time, suggesting differential effects on fields due to the arbitrary nature of the various reliefs and allowances.

In sum, the conclusions of Figure 20.1, that the tax system fails in the aims of both neutrality and progressivity, hold also for almost all of the different systems so far applied to North Sea oil production. The exception is the current system facing new developments. Here there is some progressivity. In fact many of the new marginal fields are unlikely to be profitable enough to pay any PRT, so that the only tax they will face will be corporation tax.

5. Conclusion

This paper has argued that, although there is a general presumption in favour of a neutral tax system for North Sea oil production, there may be occasions on which a government would want to improve on the market outcome. However,

Table 20.2. Tax rates under different systems (%)

	1975 Oil Taxation Act	1979 PRT @ 60%, uplift/oil allowance halved	1980 PRT @ 70%, advance payment	1980 with SPD	1981 with payback	1982 PRT @ 75%, monthly payments with PRT	1987 No royalty, APRT, or SPD
Most profitable 5 fields	58	67	71	72	73	77	71
Least profitable 5 fields	71	72	71	73	75	73	42
All fields	56	61	62	63	66	67	53
Standard deviation	13.9	14.7	15.3	15.5	16.6	17.4	20.6

Notes: Each figure represents an average tax rate for a group of fields.
Each tax system is modelled as if that system were faced for the whole of each field's life.
Source: Devereux and Morris (1983).

it is possible that such improvements could be achieved directly, rather than indirectly through the tax system. The enormous economic rents earned by some oilfields suggest that the government could increase its revenue by introducing progressivity into the tax system.

Successive governments have aimed for such a neutral yet progressive system. However, there has been continuing conflict between the desire to maximize revenue and the desire to avoid discouraging development. The resulting tax system has reflected this conflict. Instead of settling on a simple, neutral, and progressive tax, the system has elements designed to raise high revenue which have then been counteracted by *ad hoc* measures to encourage development. These measures have then on occasion been nullified by additional taxes designed to increase revenue, and so on.

The history of North Sea taxation provides a powerful case for aiming for a neutral system. This is that very quickly taxes become so complex that it is very difficult to know what the incentive effects of reform will be. Any reform is likely to have unintended consequences which might be beneficial only by chance. The evidence from the North Sea tax regime is that such consequences have not been beneficial. As a result of too many reforms, North Sea taxes are neither neutral nor progressive. Any change to the market outcome that the government desires can probably be more easily achieved outside the tax system.

References

Allsopp, C. J. (1985), 'Monetary and fiscal policy in the 1980s', *Oxford Review of Economic Policy*, **1**, 1, 1–20.
——(1986), 'The economic impact of North Sea oil in the United Kingdom', in J. R. Sargent (ed.), *Foreign Macroeconomic Experience: A Symposium*, Toronto: University of Toronto Press.
——and Mayes, D. (1985), 'Demand management in practice', in D. Morris (ed.), *The Economic System in the UK*, third edition, Oxford: Oxford University Press.
Artis, M. J., Bladen-Hovell, R., Karakitsos, E., and Dwolatzky, B. (1984), 'The effects of economic policy: 1979–82', *National Institute Economic Review*, 108, 54–67.
Atkinson, A. and Stiglitz, J. (1980), *Lectures on Public Economics*, Maidenhead: McGraw-Hill.
Atkinson, R. (1985), 'The energy policy mess', memorandum, London: Bow Group.
Averch, H. and Johnson, L. (1962), 'Behaviour of the firm under regulatory constraint', *American Economic Review*, **52**, 1052–69.
Bailey, E. E. (1986), 'Price and productivity changes following deregulation: the US experience', *Economic Journal*, **96**, 1–17.
Bank of England (1980), 'The North Sea and the United Kingdom economy: some longer term perspectives and implications', *Bank of England Quarterly Bulletin*, **20**.
——(1981), 'Factors underlying the recent recession', paper presented to the Panel of Academic Consultants 15.
——(1982), 'North Sea oil and gas: a challenge for the future', *Bank of England Quarterly Bulletin*, **22**.
Bates, R. W. and Fraser, N. M. (1974), *Investment Decisions in the Nationalised Fuel Industries*, Cambridge: Cambridge University Press.
Baumol, W. J. (1967), 'Reasonable rules for rate regulation: plausible policies for an imperfect world', in A. Phillips and O. E. Williamson (eds.), *Prices: Issues in Theory, Practice, and Public Policy*, Philadelphia: University of Pennsylvania Press.
——, Panzar, J. C., and Willig, R. D. (1982), *Contestable Markets and the Theory of Industry Structure*, San Diego: Harcourt Brace.
Bending, R. and Eden, R. (1984), *UK Energy: Structure, Prospects and Policies*, Cambridge: Cambridge University Press.
Blackaby, F. (ed.) (1979), *De-industrialisation*, London: National Institute of Economic and Social Research/Heinemann.
Bond, S. R., Devereux, M. P., and Saunders, M. (1987), *North Sea Taxation for the 1990s*, Report Series 27, London: Institute for Fiscal Studies.
Bongaerts, J. C. and Kraemer, R. A. (1986), 'The law and economics of the electricity and small scale combined heat and power generation', paper presented at the third annual meeting of the European Association for Law and Economics, Oxford, September.
Bradley, I. and Price, C. (1987), 'Regulation through an average revenue constraint', University of Leicester Discussion Paper 61.

References

Bradley, I. and Price, C. (1988), 'The economic regulation of private industries by price constraints', *Journal of Industrial Economics*, September.
Brittan, S. (1986), 'Privatisation: a comment on Kay and Thompson', *Economic Journal*, **96**, 33-8.
Brown, G. and Sibley, D. S. (1986), *The Theory of Public Utility Pricing*, Cambridge: Cambridge University Press.
Buiter, W. H. and Miller, M. (1981), 'Thatcherism: the first two years', *Brookings Papers on Economic Activity*, **2**, 305-79.
Burns, A., Newby, M., and Winterton, J. (1985), 'The restructuring of the British coal industry', *Cambridge Journal of Economics*, **9**, 93-110.
Byatt, I., Hartley, N., Lomax, R., Powell, S., and Spencer, P. (1982), 'North Sea oil and structural adjustment', HM Treasury Working Paper 22, London: HMSO.
Cairncross, A. (1986), *Years of Recovery*, London: Methuen.
Cambridge Economic Policy Group (1980), *Economic Policy Review*, Aldershot: Gower Press.
Central Electricity Generating Board (1982a), statement of case to the Sizewell B Public Inquiry, London.
——(1982b), proof of evidence to the Sizewell B Public Inquiry, CEGB/P/4, 'The need for Sizewell B' (and addenda, in particular CEGB/P/4, Add. 6), London.
——(1982c), proof of evidence to the Sizewell B Public Inquiry, CEGB/P/5, 'Scenarios and electricity demand', London.
——(1982d), proof of evidence to the Sizewell B Public Inquiry, CEGB/P/6, 'Fossil fuel supplies', London.
——(1982e), proof of evidence to the Sizewell B Public Inquiry, CEGB/P/8, 'Cost and performance estimates for new generating stations', London.
Centre Européen d'Entreprise Publique (1984), *Public Enterprise in the EEC*, Brussels.
Child Poverty Action Group (1986), *Welfare Rights Handbook*, London.
Commerzbank (1985), *Wem Gehört Zu Wer*, Hamburg.
Commission of the EEC (1984a), *Energy Policies of Member States*, Brussels.
——(1984b), *The Application of the Community's Energy Pricing Principles in Member States*, Brussels.
Commission on Energy and the Environment (1981), *Coal and the Environment*, 'the Flowers Report'.
Committee of Public Accounts (1972-3), *Parliamentary Papers*, **19**, 1.
Coopers and Lybrand Associates Ltd (1982), *Report on the Review of the Bulk Supply Tariff*, London: Department of Energy.
Corti, G. (1976), 'Electricity industries and the problem of size', *National Westminster Bank Quarterly Review*, August, 8-16.
Council of Europe (1985), 'Financial resources for local and regional authorities', Study Series—Local and Regional Authorities in Europe, 34.
Crew, M. A. and Kleindorfer, P. R. (1979), *Public Utility Economics*, London: Macmillan.
Dam, K. W. (1976), *Oil Resources: Who Gets What How?*, Chicago: Chicago University Press.
Dasgupta, P. (1986), 'Positive freedom, markets and welfare', *Oxford Review of Economic Policy*, **2**, 2, 25-36.
Davis, E. H. (1984), 'Express coaching since 1980: liberalisation in practice', *Fiscal Studies*, **5**, 1, 76-86.

References

Davis, E.H. and Kay, J. A. (1985), 'Extending the VAT base: problems and possibilities', *Fiscal Studies*, **6**, 1, 1-16.
Deloitte, Haskins, and Sells (1983), *British Gas Efficiency Study*, London.
Department of Energy (1982), 'Proof of evidence for the Sizewell "B" Public Inquiry', unpublished mimeo.
——(1983a), *Coal and the Environment*, Cmnd 8877, London: HMSO.
——(1983b), 'Department of Energy paper on energy projections methodology', unpublished mimeo.
——(1986), *Authorisation Granted and Directions Given by the Secretary of State for Energy to the British Gas Corporation under the Gas Act 1986,* London: HMSO.
——(1987), *Sizewell B Public Inquiry: Report by Sir Frank Layfield*, London: HMSO.
——(1988), *Privatising Electricity*, Cm 322, London: HMSO.
Department of the Environment (1974), *Coal Industry Examination: Final Report 1974.*
de Ru, H. J. (1985), 'Public enterprises in the Netherlands', *Annals of Public and Cooperative Economy*, **56**, 3, 313-43.
Devereux, M. P. (1986), *Oil Prices and Budget Strategy*, Report Series 21, London: Institute for Fiscal Studies.
——and Morris, C. N. (1983), *North Sea Oil Taxation*, Report Series 6, London: Institute for Fiscal Studies.
——and Saunders, M. H. (1988), 'The IFS North Sea tax model', mimeo, Institute for Fiscal Studies.
Diamond, P. A. and Mirrlees, J. A. (1971), 'Optimal taxation and public production I: productive efficiency', *American Economic Review*, **61**, 8-27.
Dixit, A. K. (1980), 'The role of investment in entry deterrence', *Economic Journal*, **90**, 95-106.
Domberger, S. and Piggott, J. (1986), 'Privatisation policies and public enterprise: a survey', *Economic Record*, **62**, 145-62.
Dowlatabadi, H. and Evans, N. (1986), 'Electricity trade in Great Britain—economic prospects and future uncertainty', *Energy Policy*, **14**, 35-45.
Dupuit, J. (1844), 'De la mesure de l'utilité des travaux publics', *Annales des Ponts et Chaussées*, 2ème series, **8**. Reprinted in English in K. J. Arrow and T. Scitovsky (eds.), *Readings in Welfare Economics*, Allen & Unwin, London, 1969.
Edison Electric Institute (1978), *Statistical Year Book*.
Edwards, R. S. and Roberts, R. D. V. (1971), *Status, Productivity and Pay*, London: Macmillan.
Electricité de France (1977), 'Direction de la production et du transport thermique', *Rapport d'Activité 1977.*
Electricity Consumers Council (1985), *Debt Collection, Disconnections and Electricity Consumers: Report on the Operation of the Code of Practice*, London.
Electricity Council (1981), *Review of the Structure of the Bulk Supply Tariff*, December, London.
——and British Gas (1982), 'Disconnection policy: a code of practice', London.
Energy Committee (1981), *The Government's Statement on the New Nuclear Power Programme*, Session 1980-1, HC 114-I, London: HMSO.
——(1983), *Combined Heat and Power*, HC 314, London: HMSO.
——(1984), *Electricity and Gas Prices*, First Report, Session 1983-4, HC 276, London: HMSO.

Energy Committee (1985), *The Development and Depletion of the United Kingdom's Gas Resources*, Seventh Report, Session 1984-5, HC 76, London: HMSO.

——(1986a), *Regulation of the Gas Industry*, First Report, Session 1985-6, HC 15, London: HMSO.

——(1986b), *Memoranda laid before the Energy Committee on the Coal Industry relating to the First Report of the Energy Committee, Session 1986-87 (HC 165), volume I*, Session 1985-6, HC 196-I, London: HMSO.

——(1986c), *CHP—Lead City Schemes*, Sixth Report, Session 1985-6, HC 488, London: HMSO.

Eurostat (1980), *Coal Monthly Bulletin*, 5.

Evans, N. L. (1981), 'Electricity supply modelling: theory and case study', Energy Discussion Paper 14, Energy Research Group, University of Cambridge.

——(1984), 'The Sizewell decision: a sensitivity analysis', *Energy Economics*, 6, January.

Fforde, J. S. (1983), 'Setting monetary objectives', *Bank of England Quarterly Bulletin*, 23, 200-8.

Forsyth, P. J. and Kay, J. A. (1980a), 'The economic implications of North Sea oil revenues', *Fiscal Studies*, 1, 3, 1-28.

——and——(1980b), 'The economic implications of North Sea oil revenues', Institute for Fiscal Studies Working Paper 10.

Foster, C. D., Jackman, R., and Perlman, M. (1980), *Local Government Finance in a Unitary State*, London: Allen and Unwin.

Galenson, A. (1984), 'Investment incentives for industry: some guidelines for developing countries', World Bank Staff Working Paper 669.

Garnaut, R. and Clunies Ross, A. I. (1983), *The Taxation of Mineral Rents*, Oxford: Oxford University Press.

Gaskin, M. (1978), *The Impact of North Sea Oil on Scotland*, London: HMSO.

Gibson, M. and Price, C. (1986), 'Standing charge rebates', *Energy Policy*, 14, 262-71.

Glyn, A. (1984), *The Economic Case Against Pit Closures*, Sheffield: National Union of Mineworkers.

Goodhart, C. A. E. (1986), 'The financial innovation and monetary control', *Oxford Review of Economic Policy*, 2, 4, 79-102.

Gravelle, H. S. E. (1984), 'Bargaining and efficiency in public and private sector firms', in M. Marchand, P. Pestieau, and H. Tulkens (eds.), *The Performance of Public Enterprises*, Amsterdam: North-Holland.

Gregory, R. G. (1976), 'Some implications of the growth of the mineral sector', *Australian Journal of Agricultural Economics*.

Haldane, G. (1938), 'Power', in M. Cole and C. Smith, *Democratic Sweden*, London: George Routledge.

Halvorsen, R. (1976), 'Energy substitution in US manufacturing', unpublished mimeo.

Hammond, E. M., Helm, D. R., and Thompson, D. J. (1985), 'British Gas: options for privatisation', *Fiscal Studies*, 6, 4, 1-20.

——, ——, and ——(1986), 'Electricity tariffs and the 1983 Energy Act', Institute for Fiscal Studies Working Paper 86.

Hannah, L. (1979), *Electricity Before Nationalisation*, London: Macmillan.

Harlow, C. (1977), *Innovation and Productivity under Nationalisation*, London: George Allen and Unwin.

Hawdon, D. and Tomlinson, M. (1982), 'Energy demand models in the USA and UK', Surrey Energy Economic Discussion Paper 8.

Helm, D. R. (1986), 'The economic borders of the state', *Oxford Review of Economic Policy*, **2**, 2, i–xxiv.
—— (1987a), 'RPI – X', *Public Money*, **7**, 1, 47–51.
—— (1987b), 'Nuclear power and the privatisation of electricity generation', *Fiscal Studies*, **8**, 4, 69–73.
—— (1988), 'Mergers, take-overs, and the enforcement of profit maximization', in J. A. Fairburn and J. A. Kay (eds.), *Mergers and Merger Policy*, Oxford: Oxford University Press.
Henney, A. (1985), 'Regulation of the gas industry', in Select Committee on Energy, *Regulation of the Gas Industry*, First Report, Session 1985–6, HC 15, London: HMSO.
—— and Thompson, D. J. (1986), 'The role of public service commissions in facilitating the development of combined heat and power generation in the US', memorandum of evidence in Select Committee on Energy, *CHP—Lead City Schemes*, Sixth Report, Session 1985–6, HC 488, London: HMSO.
Hicks, U. (1938), *The Finance of British Government*, Oxford: Oxford University Press.
Himmelmann, G. (1985), 'Public enterprise in the FRG', *Annals of Public and Cooperative Economy*, **56**, 3, 365–93.
HM Treasury (1961), *Financial and Economic Obligations of the Nationalised Industries*, Cmnd 1337, London: HMSO.
—— (1967), *Nationalised Industries: A Review of Economic and Financial Objectives*, Cmnd 3437, London: HMSO.
—— (1978), *The Nationalised Industries*, Cmnd 7131, London: HMSO.
—— (1986), *Accounting for Economic Costs and Changing Prices*, 'the Byatt Report', London: HMSO.
Hotelling, H. (1938), 'The general welfare in relation to problems of taxation and of railway and utility rates, *Econometrica*, **6**.
Howell, D. (1980), announcement in House of Commons, reported in *Hansard* for 16 January.
—— (1982), announcement at Conservative Party Conference, 6 October.
Hughes, J. and Moore, R. (1972), *A Special Case? Social Justice and the Miners*, Harmondsworth: Penguin.
Hull, I. C. (1980), *The Evaluation of Risk in Business Investment*, London: Pergamon Press.
Hyden, L. (1985), 'Industrial cogeneration in Sweden', paper presented to the Benelux Association of Energy Economists European Conference, September.
International Energy Agency (1985), *Electricity in IEA Countries*, Paris.
—— (1986), *Energy Policies and Programmes of IEA Countries*, Paris.
Jaffer, S. M. and Thompson, D. J. (1986), 'Express coaching: an analysis of contestability in practice', Institute for Fiscal Studies Working Paper 79.
James, P. (1982), *The Future of Coal*, London: Macmillan.
Jones, I. S. (1983), proof of evidence to the Sizewell B Public Inquiry on behalf of the Electricity Consumers Council, EEC/P/4, 'Fossil fuel price assumptions', London: Electricity Consumers Council.
—— (1985), 'Distortions in electricity pricing in the UK: a comment', *Oxford Bulletin of Economics and Statistics*, **47**, 275–86.
Jones, P. (1983), *The Future of Coal*, London: Macmillan.
Joskow, P. L. and MacAvoy, P. W. (1975), 'Regulation and the financial condition of

the electric power companies in the 1970s', *American Economic Review*, **65**, 295-301.
Joskow, P. L. and Rozanski, G. A. (1979), 'The effects of learning by doing on nuclear plant operating reliability', *Review of Economics and Statistics*, **61**, May, 161-8.
——and Schmalensee, R. (1983), *Markets for Power: An Analysis of Electric Utility Deregulation*, Cambridge, Mass: MIT Press.
——and——(1985), 'The performance of coal-burning electric generating units in the United States: 1960-1980', MIT Department of Economics Working Paper 379.
——and——(1986), 'Incentive regulation for electric utilities', *Yale Journal on Regulation*, **4**, 1-50.
Kahneman, D., Slovic, A., and Tversky, A. (1983), *Judgement under Uncertainty*, Cambridge: Cambridge University Press.
Kay, J. A. and Thompson, D. J. (1986), 'Privatisation: a policy in search of a rationale', *Economic Journal*, **96**, 18-32.
——and——(1987), 'Policy for industry', in R. Dornbusch and R. Layard (eds.), *Performance of the British Economy*, Oxford: Clarendon Press.
Kemp, A. G. and Rose, D. (1982), 'The reform of petroleum taxation of the UK Continental Shelf', University of Aberdeen Department of Political Economy, North Sea Study Occasional Paper 15.
Kemsley, W. F. F., Redpath, R. U., and Holmes, M. (1980), *Family Expenditure Survey Handbook*, London: HMSO.
Kepner, J., King, J., and Edmunds, T. (1985), 'Cogeneration development in Texas: one sided competition', paper presented at Rutgers University Advanced Workshop in Public Utility Economics and Regulation, 29-31 May.
Klein, B. M. (1977), *Dynamic Economics*, Cambridge, Mass: Harvard University Press.
Komanoff, C. (1981), *Power Plant Cost Escalation*, New York: Van Nostrand Rheinhold.
Krangede Power Pool (KGS) (1986), *The Swedish Electricity Supply—its Organisation and Cooperation/Competition between the Power Utilities*, mimeo, Stockholm.
Laidler, D. (1985), 'Monetary policy in Britain: successes and shortcomings', *Oxford Review of Economic Policy*, **1**, 1, 35-43.
Landon, J. H. (1985), 'Practical problems with incentive regulation', published privately for a NERA conference on electric utility, 12-15 February 1985, in *Surviving an Era of Changing Regulation*.
Lerner, A. C. (1944), *The Economics of Control*, London: Macmillan.
Littlechild, S. C. (1979), 'Controlling the nationalised industries: *quis custodiet ipsos custodes?*', University of Birmingham Discussion Paper 56.
——(1981), *The Regulation of British Telecom*, Department of Trade and Industry, London: HMSO.
——(1983), *Regulation of British Telecommunications' Profitability*, London: Department of Industry.
——(1986), *The Fallacy of the Mixed Economy*, second edition, London: Institute of Economic Affairs.
Lucas, N. (1985), *West European Energy Policies—A Comparative Study*, Oxford: Clarendon Press.
Marglin, S. A. (1963), *Approaches to Dynamic Investment Planning*, Amsterdam: North-Holland.
Marshall, A. (1890), *Principles of Economics*, London: Macmillan.

Mayer, C. P. (1985), 'Recent developments in industrial economics and their implications for policy', *Oxford Review of Economic Policy*, **1**, 3, 1-24.
——and Meadowcroft, S. A. (1985), 'Selling public assets: techniques and financial implications', *Fiscal Studies*, **6**, 4, 42-56.
Millward, R. (1982), 'The comparative performance of public and private enterprise', in Lord Roll (ed.), *The Mixed Economy*, London: Macmillan.
Molyneux, R. and Thompson, D. J. (1987), 'Nationalised industry performance: still third-rate?', *Fiscal Studies*, **8**, 1, 48-82.
Monopolies and Mergers Commission (1981), *Central Electricity Generating Board: A Report on the Operation by the Board of its System for the Generation and Supply of Electricity in Bulk*, HC 315, London: HMSO.
——(1983a), *National Coal Board: A Report on the Efficiency and Costs in the Development, Production and Supply of Coal by the NCB*, Cmnd 8920, London: HMSO.
——(1983b), *Yorkshire Electricity Board: A Report on the Efficiency and Costs of the Board*, Cmnd 9014, London: HMSO.
——(1984), *South Wales Electricity Board: A Report on the Efficiency and Costs of the Board*, Cmnd 9165, London: HMSO.
——(1985), *The Revenue Collection Systems of Four Area Electricity Boards*, Cmnd 9427, London: HMSO.
——(1988), *Gas: A Report on the Matter of the Existence or Possible Existence of a Monopoly Situation in relation to the Supply in Great Britain of Gas through Pipes to Persons other than Tariff Customers*, Cm 500, London: HMSO.
National Board for Prices and Incomes (1968), *Electricity Supply Industry National Guidelines Covering Productivity Payments*, Report 79, Cmnd 3726, London: HMSO.
——(1970), *Costs and Efficiency in the Gas Industry*, Report 155, Cmnd 4458, London: HMSO.
National Economic Development Office (NEDO) (1976), *A Study of UK Nationalised Industries: Their Role in the Economy and Control in the Future*, London: NEDO.
——(1977), 'Relationships of government and public enterprises in France, West Germany and Sweden', Background Paper 2 for *A Study of UK Nationalised Industries*, London: NEDO.
National Gas Consumers Council (1985), *Fuel Debts and Hardship: The Working of the Revised Code of Practice*, London.
Nelson, J. R. (ed.) (1964), *Marginal Cost Pricing in Practice*, Englewood Cliffs, New Jersey: Prentice-Hall.
Newbery, D. M. G. (1985), 'Pricing policy', in R. Belgrave and M. Cornell (eds.), *Energy Self-Sufficiency for the UK?*, Aldershot: Gower Press.
——(1986), 'The privatisation of British Gas and possible consequences for the European gas market', Centre for Economic Policy Research Discussion Paper 101.
Page, S. A. B. (1977), 'The value and distribution of the benefits of North Sea oil and gas', *National Institute Economic Review*, 82, 41-58.
Parris, H., Pestieau, P., and Saynor, P. (1987), *Public Enterprise in W Europe*, London: Croom Helm.
Part Committee (1981), *The Taxation of North Sea Oil*, London: Institute for Fiscal Studies.

Pfeffer, Lindsay, and Associates Inc. (1986), 'Energy policy issues in PURPA implementation', paper prepared for the US Department of Energy Office of Coal and Electric Policy, March.
Phlips, L. (1983), *The Economics of Price Discrimination*, Cambridge: Cambridge University Press.
Pindyck, R. S. (1979), *The Structure of World Energy Demand*, Cambridge, Mass: MIT Press.
Posner, M. V. (1973), *Fuel Policy*, London: Macmillan.
Powell, S. and Horton, G. (1985), 'The economic effects of lower oil prices', HM Treasury Working Paper 34.
Price Commission (1979a), *British Gas Corporation—Gas Prices and Allied Charges*, HC 165, London: HMSO.
——(1979b), *Area Electricity Boards—Electricity Prices and Certain Allied Charges*.
Pryke, R. (1981), *The Nationalised Industries: Policies and Performance since 1968*, Oxford: Martin Robertson.
Public Money (1984), 'Energy prices: economic concepts and accounting realities', 4, 1, 41–6.
Public Utilities Commission of the State of California (1985), Decision 85-04-0475, 17 April, interim opinion.
——(1986), 'Summary of cogeneration and small power production projects', as of 31 March.
Rawlinson, R. (1976), 'Producing coal', *Colliery Guardian Annual Review*.
Rees R. (1968), 'Second best rules for public enterprise pricing', *Economica*, 260–73.
——(1974), 'The economics of investment analysis', Civil Service College Occasional Paper 17, London: HMSO.
——(1983), *Public Money*, 2, 4, 13–17.
——(1984a), 'A positive theory of the public enterprise', in M. Marchand, D. Pestieau, and H. Tulkens (eds.), *The Performance of Public Enterprises*, Amsterdam: North-Holland.
——(1984b), 'The public enterprise game', *Economic Journal Conference Supplement*, 94, 109–23.
——(1984c), *Public Enterprise Economics*, second edition, London: Weidenfeld & Nicolson.
——(1985a), 'The theory of principal and agent', parts 1 and 2, *Bulletin of Economic Research*, January and May.
——(1985b) 'Principal agent theory and public control of production', University College Cardiff Discussion Paper 8514.
——(1986a), 'Incentive compatible public sector discount rates', *Journal of Public Economics*.
——(1986b), 'Is there an economic case for privatisation?', *Public Money*, 5, 4, 19–26.
Reid, G. L., Allen, K., and Harris, D. J. (1973), *The Nationalized Fuel Industries*, London: Heinemann.
Rhys, J. M. W. (1980), 'Oil, exchange rates and the UK economy', *The Business Economist*.
Robinson, C. (1979), proof of evidence to the public inquiry into the proposed NE Leicestershire coalfield, P50, 'The energy future and the Vale of Belvoir'.
——(1981), 'The errors of North Sea policy', *Lloyds Bank Review*, July.

Robinson, C. (1985), 'Coal policy in Britain', *Economic Review*, March.
——and Marshall, E. (1981), *What Future for British Coal?*, Hobart Paper 89, London: Institute of Economic Affairs.
——and——(1983), 'The coal industry and coal policy in Britain', in House of Lords Select Committee on the European Communities, *European Community Coal Policy*, Session 1983-4, Tenth Report.
——and——(1985), *Can Coal be Saved?*, Hobart Paper 105, London: Institute of Economic Affairs.
——and Morgan, J. (1978), *North Sea Oil in the Future*, London: Trade Policy Research Centre/Macmillan.
Robinson, D. and Sharpe, T. (1986), 'Hits and myths about economic regulation in the United States', paper presented at the third annual meeting of the European Association for Law and Economics, Oxford, September.
Sappington, D. E. M. and Stiglitz, J. E. (1985), 'Information and regulation', paper presented at the National Science Foundation/Carnegie Mellon University Conference on Regulation, September.
Schmalensee, R. (1979), *The Control of Natural Monopolies*, Lexington, Mass: Lexington Books.
——and Golub, B. W. (1984), 'Estimating effective concentration in deregulated wholesale electricity markets', *Rand Journal of Economics*, Spring, 12-26.
Searle Barnes, R. G. (1969), *Pay and Productivity Bargaining: A Study of the Effect of National Wage Agreements in the Nottinghamshire Coalfield*, Manchester: Manchester University Press.
Select Committee on Nationalised Industries (1968-9), *National Coal Board*, Session 1968-9, London: HMSO.
——(1977-8), *Reports and Accounts of the Energy Industries*, Session 1977-8, London: HMSO.
Sen, A. K. (1982), *Commodities and Capabilities*, Amsterdam: North-Holland.
——(1984), 'Poor, relatively speaking', *Oxford Economic Papers*, 35, 153-69.
Sharpe, L. J. (1981), 'Is there a fiscal crisis in western European local government?', in L. J. Sharpe (ed.), *The Local Fiscal Crisis in Western Europe*, London: Sage.
Sherman, R. and Visscher, M. (1982), 'Rate of return regulation and two-part tariffs', *Quarterly Journal of Economics*, 97, 27-42.
Shleifer, A. (1985), 'A theory of yardstick competition', *Rand Journal of Economics*, 16, 319-27.
Slater, M. D. E. and Yarrow, G. K. (1983), 'Distortions in electricity pricing in the UK', *Oxford Bulletin of Economics and Statistics*, 45, 317-38.
——and——(1985), 'Distortions in electricity pricing in the UK: a reply', *Oxford Bulletin of Economics and Statistics*, 47, 287-92.
Smith, G. H. (1953), *Industry in Sweden*, Stockholm.
Spence, M. (1977), 'Entry, capacity, investment and oligopolistic pricing', *Bell Journal of Economics*, 8, 534-44.
State of New York Public Service Commission (1982), 'Consolidated Edison Company of New York Inc.—electric service provided to customers with on-site generation', Opinion 82-10, Case 27574, 12 May.
——(1985), 'Long run avoided costs—methodology and estimates', paper submitted by the Department of Public Service, Case 28962, proceeding of the Commission to establish estimates of long-run avoided costs, 10 September.

Sumner, M. (1978), 'Progressive taxation of natural resource rents', *Manchester School*, **46**, 1–16.
Sydkraft (1986), *Annual Report 1985*.
Thomas, D. (1985), 'The price of providing cheaper electricity', *Financial Times*, 29 October.
Tolley, D. L. and Budden, R. J. R. (1985), *UK Electricity Supply and CHP under the 1983 Energy Act*, London: Electricity Council.
Turot, P. (1970), *Les Entreprises Publiques en Europe*, Brussels: Centre Européen d'Entreprise Publique.
Turvey, R. (1968), *Optimal Pricing and Investment in Electricity Supply*, London: Allen & Unwin.
——and Anderson, D. (1977), *Electricity Economics*, Baltimore: John Hopkins Press.
United Nations Economic Commission for Europe, Committee on Electric Power (1969), *Legal Regime of Electricity Undertakings in the Countries Participating in the Work of the Economic Commission for Europe*, ST/ECE/EP/46.
US Department of Energy, Energy Information Administration (1978), *Steam-Electric Plant Construction Cost and Annual Production Expenses 1977*, Washington DC: US Government Printing Office.
——,——(1979), *Statistics of Privately Owned Electric Utilities in the United States 1977*, Washington DC: US Government Printing Office.
Verney, D. V. (1959), *Public Enterprise in Sweden*, Liverpool: Liverpool University Press.
Vickers, J. S. (1985), 'Strategic competition among the few—some recent developments in the economics of industry', *Oxford Review of Economic Policy*, **1**, 3, 39–62.
——and Yarrow, G. K. (1985), *Privatization and the Natural Monopolies*, London: Public Policy Centre.
Virole, J. (1986), 'Electricité de France', in V. V. Ramanadham (ed.), *Public Enterprise*, London: F. Cass.
Webb, M. G. (1984), 'Privatisation of the electricity and gas industries', in D. R. Steel and D. A. Heald (eds.), *Privatizing Public Enterprises*, London: Royal Institute of Public Administration.
Whiteman, J. C. (1985), 'North Sea oil', in D. Morris (ed.), *The Economic System in the UK*, third edition, Oxford: Oxford University Press.
Williamson, O. E. (1963), 'Managerial discretion and business behaviour', in R. Marris and A. Wood, *The Corporate Economy*, London: Macmillan.
——(1975), *Markets and Hierarchies*, New York: Free Press.
Yarrow, G. K. (1985), 'Regulation and competition in the electricity supply industry', in J. A. Kay, C. P. Mayer, and D. J. Thompson (eds.) (1986), *Privatisation and Regulation: The UK Experience*, Oxford: Clarendon Press.
——(1986), 'Privatization in theory and practice', *Economic Policy*, **2**, 324–77.

Index

Note: All references are to energy and the United Kingdom, except where otherwise stated.

acid rain 4, 143
Acts of Parliament *see* legislation
advanced gas-cooled reactor 123, 223
  see also nuclear power
advanced petroleum revenue tax 421-2, 424
'adverse selection' model 99
AGR *see* advanced gas-cooled reactor
airlines 41
Allen, K. 126*n*
Allsopp, C.: on North Sea oil 377-410, 415
alternative energy 23
Anderson, D. 215-17
APRT *see* advanced petroleum revenue tax
arbitration, independent, PURPA and 210-11
Area Boards
  electricity 13, 17, 19, 207, 213, 247; competition 158, 160, 162, 164-6, 169-72, 174, 195-6, 203; created 240; energy policy 28, 43, 45; pricing 209, 210; and regulation 193; *see also* British Electric Boards; Electricity Council
  gas 41-3, 45, 264
artificial monopoly 7, 14, 18
Artis, M. J. 397
Atkinson, A. 56*n*
Atkinson, R. 176
Atomic Energy Authority 28, 207
Australia 33, 123, 298, 326, 383
average revenue constraint 272
Averch, H. 15, 184, 252-3

background variables in electricity supply industry 222-3
Bailey, E. E. 41
balance of payments 4, 351, 357, 364-5, 394, 409
Bank of England 364, 377, 383-6, 387, 389, 392, 394, 397
bankruptcy constraint 14, 160-1
bargaining *see* unions
Bates, R. W. 215, 216*n*
Baumol, W. J. 146, 179, 180*n*
BBC 153
BEB *see* British Electricity Boards
Belgium 121, 130, 245*n*
Bending, R. 45
benefits *see* social security
benefits of oil *see* oil industry, economic

BGC *see* British Gas Corporation
Blackaby, F. 373
Bladen-Hovell, R. 397
BNOC *see* British National Oil Corporation
Bond, S. R. 413*n*
Bongaerts, J. C. 242
Bradley, I. 272, 276
British Airports Authority 16
British Coal *see* National Coal Board
British Electricity Boards 118-19, 121-4
  see also Area Boards
British Gas Corporation/British Gas/gas industry 1, 4, 7, 13, 15-17, 19-20, 156, 203
  competition 41-3, 254, 275-6, 279, 282, 283-4
  in demand model 80-1, 83-6
  and electricity supply industry 51, 53, 216-17, 220, 224, 233
  energy policy 27, 28, 42-3, 68; competition 41-3; pricing 37-8, 46-50, 51-2; tax 35-6
  forecasting 32-3
  legislation 1, 13-14, 27, 42, 45, 264, 279-85
  and MMC 281-3
  performance 92, 95*n*, 115-18
  privatization and pricing 17, 40, 151, 199, 263-77, 296, 319, 430-1; efficiency of pricing 269-71; history of pricing 264-70; and regulation 271-3, 273-7, 280-1; standing charges 273-4; storage costs 274-5; unregulated 275-6; *see also under* profits
  public enterprise performance 115-17
  regulation 7, 11, 15, 19, 279-85; competition 283-4; efficiency 282-3; prices 272-3, 275-7, 280-2
  risk analysis and optional investment 216-17, 220, 224, 233
  taxation 36, 48, 266, 281
  worldwide consumption 292
British Leyland 366
British National Oil Corporation 27, 28, 416
British petroleum 414, 417
British Steel 297, 366
British Telecom 16, 50, 95*n*
Britoil 27
Brittan, S. 41
Brown, G. 133
BSC (British Steel Corpotation) 297, 366

BST *see* bulk supply tariff
Budden, R. J. R. 207
Buiter, W. H. 394
bulk supply tariff 44, 104, 211
  and competition 165, 166–73, 175
  and regulation 195–6
Burns, A. 303*n*
Byatt, I. 384, 392, 393

Cairncross, Sir A. 239, 373
California 204, 206, 208
Canada 353
capital *see* investment
Cartel Act (Germany, 1957) 250
Central Electricity Generating Board (CEGB) 19, 247
  and coal industry 123, 144, 207, 239, 256; economics of 315–17, 319, 326–7, 340; liberalization 294, 297–8, 304, 309; risk analysis and optional investment 216–17, 220, 223, 224, 231–4; *see also under* energy policy *below*
  and competition 158, 160, 162–73 *passim*, 195–7, 203
  energy policy 25, 28, 42–5; and coal industry 3, 12, 31, 33, 45–6, 43, 192–3, 198, 205; and gas industry 51, 53; pricing 194–7, 202, 210
  and gas industry 51, 53, 214–15, 220, 224, 233
  and oil industry 220, 224, 231–4
  performance 92, 121, 123, 191, 207; MMC report on 101–8, 147–8, 158, 164, 215, 222–4
  pollution 4–5
  risk analysis and optional investment 215–36 *passim*
  see also Area Boards; electricity supply industry; nuclear power
central heating 63–5, 67, 68, 281, 317
centralized systems *see* nationalized industries
Child Poverty Action Group 66*n*
Clunies Ross, A. I. 412, 418
Coal Board *see* National Coal Board
Coal and Coal Industry Acts (1936, 1977, 1980) 295, 299–301, 306, 315
Coal Commission 295, 299, 300*n*
coal industry
  in demand model 80–1, 83–6
  economics of 11, 313–405; historical background 315–17; pit closures 289, 314–15, 320, 327–9, 331–4, 340; policy response 344–5; *see also under* competition; costs; pricing; subsidies; trade
  and electricity *see under* Central Electricity Generating Board
  energy policy 31, 46, 344–5
  forecasting use 32–3
  legislation 295, 299–301, 306, 315
  liberalizing 19, 20, 289–312; demand and supply structures 292, 294–6; methods of 297–305; privatization 296–7, 299, 301–5, 307, 311
  and MMC 296, 300, 315–16, 323, 325, 331, 333, 336, 340
  nationalized 290–4, 305, 315–19
  output *see under* productivity
  public enterprise performance 124–32; *see also* National Coal Board
Coal Industry Nationalization Act (1946) 295*n*, 299, 315
Commission on Energy and Environment (1981) 301
Common Agricultural Policy (EC) 359
competition 19–20, 27, 40
  in coal industry 291, 292, 301, 312, 313, 318–19, 327, 339–40, 344–5
  in electricity *see under* electricity supply industry
  and energy market failure 6–7
  in gas industry 41–3, 256, 277–8, 280, 283, 284–6
  and incentive regulation 185
  international 24–5
  in new market philosophy 13–18
Competition Act (1980) 6
conservation of energy 5, 23, 26–7, 92–3, 321–3
Consolidated Edison 211
constraints
  and gas industry 273–5
  and public enterprise 95–101
consumption 59–66
  household budgets 59–63, 69–72
  information and market failure 63–6
  in oil industry 350–62, 379
  prices 38, 46–7
  taxed 52
  see also demand
contract prices 38–9
Corti, G. 240*n*
cost-of-service regulation 200–1
costs/cost-benefit analysis 14, 15
  in coal industry 298, 309–10, 314–15, 317, 323–6, 328–38, 342
  in electricity supply industry 207–8; in Europe 254, 260; and regulation, 188–9, 190, 191, 202, 204; *see also* risk analysis
  in gas industry 265, 267–8, 271–5, 280–3, 285
  and incentive regulation 178–9, 185
  in oil industry 351, 372
  pricing *see* marginal cost pricing
  in public enterprise 94, 97, 98, 118, 122–4, 130–2
  social *see* social security; unemployment

see also efficiency
credibility, policy, under-investment and 200-1
Crew, M. A. 8
cross-regulation 252, 254
cross-subsidization 12, 193, 333
current cost accounting 265

Dam, K. W. 300n
Dasgupta, P. 56n
Davis, E. H. 37, 52, 176
de Ru, H. J. 240
debt, national, and oil industry 401-13
decentralized systems in European electricity supply industry 238-42, 247, 258
    regulation in 250-1, 255-6
deep-mined coal 296, 302, 305, 323-5
deferment decision: risk analysis and optional investment 227-36
deficits, financial, marginal cost pricing and 149-50
Deloitte, Haskins and Sells 282
demand 2, 9-10
    in coal industry 292, 294, 313, 321-3
    in electricity supply industry *see* risk analysis
    market failure 58
    see also consumption
demand model 77-91
    conservation 90-1
    described 77-82
    dynamic properties of 89-90
    relative price elasticities 83-6
    results, robustness of 86-8
    sectoral demand elasticities 82-3
Department of Energy (UK) 6, 19
    and coal industry 300
    and electricity supply industry 170, 222
    and forecasts 10
    and gas industry 263, 266, 277
    and performance 77-91 and pricing 47
Department of Energy (US) 122n, 207, 208
Department of Health and Social Security 67-8
deregulation 40-1
    see also Energy Act (UK); privatization; PURPA
Devereux, M.: on taxing oil extraction 5, 142, 381-4, 411-27
Diamond, P. A. 37, 39, 52, 194
Dilnot, A.: on energy policy, merit goods and social security 3, 55-72, 140, 269
disconnections 68-9
distributional argument and merit goods 56-7
diversity in electricity supply imdustry *see* Europe, electricity supply
Dixit, A. K. 161, 202
DoE *see* Department of Energy
Domberger, S. 16
domestic *see* households

Dupuit, J. 133
Dwolatzky, B. 397

Eden, R. 45
EdF *see* Electricité de France
Edison Electric Institute 121n
Edmunds, T. 208, 212
Edwards, R. S. 120n, 121n
efficiency 12
    in coal industry 309; pit closures 289, 314-15, 320, 327-9, 331-4, 340
    in electricity supply industry: incentive regulation *see under* electricity supply; and regulation 188-9, 190, 191, 202, 204
    and energy policy 24, 25, 27, 36-7, 40, 45
    in gas industry 269-71, 279, 282-3
    and merit goods 57-9
    see also costs; productivity
efficient (spot) prices 38-9, 257, 260
EFL (external financing limit) 94
Electric Power Research Institute (US) 207
Electricité de France 45, 121, 122n, 240, 249, 250, 258-9, 260
Electricity Acts (1909, 1911, 1926, 1947) 158, 239-40
Electricity Council 26, 68, 121, 158, 170, 172, 206, 213, 247
    see also Area Boards
electricity supply industry
    and coal industry *see under* Central Electricity Generating Board
    competition in: in Europe 256-9, 260; and failure of Energy Act 14, 157-77; hypotheses, competing 162-5, 174-5; potential entry 159-62; pricing 157, 158-60, 161, 163-4, 165-74, 175, 176; and regulation 192, 195-7, 202-3; in USA 212-13
    in demand model 81, 82-6
    in Europe 4, 20, 23, 25, 121-3, 133, 142, 237-89; competition 256-9, 260; organizational structure 246-8; ownership 238-45; regulation 246-56, 259-60
    incentive regulation, potential of 178-87; frictions and 178-80; limitations and lessons 182-3; optimal regulation 180-2; practical principles 183-7; *see also* regulation *below*
    legislation 158, 239-40
    public enterprise performance 116-24
    regulation 17, 19, 188-205; current problems 192-8; in Europe 248-56; privatization 198-205; public policy recently 188-92; *see also* incentive regulation *above*
    risk analysis and optional investment 6, 11, 214-36; deferment decision 227-36; economic appraisal 215-27
    as special case 25-6, 133
    taxation 188-9, 194-5, 201, 208

electricity supply industry contd
  in USA 121-2, 198-9, 201, 230, 237, 244n;
    PURPA 206-13; (comparison with Energy
    Act 206-7; success of 210-13)
  see also Central Electricity Generating Board;
    and under entry; investment; pricing; trade
employment
  in coal industry 306-10, 334-8, 341-2
  in oil industry 367-8
  in public enterprise 97, 102-3, 115-22, 124-30
  see also unemployment; unions; wages
Energy Act (Germany, 1935) 250, 251, 258
Energy Act (UK, 1983) 1, 13, 14-15, 43-5, 304n
  and electricity supply industry 192, 195-6, 236, 249, 260; failure 208-10, 213; see also electricity supply industry, competition in
Energy Committee
  on competition 284
  on failure of Energy Act 213
  on gas industry 266-8, 271, 275, 282, 284, 296
  on pricing 47, 170, 266, 267, 268, 271
  on productivity 172
  on regulation 275
  and risk analysis 215, 221
  on tax 35, 36
energy market see coal industry; electricity supply industry; energy policy; gas industry; legislation; oil industry; performance
energy policy
  after privatization 5, 10, 30-54; co-ordination of policy 52-4; competition policy 41-6; pricing and investment 35-40, 44, 50-1; regulation 46-50; theoretical debate 36-41; see also under taxation
  alternative methods of provision 69-72
  coal industry 31, 46, 344-5
  in electricity supply industry see under Central Electricity Generating Board
  in gas industry see under British Gas
  and marginal cost pricing 152-4
  and oil industry 369-70, 398-405, 407-10, 413-18
  and role of state 2, 12-13, 23-9
  see also social security and under nationalized industries
entry 14, 43-4, 46
  in electricity supply industry; barriers hypothesis 157, 163, 164; efficiency hypothesis 157, 164-5; in Europe 257-8; failure of 157-77; and regulation 192, 196-7, 202-3
  see also privatization
Europe 20
  coal industry 130, 293
  electricity supply industry 4, 20, 23, 25, 121-3, 133, 142, 237-60; competition 256-9, 260; organizational structure 246-8; ownership 238-45; regulation 248-56, 257-60
  energy policy 23, 25, 41, 42, 45
  gas industry 41, 42
  oil 359, 368, 384, 395
  performance 121-2, 130
  see also in particular France; Germany; Netherlands; Sweden
European Economic Community 23, 245n, 250, 258, 359
Evans, N. L. 215, 216, 231, 233
exchange rate
  and coal industry 314, 339
  and oil industry 349-52, 356-63, 365, 369, 372-3, 375, 383, 392-5, 406
  and risk analysis and optional investment 231-3, 235
exhaustible resources see non-renewable resources
exports see trade
Extended Energy Survey Scheme 27
external financing 94, 169-70
externalities and marginal cost pricing 143-4

'fair' entry tariffs 163-4
Family Expenditure Surveys 59-65, 69-71
Federal Energy Regulatory Commission (US) 207, 211
FES see Family Expenditure Surveys
Fforde, J. S. 403
financial constraints
  in electricity supply industry 191, 194-5
  in gas industry 271-2
forecasting 6, 10, 26, 32-3
  BST 170-1
  in coal industry 289-90
  demand see demand model
  in electricity supply industry see risk analysis and optional investment
  see also demand; marginal cost pricing; public enterprise performance modelling
Forsyth, P.: on North Sea oil revenues 349-76, 377-84, 399
Foster, C. D. 247n
France
  coal industry 130
  electricity supply industry 20, 25, 33, 45, 121-2, 123, 133, 142, 238; competition 258, 259, 260; organizational structure 247; ownership 239, 240, 244; regulation 249-50, 253, 255
  nuclear power 25, 33, 45
Fraser, N. M. 215, 216n
frictions and incentive regulation 178-80
Fuel Direct Scheme 68
fuel shares equation in demand model 80-2

future energy policy 18-20
  see also forecasting
futures and insurance markets, lack of 38

Garnaut, R. 412, 418
gas industry see British Gas Corporation
Gas Acts (1948, 1972, 1986) 13, 266
Gas Bill and Proposed Authorization 278-84
Gas de France 247
Gas Levy 36, 48, 51, 266, 280
Gas Users' Council 47
Gaskin, M. 349
geographic location and BST 169
Germany
  coal industry 130, 293
  electricity supply industry 23, 231, 238, 247; competition 257-8, 259; ownership 240, 241-2, 244n, 245n; regulation 250-1, 255
  oil industry 370, 384, 395
Gibson, M. 269
Glyn, A. 328
Golub, B. W. 45
Goodhart, C. A. E. 403
government
  'failure' 152-3
  see also state
Gravelle, H. S. E. 95n
Gregory, R. G. 370, 383
gross domestic product 4, 10, 313, 322, 397
gross national product 28

Haldane, G. 243n
Halvorsen, R. 84
Hammond, E., Helm, D., and Thompson, D.:
  on competition in electricity supply 6, 14, 43, 58n, 157-77, 195, 202, 208, 237, 238n, 258; on regulation of gas industry 7, 11, 15, 19, 43, 58, 158, 174, 246, 279-85
Hannah, L. 239, 240
Harlow, C. 126n, 127n
Harris, D. J. 126n
Hartley, N. 384
Hawdon, D. 82, 84
Helm, D. 6, 152, 239, 253
  on electricity supply in Europe 3, 4, 237-60
  on energy policy, merit goods and social security 55-72, 140, 269
  see also under Hammond
Henney, A. 176: on USA 208-13
Hicks, U. 239, 245n
Holmes, M. 59n
horizontal integration in electricity supply industry 246-7, 248, 260
Horton, G. 393
Hotelling, H. 133
households
  budgets and consumption 59-63, 69-72
  demand 82, 85

Houston 208
Howe, G. 418, 421
Howell, D. 222n, 265-6, 269
Hughes, J. 129n
Hull, I. C. 231
Hyden, L. 257n
hydro power 33, 290
  see also Sweden, electricity

imports see trade
incentive regulation see under electricity supply industry
income distribution and marginal cost pricing 139-40
  see also wages
incumbent systems and electricity supply industry in Europe 238-9, 243-4, 247-8, 257, 258
  regulation in 251-2, 256
  strategic advantage and competition 160-1
indexation and rate-of-return regulation in electricity supply industry 199-200
industry, manufacturing
  and coal industry 292, 316, 325
  and electricity supply industry 209, 242, 243
  and gas industry 263, 266-7
  and oil industry 349, 352-61, 365, 367-8, 377, 383, 388-90, 409
  see also demand model
information 27
  and consumption 63-6
  and gas industry 279
  in public enterprisie 99, 100
international competition 24-5
International Energy Agency 23, 246
international trade see trade
investment/capital 5-6, 9, 24
  in coal industry 293, 313, 324-5
  in electricity supply industry: appraisal 159-60; and regulation 188-90, 191, 194, 197-8, 203
  and marginal cost pricing 137, 151
  in oil industry 363, 364-6, 372, 373
  optional see risk analysis
  in public enterprise 97-8, 101-2, 105-9
  see also capital
investors see shareholders

Jackman, R. 245n
Jaffer, S. M. 176
James, P. 256n
Japan 370, 384
Johnson, L. 15, 184, 252-3
Jones, I. 189: on risk analysis and optional investment 6, 11, 147, 215-36
Joskow, P. L. 178n, 181n, 185n, 206, 229-30, 252

Kahneman, D. 59
Karakitsos, E. 397
Kay, A. G. 412
Kay, J. 37, 52: on North Sea oil revenues 349-76, 377-84, 399
Kemsley, W. F. F. 59n
Kepner, J. 207, 212
King, J. 208, 212
Kleindorfer, P. R. 8
Komanoff, C. 233n
Kraemer, R. A. 242

labour *see* employment; unions
Laidler, D. 403, 410
Landon, J. H. 182, 184, 185n
law *see* legislation
Lawson, N.: on energy policy 1, 2, 12, 13, 23-9, 157, 237
legislation
  cartel 250
  coal 295, 299-301, 306, 315
  competition 6
  electricity 158, 239-40
  energy *see* Energy Act
  gas 1, 13-14, 27, 42, 45, 264, 279-85
  oil 1, 13-14, 27, 42, 45, 411, 418-19, 426
  public utilities (USA) 206-7, 210-13
  taxation 411, 418-19
Lerner, A. C. 133
liberalization *see under* coal industry
Littlechild, S. C. and Report (1983) 16, 93, 152, 253n
load management charges in electricity supply industry 172-3
local networks 7
  see also Area Boards
Locax, R. 384
long-run BST 168-9
long-run marginal costs 39, 44-5
  in electricity supply industry 187; and competition 163, 165, 169-70, 172-3, 175-6; in Europe 253-4, 255-6
  in gas industry 263
  rationale for 144-9
  see also marginal cost pricing
long-term contracts and PURPA 212
losses in coal industry 293, 302, 313, 323, 324, 334, 341
LRMC *see* long-run marginal costs
Lucas, N. 238n, 240, 242, 243n

MacAvoy, P. W. 185n
McGowan, F.: on electricity supply in Europe 4, 237-60
macroeconomics *see under* oil industry
Maine 212
Major, J. 67
management

in electricity supply industry 189-90, 198, 200
in gas industry 263
and incentive regulation 180-1, 182, 184
and public enterprise 94-6, 112
see also efficiency
manufacturing *see* industry
marginal cost pricing 9, 39, 93, 104-5
  in electricity supply industry: and competition 163, 165, 169-71, 172-3, 175-6; in Europe 253-4, 255-6; and regulation 189, 191, 192, 195
  and energy policy 152-4
  in gas industry 263, 265-6, 268, 272
  rationale for 10, 133-54; policy 152-4; practice 149-52; theory 134-9
  see also long-run marginal costs; short-run marginal costs
marginal costs and savings of pit closures 331-4
Marglin, S. A. 214
market
  events and macroeconomic effects of oil industry 391-8
  failure 2-7; competition 6-7; and consumption 63-6; importance of energy in economy 3-5; and merit goods 57-9; timescales implied in planning 3, 5-6; new market philosophy 12-18
  see also energy market
Marshall, A. 133
Marshall, E.: on liberalizing coal industry 19, 20, 141, 144, 289-312
Marshall, Lord 298
Mayer, C. P. 160, 280n
Mayes, D. 403
Meadowcroft, S. A. 280n
Mercur 50
merit goods *see* social security
Middle East 23
Miller, M. 394
Millward, R. 16
mining *see* coal industry
Mirrlees, J. A. 37, 39, 52, 194
mixed systems in electricity supply industry *see* decentralized systems
MMC *see* Monopolies and Mergers Commission
Molyneux, R. 9, 11, 115n
monetary policy and oil industry 403-5
Monopolies and Mergers Commission 6, 28, 174, 176, 221, 250, 255
  and CEGB 101-8, 147-8, 158, 164, 215, 222-4
  and coal 296, 300, 315-17, 323, 325, 331, 333, 336, 340
  and gas 281-3
monopoly 3, 7, 14, 18-19
  abolition *see* Energy Act; privatization

regulation of 429–31
see also nationalized industries; natural monopoly
Moore, R. 129*n*
Morgan, J. 349
Morris, C. N. 418, 426
Morrison, H. 9, 11

NACs (net avoidable costs) *see* net effective costs
National Board for Prices and Incomes 117*n*, 119*n*, 121*n*
National Coal Board/British Coal 12, 19, 28, 207, 315
  energy policy 31, 45, 48, 52, 53
  performance 123, 124–32
  pricing 141–2, 151
  see also coal industry *and under* Central Electricity Generating Board
National Gas consumers Council 68
National Loans Fund 266
national networks *see* British Gas; Central Electricity Generating Board
National Power Loading Agreement (1966) 128–9
National Union of Mineworkers 128–9, 290, 302, 304–5, 307, 308, 333–4
nationalized industries 6, 18
  coal 290–5, 299, 305, 315–19
  electricity supply in Europe 238–40, 247, 257–8; regulation in 249–50, 255
  energy policy 31–3; efficiency examination 28; pricing and investment 35–40, 44, 50–1; regulatory failure 9–12
  gas 264–9
  pricing *see* marginal cost pricing
  see also public enterprise
natural monopoly 3, 7, 18–19, 27, 37–8, 45, 265
NECs *see* net effective costs
Nelson, J. R. 131
net effective costs and net avoidable costs 214–15, 216–20, 221–7
Netherlands
  electricity supply industry 238, 240–1, 247, 257, 258; regulation 250–1, 255
  gas industry 41, 42
neutrality policy in oil industry, towards 398–405, 407–9
  taxation and 413–18
new entrants *see* entry; privatization
new market philosophy 12–18
  competition and regulation 13–18
new pits 302–3
New York 206, 210–12
Newbery, D. 412; on energy policy 5, 10, 30–54, 151, 265, 274, 417, 419
Newby, M. 303*n*
night rates 209

non-discrimination and PURPA 210–11
non-renewable resources 5
  and marginal cost pricing 151–2
  see also coal; gas; oil
normative agenda 92, 93–4
North Sea oil *see* oil industry
Norway
  electricity 23
  gas 41, 42
  oil 351, 362, 364
nuclear power 4, 28, 123, 143–4, 207, 321
  consumption, worldwide 292
  energy policy 25, 32, 33, 45, 53
  see also risk analysis and optional investment
NUM *see* National Union of Mineworkers

objective function of incumbent and electricity supply industry 160–1
OECD *see* Organization for Economic Co-operation and Development
Office of Gas Supply 13, 45, 48–9, 93, 198, 264, 276–8
Office of Telecommunications 93, 198
offsetting policy *see* neutrality
OFGAS *see* Office of Gas Supply
OFTEL *see* Office of Telecommunications
oil industry/North Sea oil 4, 23, 27, 28, 34
  consumption, worldwide 292
  in demand model 80–1, 83–6
  economic implications of 349–76; and balance of payments 351, 357, 364–5, 394, 409; depletion 363–4, 366; economic structure, effect on 352–6; employment 367–8; exchange rate and 349–52, 356–63, 365, 368, 372–3, 375, 383, 392–5, 406; size 350–2; *see also under* industry; pricing; structural change; trade
  and electricity supply industry 220, 224, 231–4
  and energy policy 369–70, 398–405, 407–10, 413–18
  investment in 363, 364–6, 372, 373
  legislation 1, 13–14, 27, 42, 45, 411, 418–19, 426
  macroeconomic impact 377–410; market events and economy 391–8; quantitative 378–83; *see also* structural change
  neutral or offsetting policy, towards 398–405, 407–9
output *see under* productivity
prices 363–5, 380–5; changes in 368–70, 378, 393; (*see also* falling *and* shocks *below*); and exchange rate 357–9, 375; falling 53–4, 275, 339, 398; linked with coal price 318–19, 327, 339–40, 344; shocks 30, 34, 186, 377, 391, 396, 397, 407; (and coal industry 291, 313, 317, 319, 340

oil industry/North Sea oil *contd*
  prices *contd*
    and conservation 26; and demand model 86–7; and electricity supply industry 190, 201; and gas industry 36)
  taxation 5, 327, 357, 374–6, 381–2, 386, 395–8, 399, 405–6, 411–27; British experience and effects 418–25; optimal 412–18
Oil and Gas (Enterprise) Act (1982) 1, 13–14, 27, 42, 45
Oil Taxation Act (1975) 411, 418–19, 426
OPEC *see* Organization of Petroleum Exporting Countries
Opencast Coal Act (1958) 295n, 300–1
Opencast Executive 295–6, 300
opencast mining 295–6, 300–8
optimal oil taxation 412–18
optional investment *see* risk analysis and optional investment
Organization for Ecomomic Co-operation and Development 375, 394, 395
Organization of Petroleum Exporting Countries 34, 327, 393, 396, 404
organizational structure of electricity supply industry in Europe 246–8, 259–60
outcomes and public enterprise 95, 101–8
output *see* productivity
ownership
  in coal industry 297–8
  in electricity supply industry 247–8, 255–6, 259; in Europe 238–45
  see also nationalized industries; privatization

Pacific Gas and Electric Company (US) 208, 210
Page, S. A. B. 349
Panzar, J. C. 146
Pareto-efficiency/optimum 92, 134–42
Parris, H. 245n
partial adjustment model 78–9
partial equilibrium justification in marginal cost pricing 141–3
peak-load pricing 104, 147, 150–1, 168, 268, 272–3
penalties, bounds on 186–7
performance
  measure and incentive regulation 184–5
  see also demand model; public enterprise
Perlman, M. 247n
Pestieau, P. 247n
petuoleum revenue tax 35, 36, 49–50, 374–6, 419–26
Phlips, L. 270
Piggott, J. 16
Pindyck, R. S. 81, 82, 84
planning, public enterprise 99–101, 105–8
plant-related variables in electricity supply industry 222–3

pluralism in electricity supply industry *see* Europe, electricity supply
Poland 123, 130
policy *see* energy policy; regualtion
political influence 152–3, 306
pollution 4, 143, 302
positive theory of public enterprise 94–101
Posner, M. V. 9, 31–3
poverty *see* social security
Powell, S. 384, 393
PPTs *see* private purchase tariffs
pressurized water reactor *see* nuclear power
Price, C.: on gas privatization 16, 17, 152, 263–78
Price Commission 117n, 123n, 124n
pricing 5, 7, 9–10, 13–15, 24, 28
  in coal industry 291, 298, 306, 313, 317–23, 334–44; and regulation in electricity supply industry 192–4, 197–8, 201–2
  cost *see* marginal cost pricing
  in demand model 78–9, 83–6
  in electricity supply industry 209–10; competition in 157, 158–60, 161, 163–4, 165–74; in Europe 253, 257; and regulation 188–9, 192–203 *passim*; *see also* risk analysis and optional investment *and under* marginal cost pricing
  in gas industry *see under* British Gas
  and incentive regulation 179
  in oil industry *see under* oil
  in public enterprise 92–4, 98, 104–5, 110–12, 118, 122–4, 130–2
  see also under marginal cost pricing
principal-agent model (agency theory) 8, 99, 181–2, 183–4, 185–6
private coal mines 293–4, 298–9
private electricity production 157, 159, 161–2, 206–9
  and regulation 192, 196, 201
  in USA *see* PURPA
  see also privatization
private purchase tariffs 44, 196, 260
  and competition 158–60, 161, 163–5, 171–2, 175–6
privatization 18, 108–12
  in electricity supply industry: and regulation, 198–205
  see also decentralized systems; private; *and under* British Gas Corporation; coal industry; energy policy
producers, prices for 38
productivity
  in coal industry 289–90, 291, 295–7, 300–2, 311, 313, 320–2, 325, 338–44
  and marginal cost pricing 137
  in oil industry 350–63, 364, 376–88, 407
  in public enterprise 94, 97, 102–3, 115–22, 124–30

see also efficiency
profits
  alternative to *see* marginal cost pricing
  in electricity supply industry 200–1
  in gas industry 264, 265, 270–1, 275–6, 280, 285
  and incentive regulation 181, 184
  lack of *see* losses
  in public enterprise 94, 98, 100–1, 108, 114, 122–4, 130–2
projections *see* forecasting
PRT *see* petroleum revenue tax
Pryke, R. 9
  on public sector performance 11, 115–32
public enterprise performance
  modelling 11, 92–112; CEGB: predictions and performance 101–8; positive theory of 94–101; privatization 92, 93, 108–12
  sectoral (1968-78) 6, 11, 115–32; coal 124–32; costs, prices and profits 118, 122–4, 130–2; electricity 118–24; employment and productivity 115–18, 118–22, 124–30; gas 115–18
  see also nationalized industries; regulation; *and under* pricing; productivity; profits
public sector borrowing and oil industry 401–3, 410
PURPA (Public Utilities Regulatory Policies Act, USA) 206–7, 210–13
PWR (pressurized water reactor) *see* nuclear power

Ramsey prices 268, 270, 274
rate-of-return pricing/regulation 15, 51, 109, 191, 271
  in electricity supply industry 201, 204, 252, 253, 260
  indexation and 199–200
  see also yardstick
Rawlinson, R. 127*n*
real price of energy (RPE) 78–9
realism 109
rebate scheme, gas 269
Redpath, R. U. 59*n*
Rees, R. 133, 220*n*, 249*n*, 430
  on management 190
  on modelling performance 11, 14, 92–112, 263–4, 268, 416
  on output maximization 10, 161
  on short time horizon 169
regional differences in pricing of gas 267–8, 272, 282
regulation 93
  failure 7–12, 15–16, 18–20; in energy market 7–8; in nationalized industries 9–12
  monopoly 429–31
  in new market philosophy 13–18

  see also under British Gas Corporation; electricity; public enterprise
regulatory capture 254, 278
Reid, G. L. 126*n*
rent taxes 34–5, 49, 51
  see also Gas Levy; oil industry taxation
required rate of return 51, 109, 191
reserves, oil 350, 363–4, 366
rewards, bounds on 186–7
Rhys, J.: on North Sea oil 377–410, 415
risk analysis and optional investment 6, 11, 214–36
  deferment decision 227–36
  economic appraisal 215–27
risks, incentives weakened by 185–6
Roberts, R. D. V. 120*n*, 121*n*
Robinson, B.: on economics of coal 11, 141, 142, 144, 312–45
Robinson, C. 349: on liberalizing coal industry 19, 20, 141, 142, 289–312
Robinson, D. 245*n*
Rockefeller family 3
Rooke, Sir D. 263, 266, 276
Rose, D. 412
Rozanski, G. A. 229–30
RPE (real price of energy) 78–9
RPI-X pricing 16–17, 18, 20, 93, 253, 260, 271, 274, 275, 429–32
RRR *see* required rate of return

Sappington, D. E. M. 181*n*
Saudi Arabia 380
Saunders, M. 413*n*, 422
Saynor, P. 245*n*
SCE (Select Committee on Energy) *see* Energy Committee
Schmalensee, R. 45, 206, 252
  on incentive regulation 178–87
Scotland 20, 122, 295, 309
Searle Barnes, R. G. 128
'second-best' problem in marginal cost pricing 141–3
Select Committees
  on Energy *see* Energy Committee
  on Nationalized Industries 121*n*, 124*n*
self-interest and incentive regulation 181–2
Sen, A. K. 56*n*, 57*n*
sensitivity tests in risk analysis and optional investment 227, 231–4
service charges 195–6
Severn Barrage 162, 164
'shadow' wages and prices 143–4
shareholders
  in coal industry 301
  in electricity supply industry: and regulation 198, 200, 201
  and incentive regulation 182, 184
  see also privatization

Sharpe, L. J. 245n
Sharpe, T. 244n
Sheldon, R. 418
Sherman, R. 16
Shleifer, A. 17, 185n, 202, 254
short-run marginal costs 9, 44–5
 in electricity supply industry 189, 256; and competition 163, 173, 175, 176
 rationale for 144–9
 see also marginal cost pricing
Sibley, D. S. 133
Sizewell Inquiry (1987) 25, 141, 143, 147, 215, 221, 224–31, 235–6
Slater, M. 169, 189
 on marginal cost pricing 10, 133–54, 263, 265
Slovic, A. 59
small-scale coal mines 295–6, 300–1
small-scale private electricity generation 157, 159, 162, 206–9
Smith, A. 154
Smith, G. H. 243n
social costs see social security; unemployment
social security 3, 12
 and coal industry 314–15, 327, 332
 gas rebates 269
 and marginal cost pricing 140
 and merit goods 55–73
South Africa 33, 298, 326
Soviet Union 23, 41, 130
Spain 245n
spare capacity and electricity supply industry 160–1
SPD see supplementary petroleum duty
Spence, M. 160
Spencer, P. 380, 384
SPK (Sweden) 251, 256
spot prices see efficient prices
SRMC see short-run marginal costs
standing charges, gas 268, 273–4
state see energy policy; legislation; nationalized industries; privatization; regulation; social security
Status Agreement (1964–5) 120
Stiglitz, J. 56n, 181n
storage costs in gas industry 274–5
structural change in oil industry 349, 354–5, 368–9, 383–91
subsidies 12, 269
 in coal industry 306–7, 314–15, 327–35
Sumner, M. 418
Sunday Times 34–5
supplementary petroleum duty 420–1, 423, 424
suppliers, prices for 38
supply 2, 3, 10
 in coal industry 292, 295–6
 market failure 58–9

model see public enterprise performance modelling
 see also productivity
surpluses, financial and marginal cost pricing 149–50
Sweden
 acid rain 4
 electricity supply industry 238; competition 258, 259, 260; organizational structure 247–8; ownership 243–4, 247n; regulation 251–2, 256
Switzerland 368
Sydcraft (Sweden) 243, 244n, 248

tariffs see pricing
taxation 12, 52
 in electricity supply industry 188–9, 194–5, 201, 208
 and energy policy 30–1, 34–5, 37, 52, 55, 70, 73
 in gas industry 36, 48, 51, 266, 281
 legislation 411, 418–19
 see also under oil industry
TDR see test discount rate
technical variables in electricity supply industry 222–3
test discount rate 39, 189, 190–1
Texas 206, 207, 208, 210, 212
Thatcher, M. 265
Thompson, D. 9, 11, 115n 173
 on USA 206–13
 see also under Hammond
Times, The 53
time-scales implied in planning 3, 5–6
Tolley, D. L. 207
Tomlinson, M. 82, 84
total useful energy equation in demand model 78, 79–80
trade, international 4, 33, 38, 42, 46
 in coal 295, 297–8, 314, 326–7, 339
 in electricity supply industry 258–9, 260
 in oil industry 350, 352–60, 369, 378–9, 382, 385–90
transport costs 35, 298, 316, 326
Treasury 51, 52, 302
 see also White Papers
Turot, P. 243n
Turvey, R. 133, 215–17
Tversky, A. 59

UDM see Union of Democratic Mineworkers
under-investment in electricity supply industry 200–1, 202
unemployment, and coal industry 314, 327, 329, 332, 342–3
 see also employment
Union of Democratic Mineworkers 290, 307

Index

unions and labour relations 4, 128–9, 290, 302, 304–5, 307, 308, 333–4
  in public enterprise 94–6, 109–10
unitary centralized systems *see* nationalized industries
United States 15, 133, 277
  airlines 41
  coal industry in 33, 84, 298, 326
  oil industry in 3, 34, 84
  pricing 84
  Treasury Bonds 400
  see also under electricity supply industry
unregulated markets in gas industry 273
use-of-system charges in electricity supply industry 173–4

value added tax 37, 52, 73, 396, 406
  zero-rating 55, 70
Vattenfall (Sweden) 243–5, 245*n*, 248, 251–2, 256, 259
Venice Declaration (1980) 23
Verney, D. V. 252
vertical intergration in electricity supply industry 246–7, 248, 260
viability and incentive regulation 182
Vickers, J. S. 158, 160, 176, 189, 194, 238*n*, 254
Virole, J. 239
Visscher, M. 16

wages
  in coal industry 313, 323–4, 337, 344
  in oil industry 370

  in public enterprise 97–8, 102–3
  see also employment
weather, exceptionally severe 67
Webb, M. G. 158, 177
welfare *see* social security
White Paper (1961) 9–10
White Paper (1967) on nationalized industries 9–10
  competition 164
  energy policy 31, 39, 188–90, 192, 197
  performance 152–3
  pricing 265
White Paper (1978) on nationalized industries 9–10, 11
  competition 166, 169
  energy policy 190–1, 192, 197
  performance 146, 153
White Paper (1983) 301
White Paper (1988) on privatizing electricity 237
Whiteman, J. C. 384
Wilberforce Inquiry 130
Williamson, O. E. 10, 246
Willig, R. D. 146
Winterton, J. 303*n*
Wood Mackenzie & Co.: model 349, 371–3
work study incentive payment (WSIP) 117

yardstick regulation 17, 185, 203–5, 252, 254
Yarrow, G. 16, 148, 158, 169, 175–7, 238*n*, 254
  on regulation in electricity supply 17, 19, 188–205